Actuators in Robotic Control—2nd Edition

Actuators in Robotic Control—2nd Edition

Guest Editor

Chih Jer Lin

 Basel • Beijing • Wuhan • Barcelona • Belgrade • Novi Sad • Cluj • Manchester

Guest Editor
Chih Jer Lin
Graduate Institute of
Automation Technology
National Taipei University of
Technology
Taipei
Taiwan

Editorial Office
MDPI AG
Grosspeteranlage 5
4052 Basel, Switzerland

This is a reprint of the Special Issue, published open access by the journal *Actuators* (ISSN 2076-0825), freely accessible at: https://www.mdpi.com/journal/actuators/special_issues/robot_actuators.

For citation purposes, cite each article independently as indicated on the article page online and as indicated below:

Lastname, A.A.; Lastname, B.B. Article Title. *Journal Name* **Year**, *Volume Number*, Page Range.

ISBN 978-3-7258-2989-7 (Hbk)
ISBN 978-3-7258-2990-3 (PDF)
https://doi.org/10.3390/books978-3-7258-2990-3

© 2025 by the authors. Articles in this book are Open Access and distributed under the Creative Commons Attribution (CC BY) license. The book as a whole is distributed by MDPI under the terms and conditions of the Creative Commons Attribution-NonCommercial-NoDerivs (CC BY-NC-ND) license (https://creativecommons.org/licenses/by-nc-nd/4.0/).

Contents

Chih-Jer Lin and Ting-Yi Sie
Design and Experimental Characterization of Artificial Neural Network Controller for a Lower Limb Robotic Exoskeleton
Reprinted from: *Actuators* **2023**, *12*, 55, https://doi.org/10.3390/act12020055 1

Chih-Jer Lin, Hsing-Cheng Wang and Cheng-Chi Wang
Automatic Calibration of Tool Center Point for Six Degree of Freedom Robot
Reprinted from: *Actuators* **2023**, *12*, 107, https://doi.org/10.3390/act12030107 26

Ling Zhang, Shiqing Lu, Mulin Luo and Bin Dong
Optimization of the Storage Spaces and the Storing Route of the Pharmaceutical Logistics Robot
Reprinted from: *Actuators* **2023**, *12*, 133, https://doi.org/10.3390/act12030133 46

Shengqiao Hu, Houcai Liu, Huimin Kang, Puren Ouyang, Zhicheng Liu and Zhengjie Cui
High Precision Hybrid Torque Control for 4-DOF Redundant Parallel Robots under Variable Load
Reprinted from: *Actuators* **2023**, *12*, 232, https://doi.org/10.3390/act12060232 62

Yi-Rong Li, Wei-Yuan Lien, Zhi-Hong Huang and Chun-Ta Chen
Hybrid Visual Servo Control of a Robotic Manipulator for Cherry Tomato Harvesting
Reprinted from: *Actuators* **2023**, *12*, 253, https://doi.org/10.3390/act12060253 87

Xun Liu, Yan Xu, Xiaogang Song, Tuochang Wu, Lin Zhang and Yanzheng Zhao
An Accurate Dynamic Model Identification Method of an Industrial Robot Based on Double-Encoder Compensation
Reprinted from: *Actuators* **2023**, *12*, 454, https://doi.org/10.3390/act12120454 104

Nikola Knežević, Miloš Petrović and Kosta Jovanović
Cartesian Stiffness Shaping of Compliant Robots—Incremental Learning and Optimization Based on Sequential Quadratic Programming
Reprinted from: *Actuators* **2024**, *13*, 32, https://doi.org/10.3390/act13010032 121

Chun-Hsiang Hsu and Jih-Gau Juang
Using a Robot for Indoor Navigation and Door Opening Control Based on Image Processing
Reprinted from: *Actuators* **2024**, *13*, 78, https://doi.org/10.3390/act13020078 139

Afrah Jouili, Boumedyen Boussaid, Ahmed Zouinkhi and M. N. Abdelkrim
Fault Detection of Multi-Wheeled Robot Consensus Based on EKF
Reprinted from: *Actuators* **2024**, *13*, 253, https://doi.org/10.3390/act13070253 157

Ranko Zotovic-Stanisic, Rodrigo Perez-Ubeda and Angel Perles
Comparative Study of Methods for Robot Control with Flexible Joints
Reprinted from: *Actuators* **2024**, *13*, 299, https://doi.org/10.3390/act13080299 178

Yao Wu, Biao Tang, Shuo Qiao and Xiaobing Pang
Bionic Walking Control of a Biped Robot Based on CPG Using an Improved Particle Swarm Algorithm
Reprinted from: *Actuators* **2024**, *13*, 393, https://doi.org/10.3390/act13100393 195

Article

Design and Experimental Characterization of Artificial Neural Network Controller for a Lower Limb Robotic Exoskeleton

Chih-Jer Lin * and Ting-Yi Sie

Graduate Institute of Automation Technology, National Taipei University of Technology, Taipei 10608, Taiwan
* Correspondence: cjlin@mail.ntut.edu.tw

Abstract: This study aims to develop a lower limb robotic exoskeleton with the use of artificial neural networks for the purpose of rehabilitation. First, the PID control with iterative learning controller is used to test the proposed lower limb robotic exoskeleton robot (LLRER). Although the hip part using the flat brushless DC motors actuation has good tracking results, the knee part using the pneumatic actuated muscle (PAM) actuation cannot perform very well. Second, to compensate this nonlinearity of PAM actuation, the artificial neural network (ANN) feedforward control based on the inverse model trained in advance are used to compensate the nonlinearity of the PAM. Third, a particle swarm optimization (PSO) is used to optimize the PID parameters based on the ANN-feedforward architecture. The developed controller can complete the tracking of one gait cycle within 3.6 s for the knee joint. Among the three controllers, the controller of the ANN-feedforward with PID control (PSO tuned) performs the best, even when the LLRER is worn by the user and the tracking performance is still very good. The average Mean Absolute Error (MAE) of the left knee joint is 1.658 degrees and the average MAE of the right knee joint is 1.392 degrees. In the rehabilitation tests, the controller of ANN-feedforward with PID control is found to be suitable and its versatility for different walking gaits is verified during human tests. The establishment of its inverse model does not need to use complex mathematical formulas and parameters for modeling. Moreover, this study introduces the PSO to search for the optimal parameters of the PID. The architecture diagram and the control signal given by the ANN compensation with the PID control can reduce the error very well.

Keywords: pneumatic artificial muscles (PAMs); neural network control; artificial neural network; iterative learning controller; lower limb robotic exoskeleton robot

1. Introduction

To perform task-oriented rehabilitation treatment for patients, a variety of robot systems for different purposes and of rehabilitation parts have been developed. The goals of robot systems are to perform specific movements to stimulate the patient's movement plasticity. To achieve the recovery of motor function or minimize the functional deficit of patients, many types of lower extremity rehabilitations have been proposed. The lower extremity rehabilitation system can be mainly classified into the following: (1) Treadmill gait trainer, (2) Footboard-based gait trainer, (3) Ground gait trainer, (4) Fixed gait trainer and (5) Ankle rehabilitation system [1]. Traditional therapies usually focus on treadmill training to improve the functional mobility [2]. This rehabilitation technique is known as partial body-weight support treadmill training. The therapists are required to assist the patient in walking on the treadmill with the legs and hips assisted when the patient's body weight is carried by hanging load belts. For example, the robotic orthosis Lokomat is an automated treadmill training system, which consists of a treadmill and a suspension system to provide the body-weight unloading [3]. The Lokomat consists of a robotic gait orthosis and an advanced body weight support system which is combined with a treadmill. It uses computer-controlled motors for each of its hip and knee joints and the drives are precisely synchronized with the speed of the treadmill to ensure that the speed of the gait orthosis

and treadmill match. The LokoHelp (LokoHelp Group) is an electromechanical device developed for improving gait after brain injury and it is placed on a treadmill parallel to the walking direction to drive the patient to walk [4]. ReoAmbulator (Motorika Ltd., U.S.A., marketed in the USA as the "AutoAmbulator"), is another body-weight-supported treadmill robotic system and it is located in the front of the treadmill and has a protruding link to support the lower limb mechanism [5]. The mechanical lower limb is tied to the patient's leg and there is also a safety strap on the top to support the patient's weight.

In recent years, with the development of neural network related research, many applications have emerged. In the field of controllers, many researchers have been developing systems to make the system more intelligent and able to adapt to complex control principles. Among them, the architecture driven by pneumatic artificial muscles (PAMs) has been a major subject of nonlinear control for many years. Among many PAMs, McKibben Muscle is more commonly used and widely known. It is a type of Braided muscle, and is composed of an air-tight elastic in the middle. The tube is the center, and the elastic tube is surrounded by a braided mesh. When the inner tube is pressurized and inflated, it expands and squeezes the braided mesh. This driving method enables PAMs to have the characteristics of small size, light weight and high output, which is very suitable for the field of rehabilitation robot driving. PAMs have been applied to the development of powered lower limb exoskeletons. For example, Beyl et al. presented a performance evaluation result of a powered knee exoskeleton [6]. The control of PAM-driven systems has proven difficult due to the nonlinear nature of the actuator and the properties of the air pressure source driving it. The model-based control strategies rely heavily on the accuracy of the model to eliminate nonlinearities. Traditional methods such as modeling hysteresis have considered as control pressure, the hysteresis phenomenon and the braided sheath initial angle. However, PAM and many PAM-driven systems generate complex nonlinear forces when pressurized [7,8]; they usually require a lot of time and effort to model the system (which usually requires empirical methods). In addition, the established system model is less resistant to environmental changes or external disturbances. Carbonell et al. [9] discussed the benefits of using three controllers in the pneumatic muscle actuator, namely robust backstepping, adaptive backstepping and sliding-mode. In the study, the tracking is well achieved by the sliding-mode and the adaptive controller. Unfortunately, properties such as PAM actuator dynamics, pneumatic/mechanical system dynamics, and payload characteristics are unknown and/or time-varying.

In many cases sliding mode control may suffer from the same problems as pure model-based control. Feedback error learning (FEL) was originally proposed by Kawato [10]. It is a method to update the feedforward controller through the output error of the feedback controller to improve the accuracy of the inverse model. There has been related discussion about FEL and nonlinear adaptive controllers [11]. However, few FEL concepts are used in the application of PAMs-actuated bidirectional (antagonistic) actuation architecture. To overcome the above-mentioned problems in PAMs modeling, Robinson et al. [12] compared three control strategies: sliding mode control, adaptive sliding mode control, and adaptive neural network (ANN) control. The results show that the ANN controller is preferable because it does not require a model of the pneumatic system or joint mechanism design, which can be difficult and time consuming to characterize, and is robust to changes in PAM actuator characteristics. In this study, a treadmill-type rehabilitation equipment was developed. The rehabilitation movements are used for two kinds of feedforward controllers, including Iterative Learning Control (ILC) and ANN feedforward controllers.

Modeling of PAM-driven rehabilitation machines has been a difficult problem in the field of rehabilitation. On the problem side, the three main challenges proposed in this study are as follows.

1. The complexity of modeling the dual PAM drive (antagonistic) actuation architecture used in this study is relatively high

2. The PAM driver used in this study is a proportional valve, which is cheaper than the pressure control valve, but will increase the complexity of the system.

3. The walking speed set in this study is relatively fast, and it is crucial to overcome the hysteresis problem of PAM, which is also a difficulty point of traditional modeling.

The data collection method used in this study directly oscillates the system through open loop control to obtain the relationship between the knee joint angle and the control command of proportional valve. In other words, we overcome the problems of 1 and 2 by using the forward-feeding ANN, and we verify the operational reliability of the system by conducting experiments on the real system.

On the technical side, there are two novelties.

1. We implemented experiments directly on our LLRER. The PSO-PID controller with a simple feedforward ANN can also obtain good tracking results by sending out the setpoint 3 sampling points ahead of the loop-oriented task.

2. We compensate the PSO-PID controller by using a queue, so that the feedforward ANN does not need the same update frequency as PSO-PID, providing a new option for future integration of other algorithms that cannot be applied to the controller due to the slow update frequency.

2. Rehabilitation System Architecture

2.1. Design of the Lower Limb Robotic Exoskeleton

Many research laboratories and companies are working on robotic exoskeletons with the intent to assist disabled individuals [10–16]. According to the structural form, lower extremity robotic exoskeletons can be classified into two types: Rigid Lower Extremity Robot Exoskeletons (RLEEX) and Compliant Lower Extremity Robotic Exoskeletons (CLEEX) [17]. Through the RLEEX research, the Human Universal Load Carrier (HULC) of Lockheed Martin [18] and the Guardian XO of Sarcos Robotics [19] in the United States have been the leaders in the development of exoskeletons. Lockheed Martin launched the HULC based on the BLEEX results and conducted a series of wearable tests with the US Army [20]. The Hybrid Assistive Limb (HAL) of the University of Tsukuba adopts a function-oriented design concept; the HAL series [21–24] for medical rehabilitation has been used in Japan and Europe and is the earliest commercial walking exoskeleton robot [25–28].

One of the most-established exoskeleton technologies for disabled assistance is the Rewalk [29]. Robotic exoskeletons can be categorized into three categories according to their purpose. The first group is human efficiency enhancement exoskeletons. The second group involves assistive devices for people with movement disorders due to stroke, spinal cord injury and muscle weakness. The third category is called therapeutic exoskeletons which are utilized for rehabilitation purposes. The first group aims to maximize the durability, stamina, and other physical abilities of persons and is also called augmentation exoskeletons. They may be employed for assisting with lifting heavy items or transporting heavy loads over long distances in manufacturing facilities, urgent relief functions, or military bases. According to the body part involved, the robotic exoskeletons can be categorized into three different categories: upper limb, lower limb and specific joint exoskeletons [14–16]. One of the most significant hurdles to be alleviated is the human-robot interaction and control. Different techniques have been presented in the literature to manage the human-robot interaction.

In this paper, the proposed lower limb robotic exoskeleton is designed for knee and hip joints. One joint has one degree of freedom and the limit of the thigh is designed in the range from −40 to 130 degrees, so that patients can wear the exoskeleton to perform squatting movements. As shown in Figure 1, a DC brushless motor is fixed on the upper side to drive the hip joint; the two PAMs are equipped on both sides to drive the knee joint as shown in Figures 2 and 3. In terms of mechanism design, we installed the PAMs on both sides of the thigh to drive the knee joint. There is a connecting piece between the hip and the back frame, it can adjust the position of the hip joint according to the user's waist circumference (up/down, left/right, front/rear). As shown in Figure 2, the hip flange face is directly connected to the thigh connecting plate and the DC motor (Maxon EC60flat) with the harmonic drive (CSG-17-100-2UH-LW) is used to drive the hip joint. Then, the thigh connecting plate is connected downward to the leg link, which is used to adjust the length of the thigh.

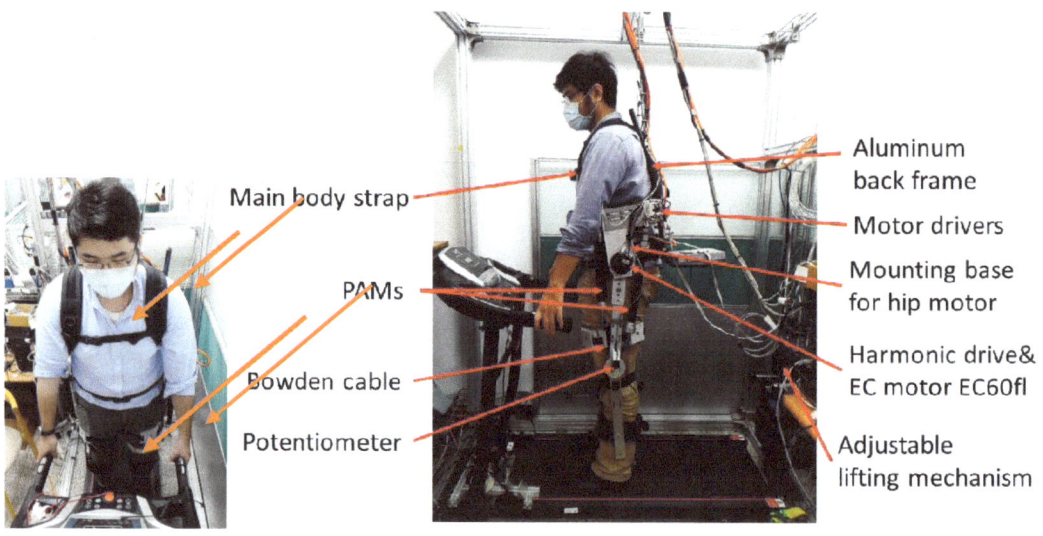

Figure 1. Overview of the lower-limb rehabilitation system.

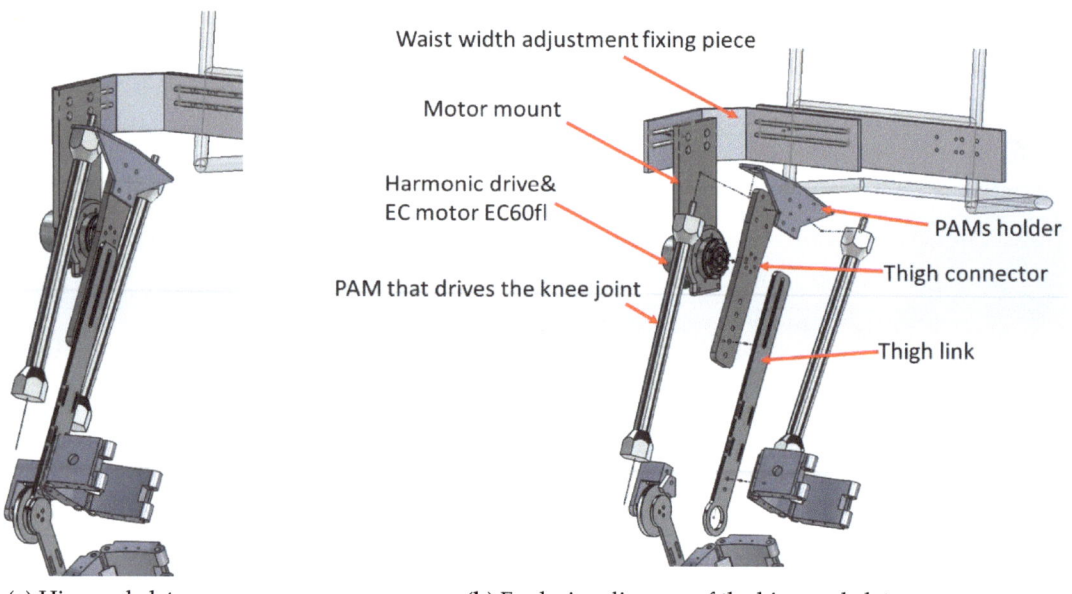

(**a**) Hip exoskeleton. (**b**) Explosion diagram of the hip exoskeleton.

Figure 2. Hip joint exoskeleton design.

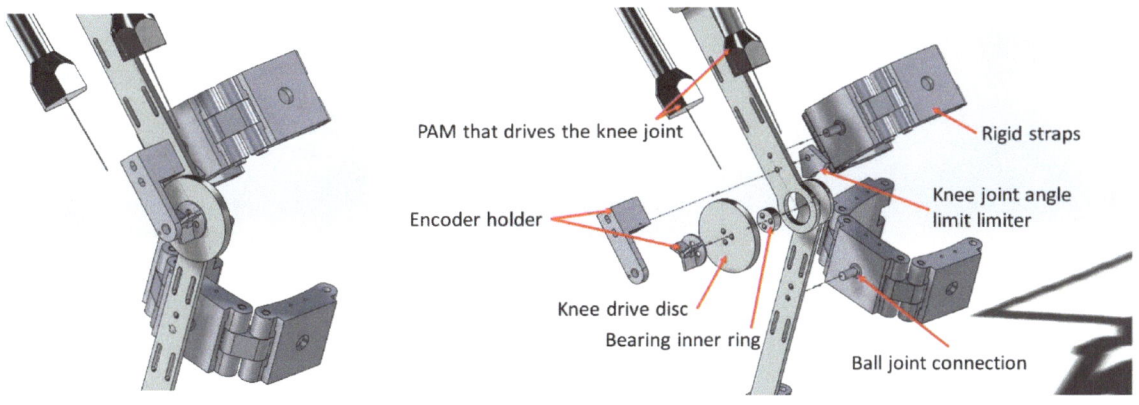

(a) Joint exoskeleton. (b) Explosion diagram of the knee exoskeleton.

Figure 3. Knee joint exoskeleton design.

As shown in Figure 3, from the design of the knee joint mechanism, the movement of the knee joint comes from the drive disc which is pulled by the two PAMs. The wire is used to maintain the tension pulled by the two PAMs and the proportional directional valve is used to control the contraction and release of the PAMs. In the knee mechanism, a limiting mechanism is used to limit the rotation angle of the knee joint and the rotation range is designed from -10 to 90 degrees. The fixing strap is fixed on the leg, as shown in Figure 3, and the connection part with the mechanism uses a ball joint, so that the rigid strap fits the shape of the subject to a certain extent, and has better rigidity than a pure cloth strap. As shown in Figure 4, the thigh length adjustment mechanism can be adjusted from the shortest distance of 37.2 cm to the longest distance of 52 cm, which can meet the thigh length range of most people. The joint part uses a potentiometer to measure the joint angle.

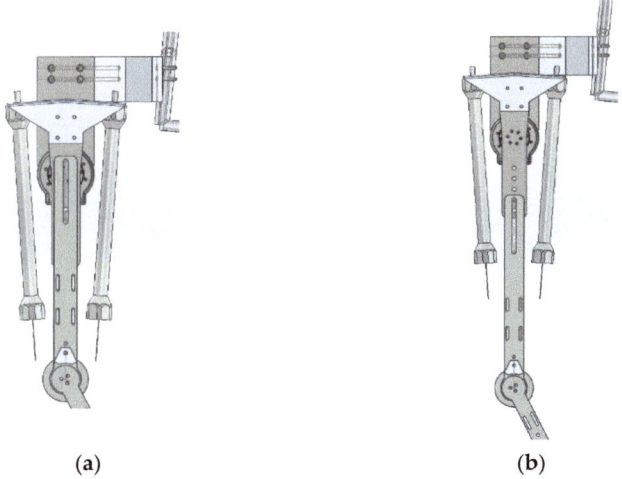

(a) (b)

Figure 4. Thigh length adjustment: (**a**) 37.2 cm; (**b**) 52 cm.

2.2. Electromechanical System of Powered Lower Limb Rehabilitation Exoskeleton Robot

This research develops a powered exoskeleton system which has two degree-of-freedom lower limb power exoskeletons, as shown in Figure 5. The hip joint uses a brushless DC motor with a reducer (Maxon EC60flat + CSG-17-100-2UH-LW) for positioning control as shown in Figure 5. The knee joint uses two PAMs to drive, with driving architecture as shown in Figure 6. The whole system of the proposed lower limb rehabilitation exoskeleton robot system (LLRER) is shown in Figure 7 and all hardware and equipment are used for the LLRER listed in Table 1. The proposed LLRER system is controlled by CompactRIO SbRIO-9631 (National Instrument) with NI 9516 modules, which are responsible for receiving the encoder signal of the motor (Maxon EC 60 flat) with feedback for the current angles of the hip joint to sbRIO-9631 for calculation. In this study, the knee joint is driven by two PAMs (Festo, Germany, MAS-20, as shown in Figure 5) and a proportional directional control valve (Festo, MPYE-5-M5-010-B) is used to control the two PAMs.

The knee joint is controlled by the bidirectional actuation via the two PAMs to exert force in two directions, respectively. The proportional directional control valve is operated by converting the voltage input signal to flow directional control signal. The valve is used to control the opening area as well as the inlet and outlet direction through the input voltage to achieve the purpose of controlling the valve. Compared with the single PAM system in which the restoring force comes from gravity or spring, this control method can generally obtain greater joint torsional rigidity, thereby achieving more accurate tracking control results. After the controller algorithm is calculated, the control signal is used to control the knee joint and the hip joint through analog output to achieve the control of the system. The airflow direction of the pressure source is controlled by the proportional directional control valve. The air pressure value and the joint angle value are feedbacked to the embedded controller. The position control PID outputs a directional valve control voltage of 0 V to 10 V, which controls the stretching and contraction of the PAM, and returns the position through the potentiometer of the knee joint as shown in Figure 5.

In terms of research and development, PAM is well known to have a relatively small volume ratio while having a high output force. In terms of safety, the shrinkage limit of PAM is about 25%, which is relatively safe, although it is difficult to model, but it has a certain degree of stretching and elasticity, so it is still popular in the field of rehabilitation, providing a certain degree of comfort for the rehabilitation. For the discussion of the controller, we also used PAM drive at the hip joint in the previous research. In the case of fast walking (1 km/h), the PAM response of the hip is not fast enough, so we developed a compound type to support the faster rehabilitation action.

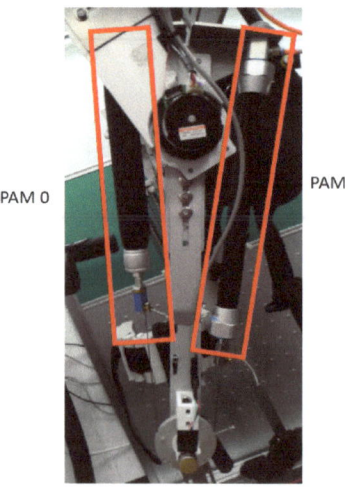
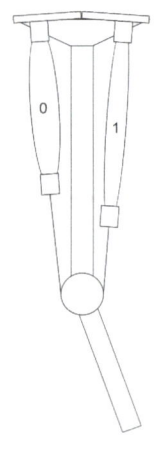

(a) Actuation for the knee joint.   (b) Mechanism of the knee joint.

Figure 5. Knee joint mechanism.

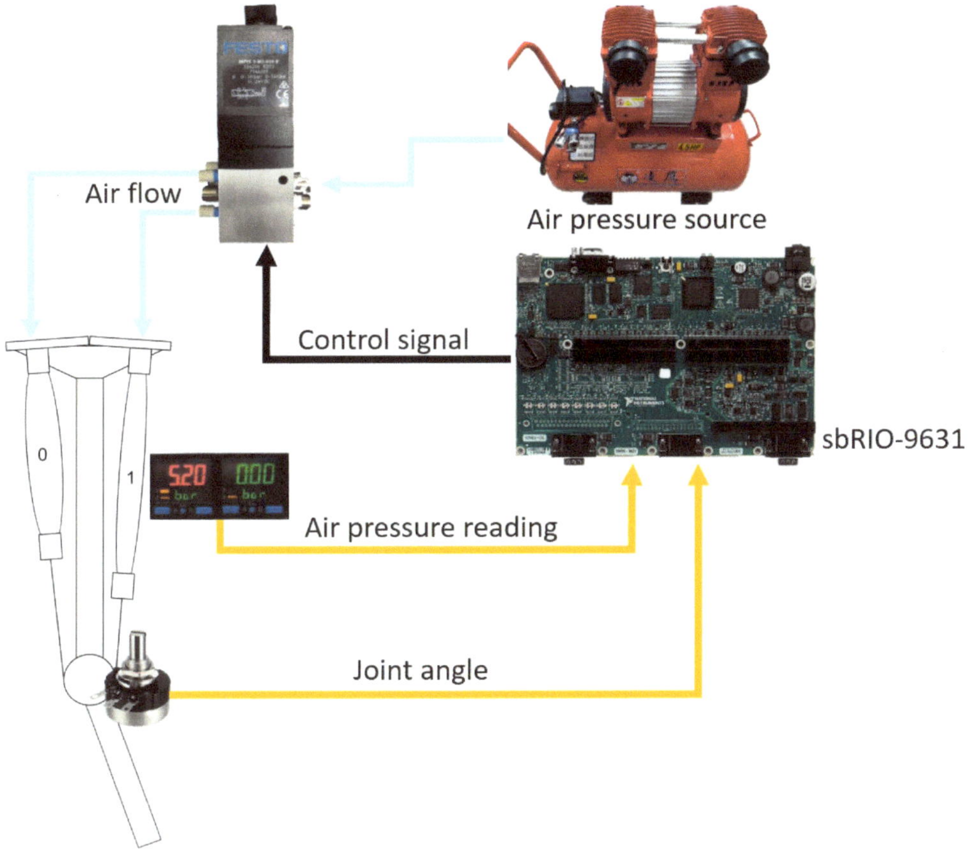

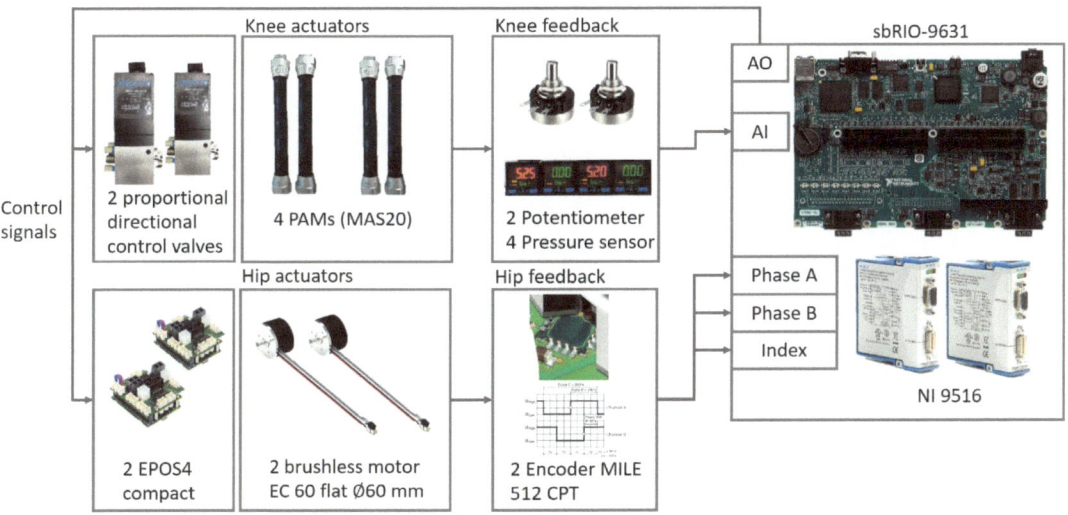

Figure 6. Knee joint control architecture.

Figure 7. Block diagram of the powered lower limb exoskeleton robot.

Table 1. Specification list of hardware and equipment for the proposed LLRES.

Item	Type	Specification
NI SBRIO-9631	Embedded controller	Analog&Digital I/O, 266 MHz CPU, 64 MB DRAM, 128 MB Storage, 1 M Gate FPGA
NI 9516	Servo Drive Interface Module	Servo, 1-Axis, Dual Encoder
MPYE-5-M5-010-b	Proportional directional control valve	Pressure range: 0~10 bar Input voltage range: 0~10 V
MAS-20-300N-AA-MC-O-ER-BG	Pneumatic Artificial Muscle	Operating pressure: 0~6 bar; Maximal permissible contraction: 25% of nominal length
Maxon EC60flat	Flat brushless DC motor	Nominal speed: 3730 rpm Nominal torque: 269 mNm
CSG-17-100-2UH-LW	Harmonic Drive; with cross roller bearing	Limit for average torque: 51 Nm Limit for Momentary torque:143 Nm
SPAB-P10R-G18-NB-K1	Air pressure sensor	Pressure range: 0~10 bar; Electrical output: 1~5 V analog voltage output

3. LLRER Controller Design

3.1. Gait Model Acquisition

To capture the tracking reference of the LLRER system, an unpowered exoskeleton system is made to capture a normal walking reference for the tracking command. As shown in Figure 8, the motion capture system is equipped with 6 sensors on the body. There are two potentiometers on the hip position and the knee joint; a 9-axis IMU (MPU9250) is fixed on the thigh hip to correct the distortion of the hip joint data caused by the back and forth shaking as walking. The sensor signals are captured by the microprocessor (Arduino Uno) for the computation as shown in Figure 9. The IMU is used to transmit the yaw angle from the waist to the hip joint to the PC through I2C; the embedded system converts the potential angular positions of the hip and knee joints into the rotation angle directly. The sampling time of this data collector is 16.3 ms, and the average precision is 0.23 degrees.

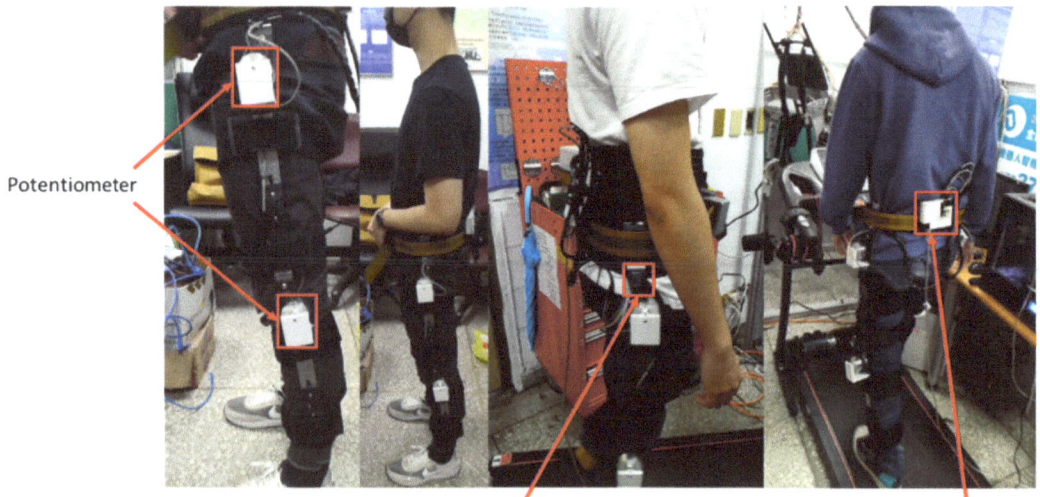

Figure 8. Wearing an unpowered exoskeleton.

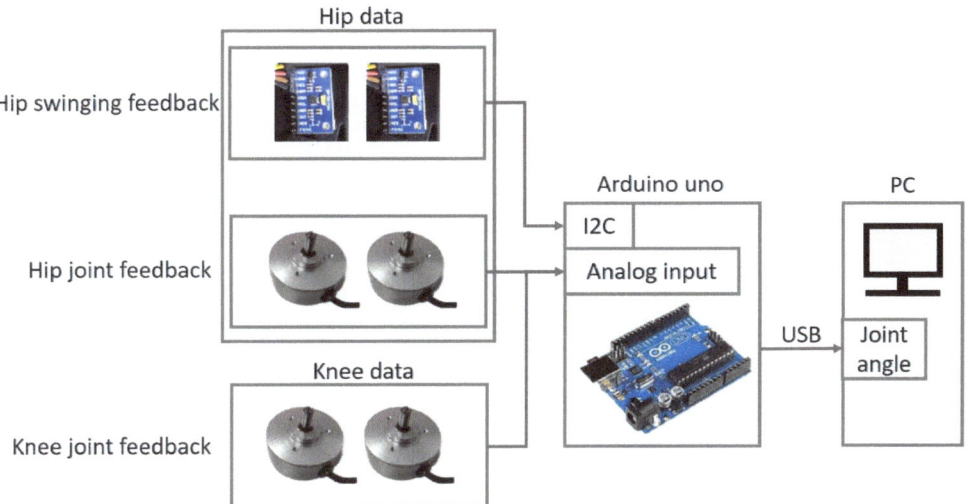

Figure 9. Unpowered exoskeleton communication architecture.

The captured angles are filtered and added to the embedded processor; then, the sorted individual gait models are as shown in Figure 10, where V1 represents the walking model at a treadmill speed of 1 km/h and V4 represents a model at the speed of 4 km/h. The data in Figure 10 is the gait motion model at different walking speeds. The gait model is obtained by averaging the trajectories of each walking speed and curve fitting the average trajectory. The gait model is resampled directly to the desired control frequency at the time of use.

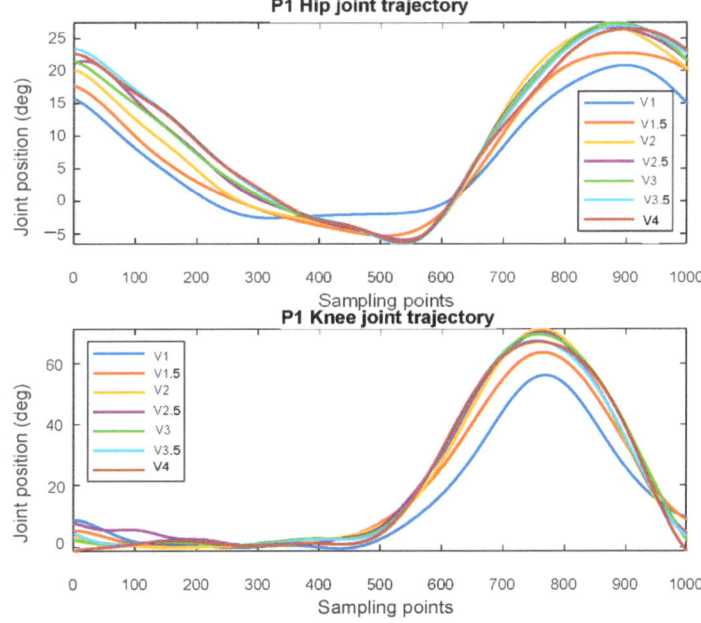

Figure 10. Gait model of the P1 subject.

3.2. Iterative Learning Control for the LLRER

The Iterative Learning Control (ILC) is an effective control method to improve the tracking error of the repetitive operation of dynamic systems; the rehabilitation gait and movements are usually repetitive movements. The ILC is different from other learning control strategies, such as adaptive control, Repetitive Control (RC) and Neural Networks. The adaptive method modifies the controller itself, while the ILC modifies the output of the controller which reduces the tracking error by changing the control signal. After adjusting the controller using the Ziegler–Nichols method, the tracking error is corrected by the ILC. The definitions of each variable are shown in Table 2. The ILC iteration is calculated in matrix form and the expected trajectory matrix Y_d is determined by the previous measurement. The definition of ILC is shown in Equation (1), where the error of this cycle (the gait cycle) $e_{k \times N}$ is the difference between the expected trajectory matrix Y_d and the real output matrix $y_{k \times N}$. Then the error is multiplied by the learning rate and compensated to the next rehabilitation $\theta_{(k+1) \times N}$.

$$\theta_{(k+1) \times N} = \theta_{k \times N} + L e_{k \times N} \tag{1}$$

$$e_{k \times N} = Y_d - y_{k \times N} \tag{2}$$

Table 2. ILC symbol table.

Notations Type	Specification
N	Tracking points per gait cycle
$Y_d = (Y_1, \ldots, Y_N)$	Desired output profile
$y_{k \times N} = (y_1, \ldots, y_N)$	Real output in the current cycle
$e_{k \times N} = (e_1, \ldots, e_N)$	Output error in the current cycle
L	Learning rate
$\theta_{k \times N} = (\theta_1, \ldots, \theta_N)$	Control signal of current cycle
$\theta_{(k+1) \times N} = (\theta_1, \ldots, \theta_N)$	Control signal of next cycle

The control system diagram is shown in Figure 11 and the ILC algorithm updates the desired control signal according to the desired gait. The ILC also compensates the change of the tracking errors, so that the controller can change the control before the change of the tracking error. At first, the ILC is applied to the hip and knee joint control to test the tracking performances. In response, the learning rate L is fixed at 0.02 and the iteration loops are performed 25 times. The same learning rate is used for both the knee and hip joints and the gait model, and then the ILC control experiments are carried out on the knee and hip joints, respectively.

The experimental initial parameters of the PID are obtained through the Z-N method. The controller performance was observed by performing multiple no-load gait experiments at five different speeds, as shown in Table 3. The PID parameters of the hip joint measured by the Z-N method are designed as P: 1.397, I: 0.004, D: 0.001 in the experiments; these PID parameters are used for both hip joints. Figure 12 shows the tracking response of the left and right hip joints using the ILC with the PID learning at a treadmill speed of 1 km/h. Comparing with the results of Figure 12, the ILC can compensate the tracking errors and the lowest tracking errors appear after 10 updates at the speed of 1 km/h. From Table 3, it can be seen that the average error is less than 1 degree and the maximum error is less than 2 degrees. In this hip tracking test, the ILC can compensate the tracking error effectively for the rehabilitation tasks.

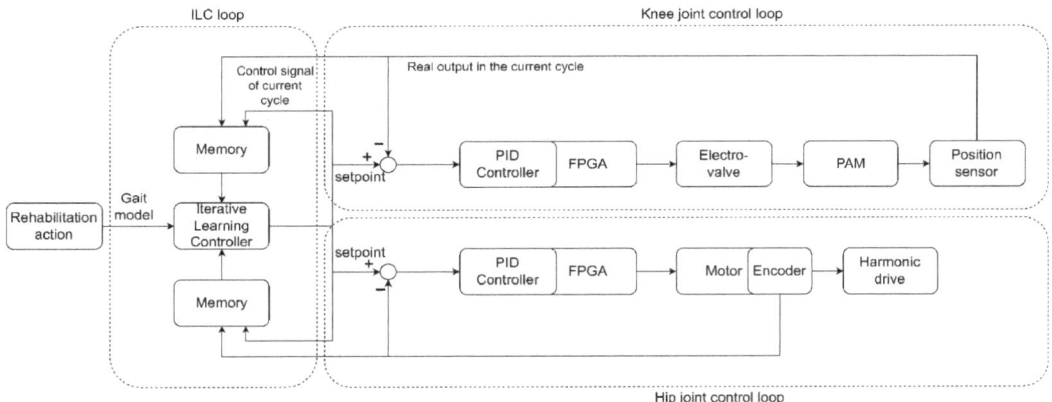

Figure 11. Control system of ILC with PID controller.

Table 3. Tracking error of hip joint using ILC with PID controller.

Treadmill Speed (km/h)	Sec/Cycle	Right Hip		Left Hip	
		MAE (°)	MAXE (°)	MAE (°)	MAXE (°)
0.12	30	0.0241	0.5910	0.0225	0.1280
0.24	15	0.0494	0.2440	0.0440	0.2030
0.53	6.8	0.1150	0.4490	0.0890	0.4560
0.85	4.25	0.3123	0.7690	0.1856	0.7460
1	2.89	0.5616	1.7750	0.4778	1.7490

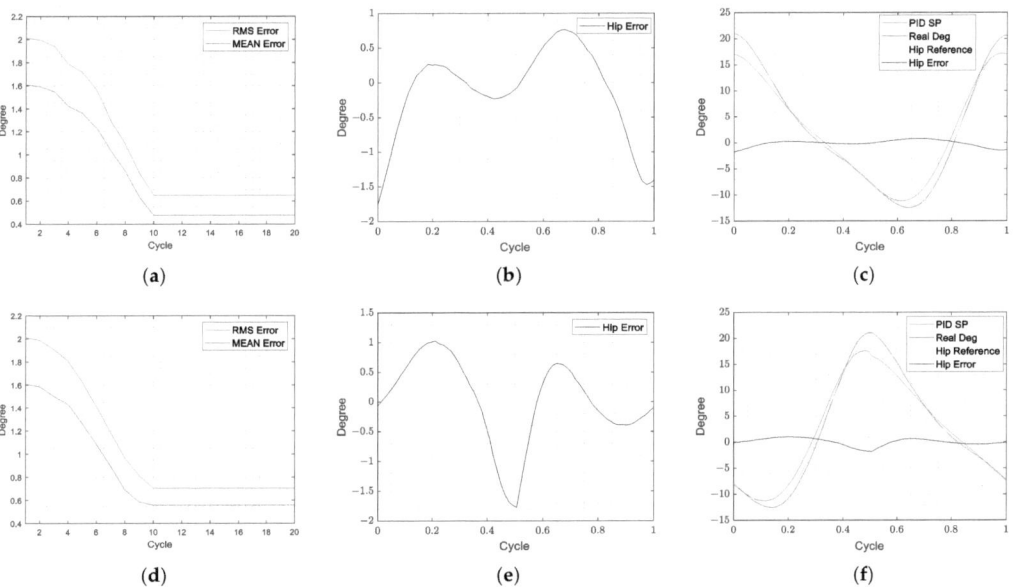

Figure 12. ILC with PID learning at a treadmill speed of 1 km/h. (**a**) (Left hip) The number of ILC iterations and the RMS/MEAN error of the trajectory; (**b**) (Left hip) The tracking error diagram of the best gait cycle; (**c**) (Left hip) The tracking response of the best gait cycle.; (**d**) (Right hip) The number of ILC iterations and the RMS/MEAN error of the trajectory; (**e**) (Right hip) The tracking error diagram of the best gait cycle; (**f**) (Right hip) The tracking response of the best gait cycle.

After the hip joint test, the knee joint of the proposed system is tested by using the PID controller. The PID parameters (P: 0.203, I: 0.006, D:0.001) are obtained by the Z-N method and the ILC structure is as shown in Figure 13. The knee joints are also tested at five rehabilitation speeds; the tracking results are shown in Table 4. According to the results, the knee joint's response is different from the hip joint, because the use of PAMs gives the system a large tracking error due to the nonlinear characteristics of the PAMs.

The tracking results of the treadmill at 0.85 km/h and 1 km/h are shown in Figure 13 to compare the tracking performance of the PAMs; the SP (setpoint) is the control position command corrected by the ILC controller, Real Deg is the actual response of the system, and knee error is the difference between the knee reference and the actual response of the system. From the experimental results of the system in Figure 13, the knee joint using the PID and ILC cannot achieve performance as the same as the hip joint at the speed of 1 km/h. As the walking speed of the system increases, the effect of the ILC controller is worse. The tracking result of 1 km/h has a large overshoot of the rehabilitation reference trajectory, especially at 0.4 and 0.7 cycles (Figure 13d) and 0.1 and 0.9 cycles (Figure 13b). This indicates that the PAM system needs to find other control methods to compensate it. After using the ILC in the hip and knee joints, it was found that the hip joint could be used with the ILC, while the knee joint needed further improvement. The next section focuses on the improvement of the knee controller.

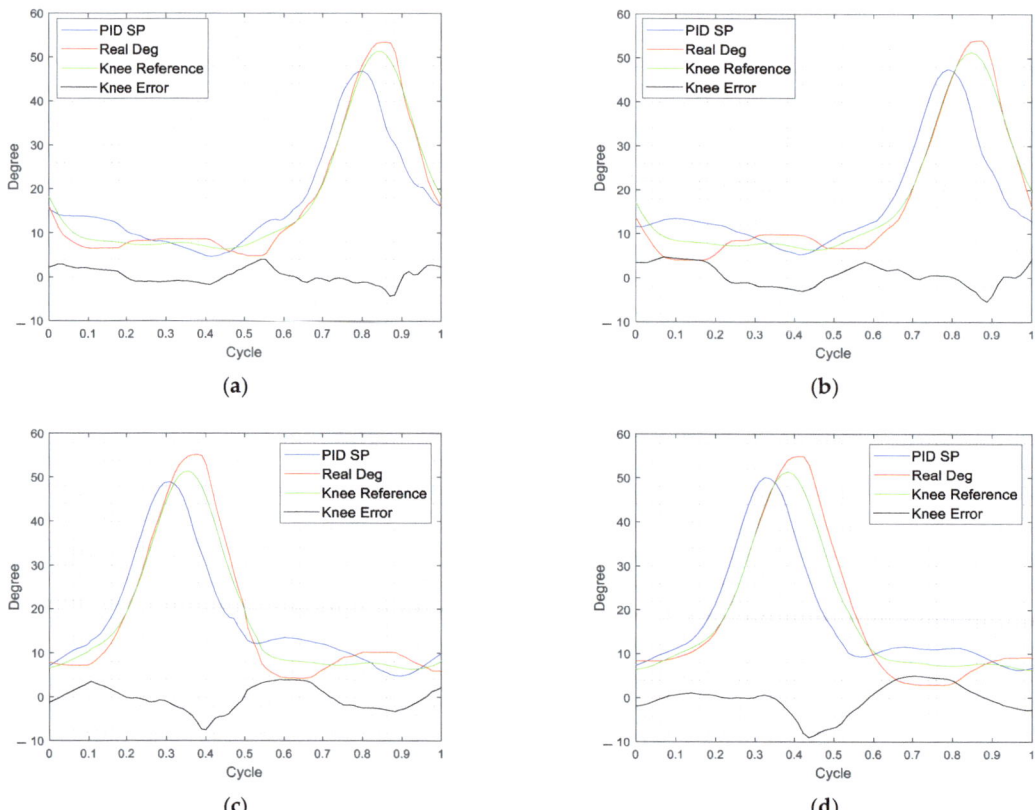

Figure 13. Figure 13. Knee ILC performance at different treadmill speeds. (**a**) (Left knee) Treadmill speed 0.85 km/h; (**b**) (Left knee) Treadmill speed 1 km/h; (**c**) (Right knee) Treadmill speed 0.85 km/h; (**d**) (Right knee) Treadmill speed 1 km/h.

Table 4. Tracking error of the knee joint using ILC with PID controller.

Treadmill Speed (km/h)	Sec/Cycle	Right Knee		Left Knee	
		MAE (°)	MAXE (°)	MAE (°)	MAXE (°)
0.12	30	0.3944	1.9910	0.4288	1.4910
0.24	15	0.9204	3.0030	0.7004	2.4390
0.53	6.8	1.1085	5.5600	0.7162	2.7850
0.85	4.25	2.3364	7.4040	1.4856	4.2670
1	2.89	2.5477	9.0250	2.1554	5.3690

4. Design of the Feedback Controller for the Knee Joint

4.1. Feedforward Artificial Neural Network (ANN) with the Inverse Model

For the network development part, we use Matlab's Neural Net Fitting app for network training, and for the training algorithm, we use Levenberg–Marquardt to update weight and bias values. After training, we integrate the network model into LabVIEW for exoskeleton control. The integration method uses LabVIEW Matlab script to call the established ANN model in the loop of the controller [30–33].

We use the data measured by the real system to train the feedforward ANN controller model in advance. We first set the control command of the proportional directional valve as a linear change in a fixed time, and measure six different time periods to complete a single system response to directional actions. There are two different movements of the knee joint: one is from the straight to the bend (forward movement), and the other is the knee from the bend to the straight (backward movement). Taking Figure 5b as an example, the forward action is PAM0 stretching and PAM1 compression, and the backward action is PAM0 compression and PAM1 stretching. We directly measure a series of system data of these two actions, such as the air pressure of PAM0 (bar) P_{A0} and air pressure of PAM1 (bar) P_{A1} and the knee joint angle θ_d. The time represents that the control signal of the directional valve is sent within 0.5, 1, 2, 3, 4, and 5 s. The corresponding system architecture is shown in Figure 11. Control signals, air pressure readings, and joint angle values are captured while moving, and are used for ANN to learn the system characteristics in advance.

$$\Delta P_{A0} = P_{A0}(n+1) - P_{A0}(n) \tag{3}$$

$$\Delta P_{A1} = P_{A1}(n+1) - P_{A1}(n) \tag{4}$$

$$\Delta \theta_d = \theta(n+1) - \theta(n) \tag{5}$$

$$\Delta cmd = cmd(n+1) - cmd(n) \tag{6}$$

where $P_{A0}(n)$ is the current air pressure (bar) value of PAM0, and $P_{A1}(n)$ corresponds to the air pressure (bar) value of PAM1. $\theta(n)$ is the current knee angle, $cmd(n)$ is the current directional valve control voltage. The data required for training ANN1 (estimating future air pressure changes) can be obtained, and the corresponding current air pressure values P_{A0} and P_{A1}, the angle change amount $\Delta \theta_d$ at the next moment, and the corresponding sampling time can be modified according to the delay time that the system needs. The corresponding output is the predicted change in air pressure in the future ΔP_{A0} and ΔP_{A1}.

To train the ANN1, we use the six experiments to capture the data. Figure 14 shows the six experiments to train the ANN1. Figure 14a,b are the time responses of the P_{A0} and P_{A1} of the knee PAMs with respect to the valve command. Figure 14c represents the knee joint angle with respect to the P_{A0} and P_{A1}. The ANN1 is trained with these data to establish the dynamic model. The training set of ANN1 is represented as TS_{ANN1}, and the purpose is to give the estimated pressure change with reference to the current system pressure for the input of an ideal angle variation. The collected ANN1 training set is about 3000 sets of input and output corresponding data.

$$TS_{ANN1} = \{P_{A0}, P_{A1}, \Delta \theta_d, \Delta P_{A0}, \Delta P_{A1}\} \tag{7}$$

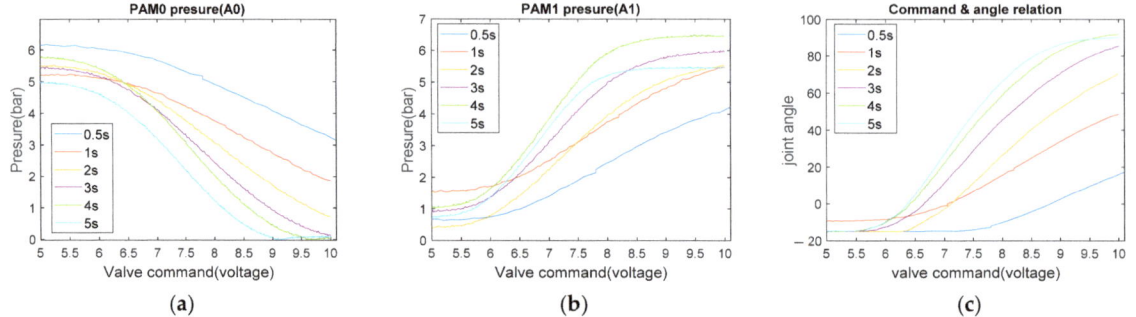

Figure 14. Forward movement system response. (**a**) PAM0 pressure changes (bar); (**b**) PAM1 pressure changes (bar); (**c**) command signals and angle relation.

The data for training ANN2 (proportional directional valve) can also be obtained from the same recorded data. The network inputs are the desired angle change $\Delta\theta_d$ and the air pressure change ΔP_{A0} and ΔP_{A1} corresponding to the angle change; the output is the corresponding directional valve control voltage change Δcmd. ANN1 training uses a fully connected network with 3 inputs, 10 hidden layers, and 2 outputs. ANN2 training uses a fully connected network with 3 inputs, 10 hidden layers, and 1 output. This weight is pre-trained and integrated with the controller, as the network is not updated immediately during operation. The collected ANN2 training set has about 1000 input and output corresponding data. The training set of ANN2 is denoted as TS_{ANN2}; the purpose is to imitate the model of the proportional directional valve, and convert the air pressure change into control commands.

$$TS_{ANN2} = \{\Delta P_{A0}, \Delta P_{A1}, \Delta\theta_d, \Delta cmd\} \tag{8}$$

Figure 15 uses the feedforward ANN in combination with the PID controller. First, ANN1 refers to the current air pressure A0 and A1 with the desired angle change $\Delta\theta_d$ to predict the expected air pressure change value. ANN2 refers to these air pressure change values and $\Delta\theta_d$ gives a compensated control command Δcmd, and the tracking trajectory of PID is also advanced by 3 sampling points. The $\Delta\theta_d$ buffer is 10 sampling points in advance, which is a limitation of program development. The prediction time of two pre-trained ANNs integrated into the controller is measured to be 200 ms. In order to make immediate compensation for control commands, it is necessary to predict 4 sets of data at a time to catch up with the time when the ANN runs the next time. It takes 200 ms to wait for 4 data input, and 200 ms to predict, so it is necessary to prepare the ANN data 8 sampling points in advance. Adding the system response delay, the final choice is 10 sampling points in advance.

In other words, the update frequency of the ANN block is 5 Hz, the PID block is 20 Hz, and the control frequency of the exoskeleton is the same as the PID at 20 Hz. ANN predicts 4 pieces of compensation data at a time and queues them at the v buffer. Because the nature of the rehabilitation action is a cyclic action, ANN's queue compensation is feasible. If the controller tracks an acyclic action, the compensation effect of this advance queue may not be ideal. The ideal situation is that there is no need for queues. Here, queues are used because of performance problems in system integration, so the asynchronous method is used.

Figure 16c is the control signal of the feedforward ANN with the inverse model using the PID controller. The ANN trained by using the pre-measured system data can obtain the same control effect. The main control variables are output by the pre-trained network and the PID control is responsible for a small amount of control. The trend of the pressure change predicted by ANN in Figure 16a,b is the same as that of the actual system and the ideal air pressure change is given before the change. The difference of the ideal air pressure will be compensated with the PID control. From the results shown in Figure 17, the tracking results are good, especially in the area where the tracking angle changes greatly (from 2.5 to 3.6 s). In this experimental result, the performance of the ANN is better than that of the ILC.

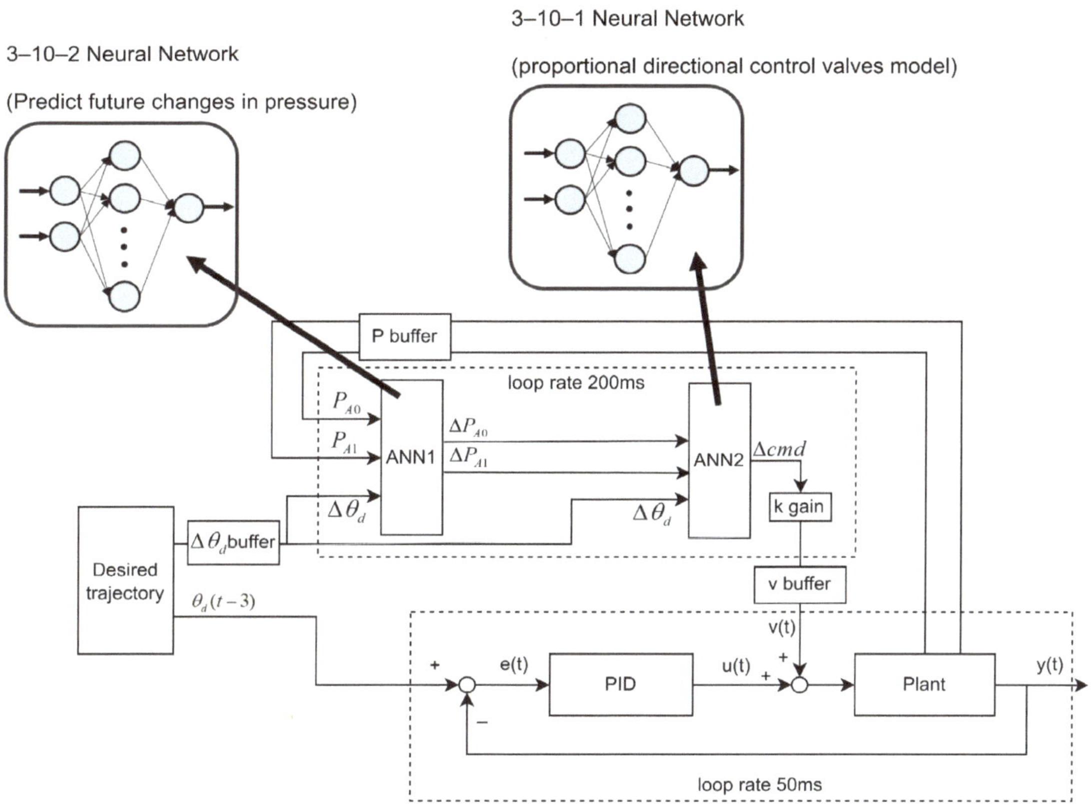

Figure 15. ANN-feedforward with PID controller.

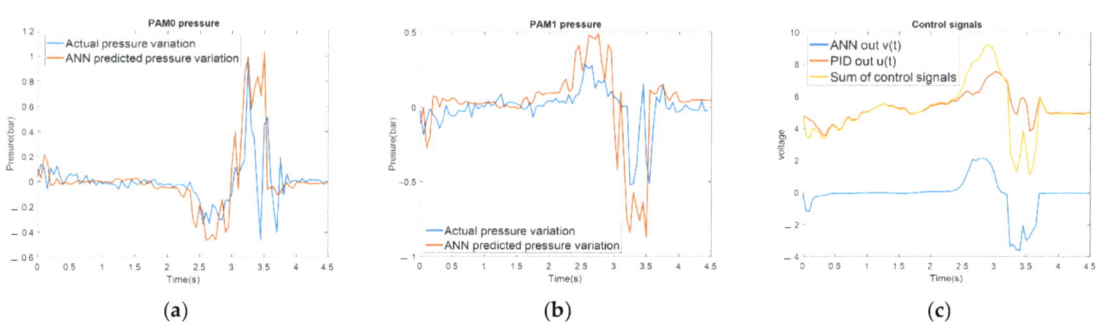

Figure 16. Controller signals. (**a**) PAM0 air pressure actual value and ANN predicted value; (**b**) PAM1 air pressure actual value and ANN predicted value; (**c**) control signal of feedforward ANN (IV) with PID.

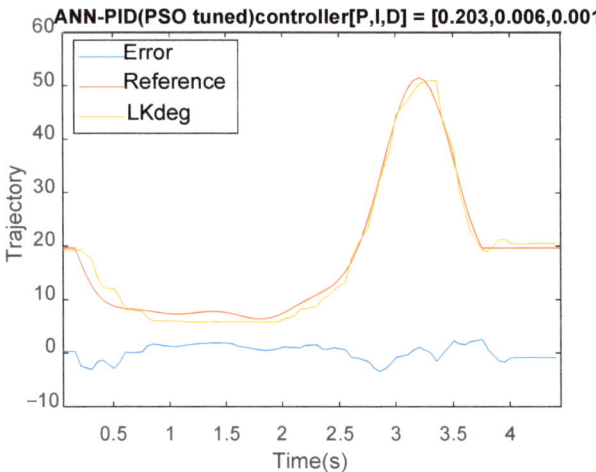

Figure 17. Feedforward ANN controller with PID system tracking results.

4.2. PSO Tuned PID with ANN Feedforward Control

After the compensation of the ANN feedforward control, the particle swarm optimization (PSO) is used to adjust the parameters of the PID controller. Since the ANN feed-forward has been trained in advance, the next step is to adjust the PID parameters to fit the ANN feedforward controller. Equations (9) and (10) are the calculation methods of the objective function, which are the minimum mean absolute error (MAE) and maximum absolute error (MAXE), respectively, where e is the tracking error of each gait cycle. To minimize MAE and MAXE, the initial individual generation uses the PID parameters obtained by the Z-N method as the initial values (P: 0.203, I: 0.006, D: 0.001); then, the upper limit of the initial population range is set as 0.8~1.2 times the original value. The objective function is set as the sum of 0.7 times MAE and 0.3 times MAXE, as shown in Equation (11). The tracking errors are calculated for each gait cycle and each group of the PID parameters is evaluated for each cycle. The PSO parameter adjustment of the PID parameters is used to test the real system for evaluation. In the PSO method, the population size (popsize) is set to 5 and 20 iterations are performed to find the best parameters. The suitable parameter of interval threshold is set as the 0.2 times of the current best parameter gbest. The update flow chart of PSO is shown in Figure 18 and the overall control flow chart is shown in Figure 19.

$$MAE = \frac{1}{n}\sum_{j=1}^{n}|f_j - y_j| = \frac{1}{n}\sum_{j=1}^{n} e_j \tag{9}$$

$$MAXE = \max_{j=1}^{n}\{|f_j - y_j|\} = \max_{j=1}^{n}\{e_j\} \tag{10}$$

$$\min F = 0.7 MAE + 0.3 MAXE \tag{11}$$

$$\Delta P_i^{new}(k+1) = w \cdot \Delta P_i(k) + c_1 \cdot r_1 \cdot (pbest_i^P - P_i(k)) + c_2 \cdot r_2 \cdot (gbest^P - P_i(k)) \tag{12}$$

$$\Delta I_i^{new}(k+1) = w \cdot \Delta I_i(k) + c_1 \cdot r_1 \cdot (pbest_i^I - I_i(k)) + c_2 \cdot r_2 \cdot (gbest^I - I_i(k)) \tag{13}$$

$$\Delta D_i^{new}(k+1) = w \cdot \Delta D_i(k) + c_1 \cdot r_1 \cdot (pbest_i^D - D_i(k)) + c_2 \cdot r_2 \cdot (gbest^D - D_i(k)) \tag{14}$$

$$P_i^{new}(k+1) = P_i(k) + \Delta P_i^{new}(k+1) \tag{15}$$

$$I_i^{new}(k+1) = I_i(k) + \Delta I_i^{new}(k+1) \tag{16}$$

$$D_i^{new}(k+1) = D_i(k) + \Delta D_i^{new}(k+1) \tag{17}$$

where Equations (12)–(17) are PSO update formulas for PID parameters; $P_i(k), I_i(k), D_i(k)$ is the position of the i-th particle and the individual in the k-th iteration, $\Delta P_i(k), \Delta I_i(k), \Delta D_i(k)$ is the velocity of the i-th particle and the individual in the k-th iteration; w is the inertia weight; r_1 and r_2 are two random numbers in the range of 0 to 1; c_1 and c_2 represent the confidence weight of the particle to itself and the group, generally set from 0 to 4; $pbest_i$ denotes the best position experienced so far by the i-th particle; $gbest$ denotes the best position experienced so far by the entire population.

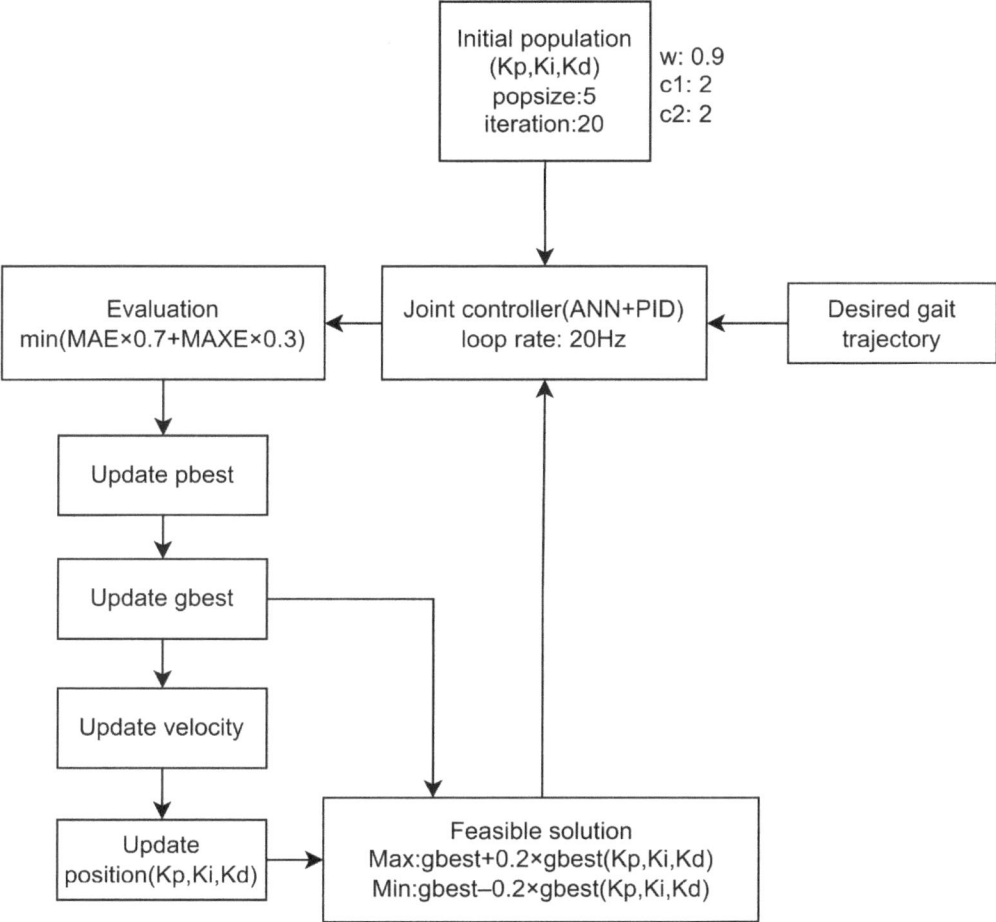

Figure 18. PSO tuned PID with ANN flow chart.

When the PSO controller iterates for 20 times, the optimal objective function changes as shown in Figure 20. Figure 21 shows the response with the PID optimization adjustment, after the PSO optimization is performed. With the comparison to Figure 17 (at 1 s), the controller after the PSO adjustment has a better performance than the original in Section 4.1. After the PSO adjustment of the parameters, the MAXE has changed from 4.4 to 3.9 with some overshoot at 3.7 s. Figure 22a shows the difference of the time response for the controllers of PID, ANN + PID, and ANN + PID (PSO adjustment). After adjusting PID parameters, the tracking error is better than the original and the overall MAE decreases as shown in Figure 22b, especially around 1 s. The control signal given by the ANN compensation with the PID can reduce the error very well.

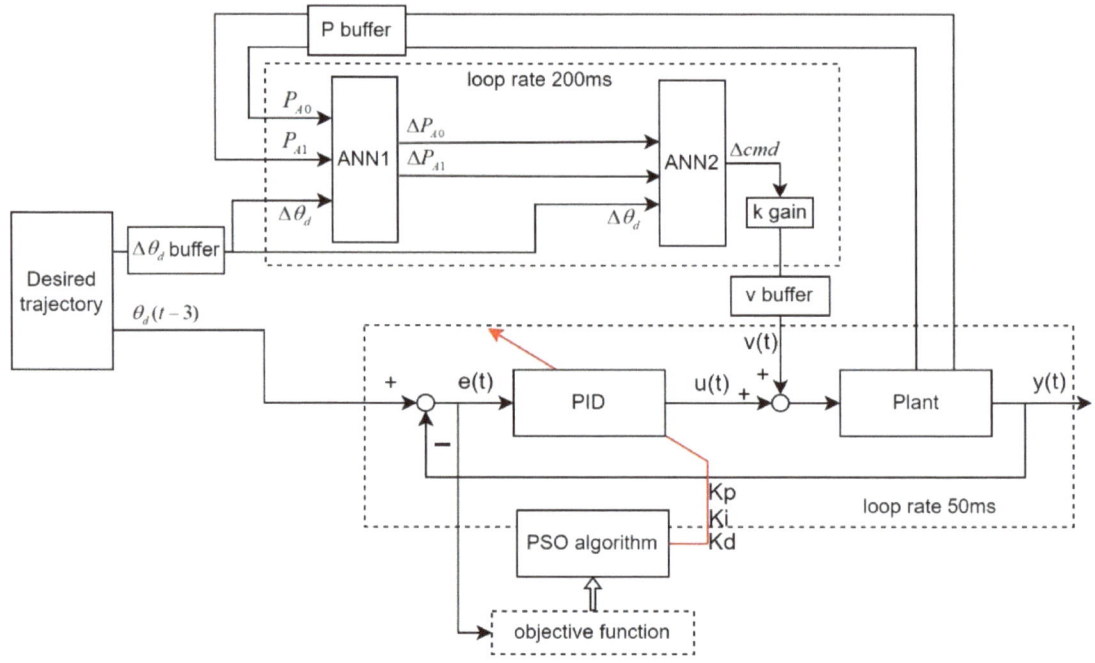

Figure 19. PSO tuned PID with ANN controller.

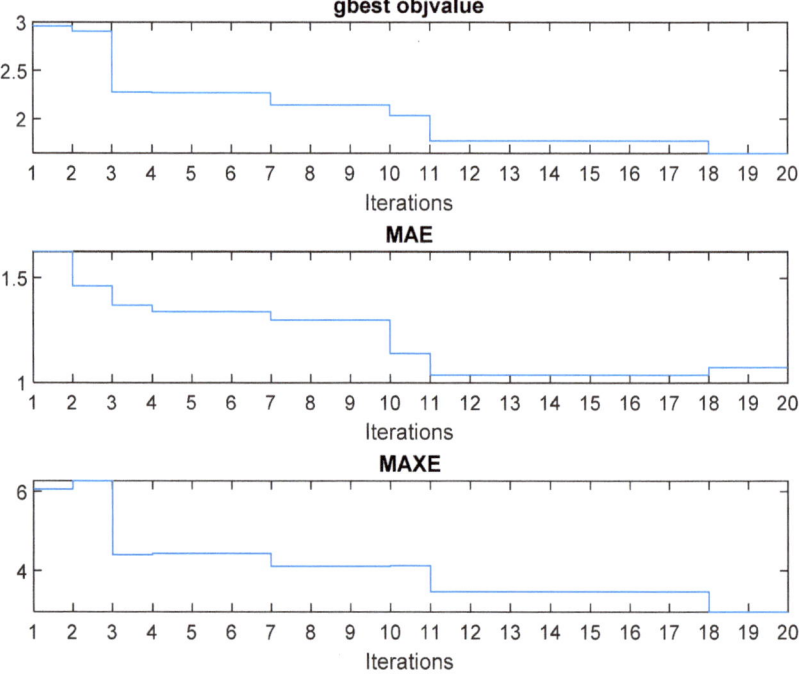

Figure 20. PSO iterative parameters.

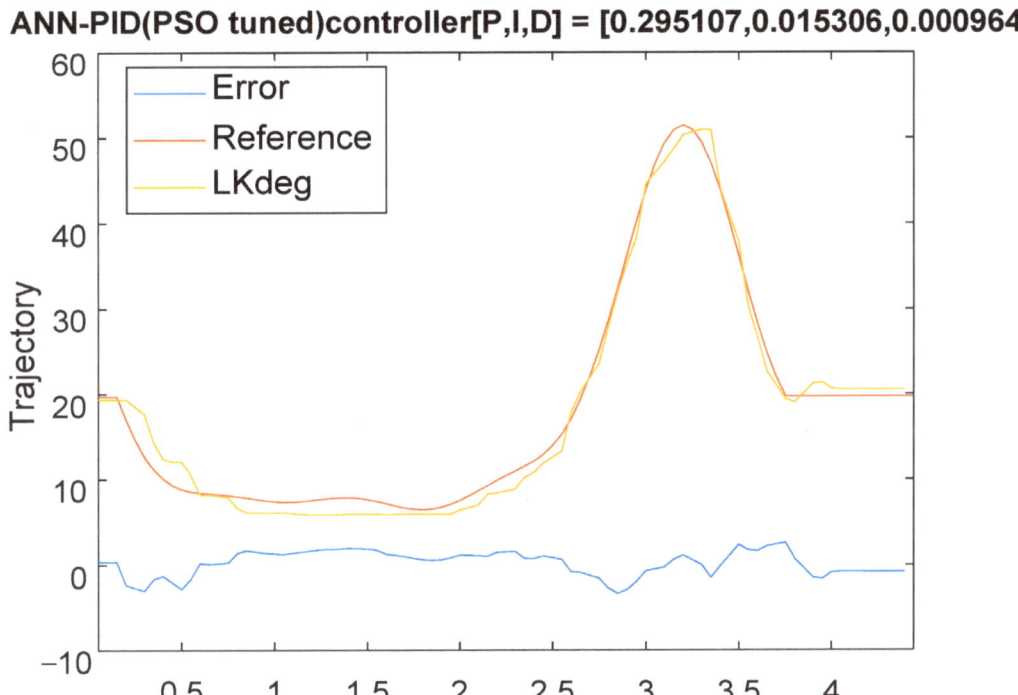

Figure 21. PSO tuned PID with ANN feedforward control results.

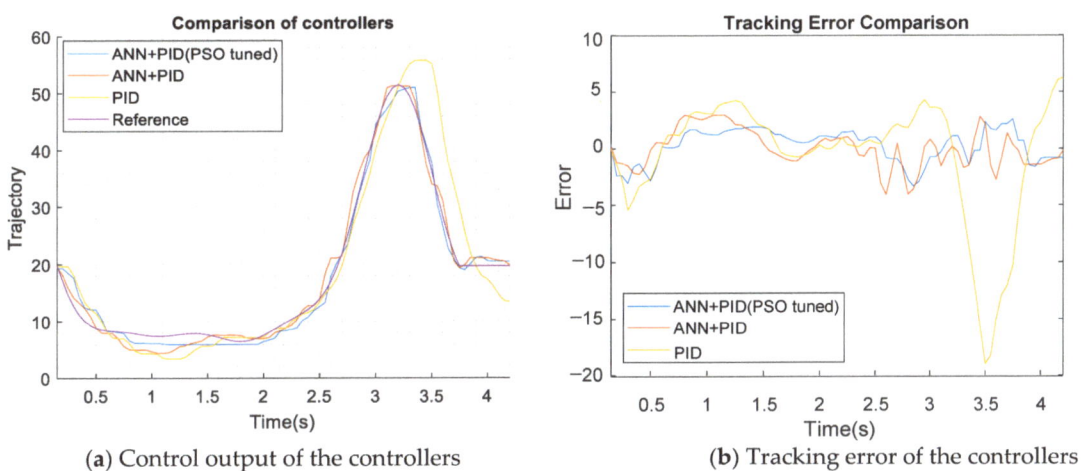

(a) Control output of the controllers (b) Tracking error of the controllers

Figure 22. PSO tuned controller performance comparison.

5. Experiment and Discussion

Previously we discussed three improvements to the knee controller. First, the ILC control architecture is used with the PID control error as the feed-forward update error, expecting to get a good control effect. Secondly, the air pressure and the angle data of different control quantities are collected and the measured data are used to train the inverse model. The trained control structure (shown in Figure 15) has compensated the tracking error as shown in Figure 17. The third method is to use the PSO to search the optimal parameters of the PID; the architecture diagram is shown in Figure 19. It can be seen that the control signal given by the ANN compensation with the PID control can reduce the error very well.

5.1. Knee Joint Controller Performance Comparison

Comparing the effects of the different knee joint controllers, Tables 5 and 6 are the comparative data of the left and right knee joints under the rehabilitation speed of 1 (km/h). According to the experimental results, the PID controller has the worst control response; the feedforward ANN with the PID controller has a better performance than the PID controller; the feedforward ANN with PID (PSO tuned) controller is the best among the three controllers. To test the performance with the subjects, the walking rehabilitation (1 km/h) is performed by the subject. The 1 km/h walking speed is converted into a knee joint cycle time of 3.6 s per cycle, which is relatively fast in the PAM control.

In this experiment, the controller structure is the same as in Figure 15. After the controllers of all joints are integrated into the same program, the operation time of the ANN block is increased from the previous 200 ms to 350 ms due to the computer performance. The buffer size is adjusted from the previous 4 to 7 (350 ms/50 ms) to keep up with the speed of the control loop (50 ms). The controller adjusts the parameters suitable for the current ANN model through PSO and then fixes the optimal parameters. The parameters of the left knee are (P:0.295107, I:0.015306, D:0.000964) and the ones of the right knee are (P: 0.465371, I: 0.017837, D: 0.000236). The control frequency is 20 Hz (sampling time 50 ms) and Figure 23a,b are the experimental result of left knee and the right knee for the PSO tuned PID with ANN feedforward controller. From the experimental results, the on-load tracking error for the proposed controller is still good. In Tables 5 and 6, the MAXE of PSO tuned PID with ANN feedforward is about 3.2 to 6.6 degrees and the MAE is lower than 2 degrees. It can be seen that this control architecture is robust for the subject interference with the system.

Table 5. Comparison table of tracking outcomes of different controller (left knee).

LK	PID		ANN (Trained IV) + PID		ANN (Trained IV) + PID (PSO Tuned)		ANN(Trained IV) + PID (PSO Tuned) with Load	
Test NO.	MAE	MAXE	MAE	MAXE	MAE	MAXE	MAE	MAXE
1	3.091	18.381	1.425	5.273	1.226	3.680	1.870	5.336
2	3.665	19.497	1.480	6.481	1.214	3.976	1.575	3.524
3	3.388	19.282	1.199	4.426	1.195	4.275	1.608	3.849
4	3.325	18.329	1.257	4.099	1.237	3.357	1.333	3.174
5	3.590	18.961	1.217	4.728	1.181	3.933	1.901	5.348

Table 6. Comparison table of tracking outcomes of different controller (right knee).

RK	PID		ANN (Trained IV) + PID		ANN (Trained IV) + PID(PSO Tuned)		ANN (Trained IV) + PID(PSO Tuned) with Load	
Test NO.	MAE	MAXE	MAE	MAXE	MAE	MAXE	MAE	MAXE
1	3.190	16.310	1.334	4.972	1.172	5.205	1.361	6.154
2	3.897	16.228	1.666	5.082	1.190	4.122	1.427	6.618
3	4.018	16.550	1.258	4.611	1.361	3.462	1.293	3.863
4	3.580	16.309	1.295	5.007	1.077	3.512	1.530	5.752
5	3.997	16.444	1.955	5.840	1.189	3.528	1.350	5.990

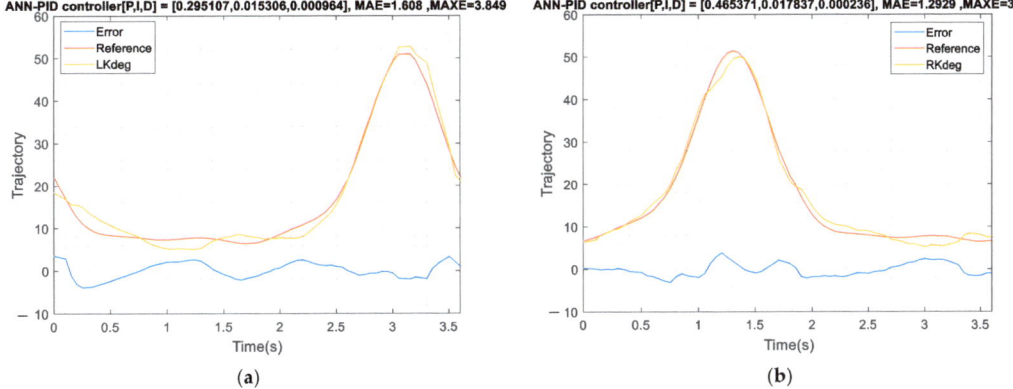

Figure 23. PID (PSO tuned)-ANN feedforward control loaded response. (**a**) Tracking result of left knee; (**b**) Tracking result of right knee.

5.2. Multi-Subject LLRER Load Experiment

In order to verify the practicability of the proposed PID (PSO tuned)-ANN feedforward controller for the knee joints, an experiment is designed with 10 subjects using the proposed rehabilitation system. In these experiments, the hip joint uses the ILC controller proposed in Section 3.2 and the knee joint uses the PID (PSO tuned)-ANN feedforward controller proposed in Section 5.1. The unique gait models are obtained with an unpowered exoskeleton system as shown in Figure 10 and then the users wear the proposed LLRER for testing. The experimental treadmill speed is set as 1 km/h and the time for one gait cycle is 3.6 s. Both MAE and MAXE are calculated in each gait cycle and the experimental data of the subjects P1 and P2 are shown in Figures 24 and 25.

From the system response of Figures 24 and 25, if the tracking model is replaced with an individual's unique gait, the control strategy proposed can still maintain a good control response. Table 7 shows ten subjects' experimental results and the experimental results show that the controller performs well in the real experiments. The ILC results for the hips show the MAE is 0.915 degrees. In the knee joint experiments using the feedforward ANN with PID (PSO tuned) controller, the average MAE is about 1.66 degrees and the experimental results are also excellent. To indicate the generality of the feedforward controller architecture, the system response data for pre-training is sufficient. The experimental results show that the concept of ANN prediction for this LLRER system is feasible.

Table 7. Rehabilitation controller performance data for 10 subjects.

Loaded Test	Treadmill Speed (1 km/h)							
	Left_Hip		Right_Hip		Left_Knee		Right_Knee	
Controller	PID + ILC		PID + ILC		PID (PSO Tuned) +ANN		PID (PSO Tuned) +ANN	
Test NO.	MAE	MAXE	MAE	MAXE	MAE	MAXE	MAE	MAXE
P1	0.782	2.135	0.797	2.097	1.989	6.939	1.951	4.665
P2	0.698	1.904	0.666	1.834	1.045	4.106	1.763	4.373
P3	1.145	3.741	1.125	3.235	1.427	4.067	2.580	6.405
P4	1.317	3.429	1.307	3.058	1.586	3.867	1.773	6.671
P5	0.351	1.390	0.350	1.407	1.970	6.615	1.106	5.544
P6	0.967	2.996	0.976	2.320	1.302	2.812	0.981	3.284
P7	0.987	3.316	0.813	3.006	2.058	5.191	1.367	4.046
P8	0.778	2.361	0.715	2.250	1.798	4.409	1.465	4.188
P9	1.299	2.777	1.315	2.800	1.615	7.935	1.299	6.226
P10	0.825	1.953	0.827	1.949	1.829	6.460	1.950	5.665
avg	0.915	2.600	0.889	2.396	1.662	5.240	1.623	5.107

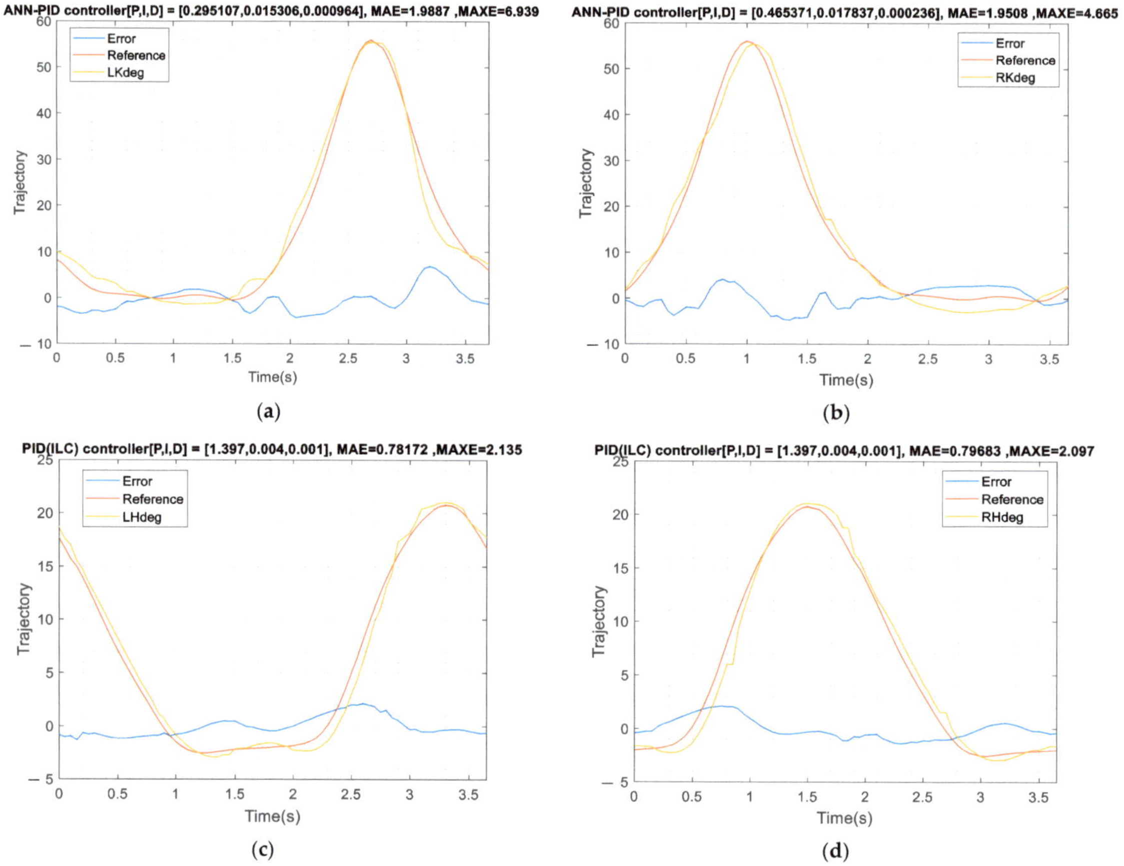

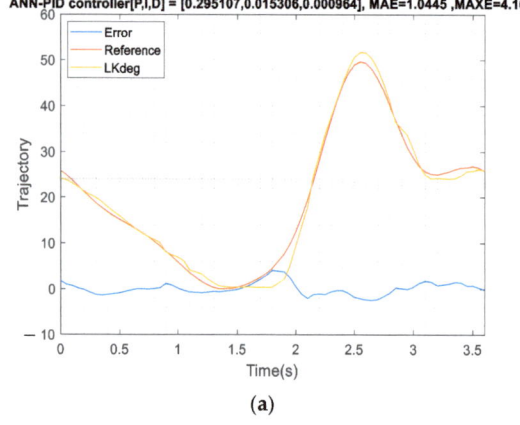

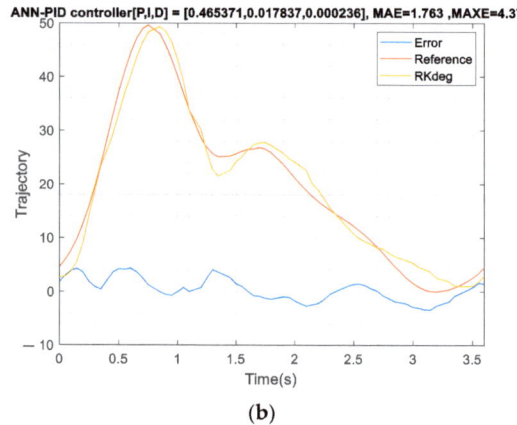

Figure 24. Subject P1 data. (**a**) Left knee tracking results; (**b**) right knee tracking results; (**c**) left hip tracking results; (**d**) right hip tracking results.

Figure 25. *Cont.*

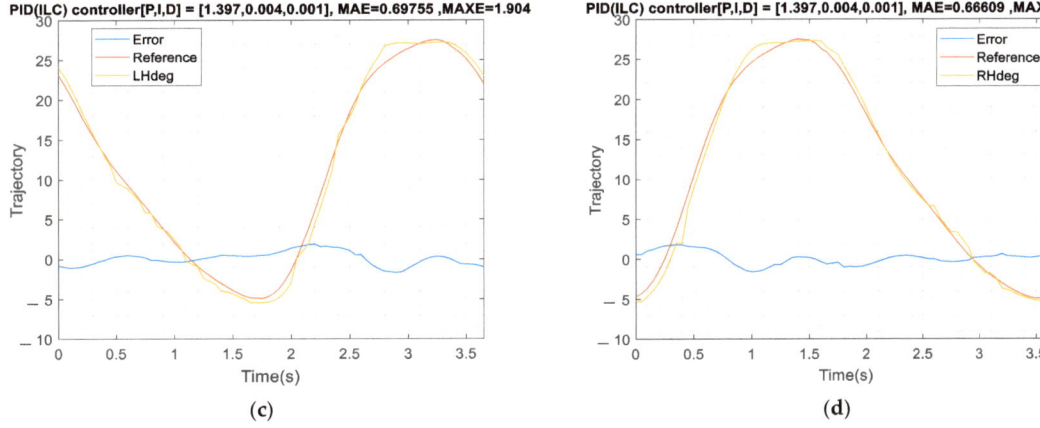

Figure 25. Subject P2 data. (**a**) Left knee tracking results; (**b**) right knee tracking results; (**c**) left hip tracking results; (**d**) right hip tracking results.

6. Conclusions

In this study, the data collection method of feed-forward ANN is simple. When the system is adjusted due to individual differences of rehabilitation patients, the knee joint only needs to swing back and forth at different speeds to complete the data collection of the new system parameters. There is potential for rapid adaptation in applications. Secondly, the queue method is used to compensate the PSO-PID controller, so that the ANN does not need to update at the same frequency as the PSO-PID, providing a new option for future controller integration. In addition, in the field of lower limb rehabilitation, there are few experimental conditions like ours. Our rehabilitation speed is relatively fast in the application of PAM. The feed-forward ANN combined with PSO-PID can make the performance of the controller on the basis of the traditional Z-N method, and it is optimized to effectively solve the well-known PAM hysteresis problem.

The lower extremity rehabilitation system provides good rehabilitation quality. A DC motor with a reducer for the hip joint and a PAMs-driven bidirectional (antagonistic) actuation for the knee joint are used for the rehabilitation task. First, the ILC algorithm based on the PID controller is used with the feedforward concept and the actual measurement shows that the DC motor of the hip mechanism works well and can provide good rehabilitation (average MAE 0.915 and 0.889 degrees); however, there are nonlinear characteristics for the knee joints and the tracking error is not good enough. Second, to compensate the tracking error of the knee joints, the feedforward concept was used to measure the actual system and the dynamic model was measured by the ANN feedforward control. The air pressure and the angle data of different control quantities are collected and the measured data are used to train the inverse model. The PID controller with the ANN feedforward shows that its response is much better than that of PID. The trained control structure has compensated the tracking error. Third, the PSO is used to search the optimal parameters of the PID and the architecture diagram. It can be seen that the control signal given by the ANN compensation with the PID control can reduce the error very well. The results with the inverse model can be trained with the experimental data without any mathematical modeling. Its versatility for different walking gaits can also be verified during human testing (average MAE 1.66 and 1.623 degrees).

Author Contributions: Conceptualization, C.-J.L.; methodology, C.-J.L.; software, T.-Y.S.; validation, C.-J.L. and T.-Y.S.; formal analysis, C.-J.L.; data curation, C.-J.L.; writing—original draft preparation, T.-Y.S.; writing—review and editing, C.-J.L. and T.-Y.S.; visualization, T.-Y.S. and C.-J.L.; supervision, C.-J.L.; project administration, C.-J.L.; funding acquisition, C.-J.L. All authors have read and agreed to the published version of the manuscript.

Funding: This research was funded by National Science Council of the Republic of China, grant number Contract No. MOST 108-2221-E-027-112-MY3. The APC was funded by Contract No. MOST 108-2221-E-027-112-MY3.

Informed Consent Statement: Informed consent was obtained from all subjects involved in the study.

Conflicts of Interest: The authors declare no conflict of interest.

References

1. Díaz, I.; Gil, J.J.; Sánchez, E. Lower-limb robotic rehabilitation: Literature review and challenges. *J. Robot.* **2011**, *2011*, 759764. [CrossRef]
2. Wernig, A.; Muller, S.; Nanassy, A.; Cagol, E. Laufband Therapy Based on 'Rules of Spinal Locomotion' is Effective in Spinal Cord Injured Persons. *Eur. J. Neurosci.* **1995**, *7*, 823–829. [CrossRef] [PubMed]
3. Colombo, G.; Joerg, M.; Schreier, R.; Dietz, V. Treadmill training of paraplegic patients using a robotic orthosis. *J. Rehabil. Res. Dev.* **2000**, *37*, 693–700. [PubMed]
4. Freivogel, S.; Mehrholz, J.; Husak-Sotomayor, T.; Schmalohr, D. Gait training with the newly developed "LokoHelp"—System is feasible for non-ambulatory patients after stroke, spinal cord and brain injury. A feasibility study. *Brain Inj.* **2008**, *22*, 625–632. [CrossRef]
5. West, G.R. Powered Gait Orthosis and Method of Utilizing Same. Patent number 6,689,075, 10 February 2004.
6. Beyl, P.; Van Damme, M.; Van Ham, R.; Vanderborght, B.; Lefeber, D. Pleated Pneumatic Artificial Muscle-Based Actuator System as a Torque Source for Compliant Lower Limb Exoskeletons. *IEEE/ASME Trans. Mechatron.* **2014**, *19*, 1046–1056. [CrossRef]
7. Yang, H.; Xiang, C.; Han, H.; Hao, L. Inverse kinematics modeling and motion control of PAM bionic elbow joint. In Proceedings of the 2015 IEEE International Conference on Robotics and Biomimetics (ROBIO), Zhuhai, China, 6–9 December 2015; pp. 1347–1352. [CrossRef]
8. Tondu, B. Modelling of the McKibben artificial muscle: A review. *J. Intell. Mater. Syst. Struct.* **2012**, *23*, 225–253. [CrossRef]
9. Carbonell, P.; Jiang, Z.P.; Repperger, D.W. Nonlinear control of a pneumatic muscle actuator: Backstepping vs. sliding-mode. In Proceedings of the 2001 IEEE International Conference on Control Applications, Mexico City, Mexico, 7 September 2001; pp. 167–172. [CrossRef]
10. Kawato, M. Feedback-error-learning neural network for supervised motor learning. *Adv. Neural Comput.* **1990**, 365–372.
11. Nakanishi, J.; Schaal, S. Feedback error learning and nonlinear adaptive control. *Neural Networks* **2004**, *17*, 1453–1465. [CrossRef]
12. Robinson, R.M.; Kothera, C.S.; Sanner, R.M.; Wereley, N.M. Nonlinear Control of Robotic Manipulators Driven by Pneumatic Artificial Muscles. *IEEE/ASME Trans. Mechatronics* **2015**, *21*, 55–68. [CrossRef]
13. Nho, H.; Meckl, P. Intelligent feedforward control and payload estimation for a two-link robotic manipulator. *IEEE/ASME Trans. Mechatronics* **2003**, *8*, 277–283. [CrossRef]
14. Liu, D.; Chen, W.; Pei, Z.; Wang, J. A brain-controlled lower-limb exoskeleton for human gait training. *Rev. Sci. Instrum.* **2017**, *88*, 104302. [CrossRef]
15. Luu, T.P.; Nakagome, S.; He, Y.; Contreras-Vidal, J.L. Real-time EEG-based brain-computer interface to a virtual avatar enhances cortical involvement in human treadmill walking. *Sci. Rep.* **2017**, *7*, 1012. [CrossRef]
16. Formaggio, E.; Massiero, S.; Bosco, A.; Izzi, F.; Piccion, F.; Del Felice, A. Quantitative EEG Evaluation during Robot-Assisted Foot Movement. *IEEE Trans. Neural Syst. Rehabil. Eng.* **2017**, *25*, 1633–1640. [CrossRef]
17. Qiu, S.; Pei, Z.; Wang, C.; Tang, Z. Systematic Review on Wearable Lower Extremity Robotic Exoskeletons for Assisted Locomotion. *J. Bionic Eng.* **2022**. [CrossRef]
18. Bogue, R. Robotic exoskeletons: A review of recent progress. *Ind. Robot.-Int. J. Robot. Res. Appl.* **2015**, *42*, 5–10. [CrossRef]
19. Kim, S.; Srinivasan, D.; Nussbaum, M.A.; Leonessa, A. Human gait during level walking with an occupational whole-body powered exoskeleton: Not yet a walk in the park. *IEEE Access* **2021**, *9*, 47901–47911. [CrossRef]
20. Kazerooni, H.; Amundson, K.; Harding, N. Device and Method for Decreasing Energy Consumption of a Person by Use of a Lower Extremity Exoskeleton. Application Publication. US Patent EP2326288A1, 26 January 2015.
21. Suzuki, K.; Mito, G.; Kawamoto, H.; Hasegawa, Y.; Sankai, Y. Intention-based walking support for paraplegia patients with robot suit HAL. *Adv. Robot.* **2007**, *21*, 1441–1469. [CrossRef]
22. Tsukahara, A.; Hasegawa, Y.; Sankai, Y. Standing-up motion support for paraplegic patient with robot suit HAL. In Proceedings of the 2009 IEEE International Conference on Rehabilitation Robotics, Kyoto, Japan, 23–26 June 2009; pp. 211–217. [CrossRef]
23. Tsukahara, A.; Kawanishi, R.; Hasegawa, Y.; Sankai, Y. Sit-to-stand and stand-to-sit transfer support for complete paraplegic patients with robot suit HAL. *Adv. Robot.* **2010**, *24*, 1615–1638. [CrossRef]
24. Kawamoto, H.; Taal, S.; Niniss, H.; Hayashi, T.; Kamibayashi, K.; Eguchi, K.; Sankai, Y. Voluntary motion support control of robot suit HAL triggered by bioelectrical signal for hemiplegia. In Proceedings of the 2010 Annual International Conference of the IEEE Engineering in Medicine and Biology 2010, Buenos Aires, Argentina, 31 August–4 September 2010; pp. 462–466. [CrossRef]
25. Hassan, M.; Kadone, H.; Suzuki, K.; Sanka, Y. Wearable gait measurement system with an instrumented cane for exoskeleton control. *Sensors* **2014**, *14*, 1705–1722. [CrossRef]
26. Nilsson, A.; Vreede, K.S.; Haglund, V.; Kawawamoto, H.; Sankai, Y.; Borg, J. Gait training early after stroke with a new exoskeleton-the hybrid assistive limb: A study of safety and feasibility. *J. Neuro-Eng. Rehabil.* **2014**, *11*, 92. [CrossRef]

27. Tsukahara, A.; Hasegawa, Y.; Eguchi, K.; Sankai, Y. Restoration of gait for spinal cord injury patients using HAL with intention estimator for preferable swing speed. *IEEE Trans. Neural Syst. Rehabil. Eng.* **2014**, *23*, 308–318. [CrossRef] [PubMed]
28. Hassan, M.; Kadone, H.; Ueno, T.; Hada, Y.; Sankai, Y.; Suzuki, K. Feasibility of synergy-based exoskeleton robot control in hemiplegia. *IEEE Trans. Neural Syst. Rehabil. Eng.* **2018**, *26*, 1233–1242. [CrossRef] [PubMed]
29. Zeilig, G.; Weingarden, H.; Zwecker, M.; Dudkiewicz, I.; Bloch, A.; Esquenazi, A. Safety and tolerance of the ReWalk™ exoskeleton suit for ambulation by people with complete spinal cord injury: A pilot study. *J. Spinal Cord Med.* **2012**, *35*, 96–101. [CrossRef] [PubMed]
30. Shi, D.; Zhang, W.; Zhang, W.; Ding, X. A Review on Lower Limb Rehabilitation Exoskeleton Robots. *Chin. J. Mech. Eng.* **2019**, *32*, 74. [CrossRef]
31. Suszyński, M.; Peta, K. Assembly Sequence Planning Using Artificial Neural Networks for Mechanical Parts Based on Selected Criteria. *Appl. Sci.* **2021**, *11*, 10414. [CrossRef]
32. Wang, Y.; Li, Y.; Song, Y.; Rong, X. The Influence of the Activation Function in a Convolution Neural Network Model of Facial Expression Recognition. *Appl. Sci.* **2020**, *10*, 1897. [CrossRef]
33. Suszyński, M.; Meller, A.; Peta, K.; Trączyński, M.; Butlewski, M.; Klimenda, F. Application of Neural Networks for Water Meter Body Assembly Process Optimization. *Appl. Sci.* **2022**, *12*, 11160. [CrossRef]

Disclaimer/Publisher's Note: The statements, opinions and data contained in all publications are solely those of the individual author(s) and contributor(s) and not of MDPI and/or the editor(s). MDPI and/or the editor(s) disclaim responsibility for any injury to people or property resulting from any ideas, methods, instructions or products referred to in the content.

Article

Automatic Calibration of Tool Center Point for Six Degree of Freedom Robot

Chih-Jer Lin [1], Hsing-Cheng Wang [1] and Cheng-Chi Wang [2,*]

[1] Graduate Institute of Automation Technology, National Taipei University of Technology, New Taipei 106, Taiwan
[2] Department of Intelligent Automation Engineering, Graduate Institute of Precision Manufacturing, National Chin-Yi University of Technology, Taichung 411, Taiwan
* Correspondence: wcc@ncut.edu.tw

Abstract: The traditional tool center point (TCP) calibration method requires the operator to use their experience to set the actual position of the tool center point. To address this lengthy workflow and low accuracy, while improving accuracy and efficiency for time-saving and non-contact calibration, this paper proposes an enhanced automatic TCP calibration method based on a laser displacement sensor and implemented on a cooperative robot with six degrees of freedom. During the calibration process, the robot arm will move a certain distance along the X and Y axes and collect the information when the tool passes through the laser during the process to calculate the runout of the tool, and then continue to move a certain distance along the X and Y axes for the second height calibration. After the runout angle is calculated and calibrated by triangulation, the runout calibration is completed and the third X and Y axis displacement is performed to find out the exact position of the tool on the X and Y axes. Finally, the tool is moved to a position higher than the laser, and the laser is triggered by moving downward to obtain information to complete the whole experimental process and receive the calibrated tool center position. The whole calibration method is, firstly, verified in the virtual simulation environment and then implemented on the actual cooperative robot. The results of the proposed TCP calibration method for the case of using a pin tool can achieve a positioning deviation of 0.074 and 0.125 mm for the robot moving speeds of 20 and 40 mm/s, respectively. The orientation deviation in the x-axis are 0.089 and −0.184 degrees for the robot moving speeds of 20 and 40 mm/s, respectively. The positioning repeatability of ±0.083 mm for the moving speed of 20 mm/s is lower than ±0.101 mm for the speed of 40 mm/s. It shows that lower moving speed can obtain higher accuracy and better repeatability. This result meets the requirements of TCP calibration but also achieves the purpose of being simple, economical, and time-saving, and it takes only 60 s to complete the whole calibration process.

Keywords: six-axis manipulator; tool center point; calibration; laser displacement sensor

Citation: Lin, C.-J.; Wang, H.-C.; Wang, C.-C. Automatic Calibration of Tool Center Point for Six Degree of Freedom Robot. *Actuators* 2023, 12, 107. https://doi.org/10.3390/act12030107

Academic Editor: Matteo Cianchetti

Received: 17 January 2023
Revised: 23 February 2023
Accepted: 23 February 2023
Published: 27 February 2023

Copyright: © 2023 by the authors. Licensee MDPI, Basel, Switzerland. This article is an open access article distributed under the terms and conditions of the Creative Commons Attribution (CC BY) license (https://creativecommons.org/licenses/by/4.0/).

1. Introduction

Robotic manipulators are widely used in industrial manufacturing, while the use of collaborative arms, in addition to industrial arms, is also increasing year by year. To realize Industry 4.0, automation has become an important indicator for factory transformation and the high accuracy manufacturing procedures are even more important [1–4]. According to the International Federation of Robotics [5], the number of manipulators operating in the factories is increasing year on year, moreover, the annual installations are increasing too. It can also be understood that the role of manipulators in the industry is becoming increasingly significant. With long operating hours, high repeatability precision, and low error rate, it has certainly improved the automatic production capacity and flexibility [6]; furthermore, lowering the production duty and equipment budget. In a nutshell, to maximize the advantage of the manipulator, increasing the precision is the key point. A robot arm is most often defined as a set of rigid linkage mechanisms connected by joints. One side is attached to an external rigid surface, called the base, and the other side can be fitted with various tools, called flanges. The end effector is based on the robot's operating position, such as the center point of the vacuum suction product or where the tip of the robot torch is actually welded, which is called the operating point, also called the tool center point (TCP) [7–9]. The traditional way of TCP calibration is mostly performed by the operator, who needs to move the robot arm to reach the actual position of the tool center point to the reference station and check its accuracy by eye, and this process is repeated six to twelve times to retrieve the most accurate TCP. In recent years, many researchers proposed different methods for TCP calibration to improve the accuracy of the robot arm. Bergström [10] provided a standard idea of using a spherical probe tool and a calibration cup to calculate the TCP. In order to define the actual tool center point, he used a soft servo to move the spherical probe tool into the calibration cup, which prevents the robot system from triggering a collision alarm, while allowing the soft servo to deactivate the proportional part of the PID position control, allowing deviations from the defined program trajectory. After loading the spherical probe tool into the calibration fixture, he then reoriented the spherical probe and recorded it at least four times. After repeating this procedure several times in different directions of the tool, the final TCP would be defined. This method is one of the contact calibration methods but it cannot be implemented on any type of machining tool and is inflexible. Guo et. al. [11] proposed a constraint method for the posture of an irregular-shaped tool in this scheme. Theoretical foundations for the four-posture calibration method of the irregular-shaped tool for dual-robot-assisted ultrasonic non-destructive testing (NDT) were presented in detail. This strategy has been successfully applied in the NDT experiment of semi-enclosed composite workpieces. Experimental results show that: the calibration method can be used to obtain the correct TCP position efficiently; the TCP orientation constraint rule can ensure the extension pole of the irregular-shaped ultrasonic probe is parallel to the axis of the semi-enclosed cylindrical workpieces; and the ultrasonic transducer axis is perpendicular to the surface of the workpiece. Fares et al. [12] studied to maximize the variance of the robot's TCP value obtained by the four-point method by using the industrial robot in a set of n points generated by a random distribution and using this set of data as the input data for the sphere fitting algorithm developed in MATLAB. Moreover, the accuracy and stability of the proposed method were subsequently validated against experimental results.

Machine vision is becoming increasingly important in scientific, industrial, smart manufacturing, and medical applications due to the tremendous development of PC-based languages, vision technologies, and algorithms. Erick et. al. [13] proposed a novel calibration system that uses position sensitive calibration, position sensitive detector, and camera and laser fixtures to calibrate the TCP. In order to calibrate the TCP, the laser pointer and the TCP must be located at several positions set by the user. Borrmann et al. [14] proposed a laser tracker-based calibration method for TCP. They designed a system using a laser tracker and two tool balls that can reflect the laser beam and installed this measurement tool on the TCP. The actual TCP can be obtained by rotating the robot arm, recording the information of the tool ball with the laser tracker, and finally calculating the homogeneous transformation matrix. The advantages of using laser trackers to measure the TCP are high accuracy and operator independence but the disadvantages are that they require additional equipment, are expensive, and require special environmental conditions. Zhang et al. [15] analyzed to solve the problems of poor accuracy stability and strong operational dependence in traditional TCP calibration methods and proposed a TCP calibration method for robot-assisted puncture surgery. It is more suitable and helpful for a physician. This paper designs a special binocular vision system and proposes a vision-based TCP calibration algorithm that simultaneously identifies the tool center point position (TCPP) and tool center point frame (TCPF). An accuracy test experiment proves that the designed special binocular system has a positioning accuracy of ± 0.05 mm. Comparison experiments show that the proposed TCP calibration method reduces the time consumption by 82%, improves the accuracy of TCPP by 65%, and improves the accuracy of TCPF by 52% compared to the traditional method. Liu et.al. [16] proposed a robot TCP automatic calibration algorithm based on binocular vision measurement. A target that can be recognized by the binocular vision sensor is attached to the robot TCP. The pose transformation between the vision sensor and the robot base is calculated by taking the binocular vision three-dimensional space measurement as the constraint and combining it with the multiple translational motions of the robot end tool. After several free rotations of the end tool of the robot, the TCP takes the measurement vector of the corresponding binocular vision sensor as the stroke to carry out the hypothetical parallel movement.

The main mission of the manipulator is to follow a specific trajectory and orientation use of the end effector to reach the target point. In the field of manipulator calibration, TCP calibration has rarely been studied. The purpose to calibrate TCP is to ensure precision and efficiency every time the tool changes automatically. However, inherent error often occurs when packaging or processing so the relation between the flange and TCP has to be calibrated. The traditional way to calibrate TCP is mainly by using a quick check for if the TCP is precisely targeting the reference point at the different postures and record these different postures by the manipulator controller. Yet, the process is time-consuming and labor-intensive; moreover, for non-specialized and experienced operators, the error will be magnified. Hereby, to retain high precision when the manipulator is processing, the correction of the TCP is the key factor, in addition, this research proposes a method that is based on the content mentioned above, focuses on automatic calibration technology by using a laser sensor to process runout, and offsets calibration after installing the tool; moreover, this method can achieve an easy, low time consuming, affordable price, and ultimately realize automatic operation.

2. Methodology

2.1. Design of the Experimental Structure

In this paper, TM5M-900 robot is used to verify the calibration experiments shown in Figure 1 [17,18]. This 6-axis robot is suitable for mobile assembly applications in the automated chemical and electronic industries. It is easy to program, highly customizable, has a radius range of 900 mm, and has a payload of 4 kg. The TCP, shown in Figure 1, is the working point used by the robot with a reference coordinate system attached to the robot's flange. Typically, the robot motion is programmed to define a path relative to the reference coordinate preset, which can be represented by various coordinate systems. In addition, a robot system can have multiple TCPs in it but it can only have one TCP active at a time. The robot base coordinates are attached to the robot base, and, in this study, the base coordinate system corresponds to the world coordinate system. The wrist frame is attached to the robot flange and the surface is on the robot's last axis and can mount tools. The center of the wrist frame is located in the flange center, the six axes of the robot are coinciding with the blue z-axis of the wrist frame, and the red axis and green axis represent the X and Y axes, respectively.

First, the z-axis of the tool coordinate systems is defined as the direction extending in the tool axis direction, and the TCP, which is also the origin of the tool coordinate system, is defined at this end effector (Figure 2a) [18].

From Figure 2, we can find that since the cylindrical tool is symmetrical in the X and Y axes, the rotation error along the z-axis is negligible. In this case, if the actual position and direction of the TCP, i.e., the coordinate system of the tool, is required, only the axial direction of the tool is required and then the direction of the coordinate system can be obtained by calculation. Once the direction of the coordinate system is obtained by calculation, the position of the coordinate system can be calculated by finding the end of the tool along the axial direction (Figure 2b).

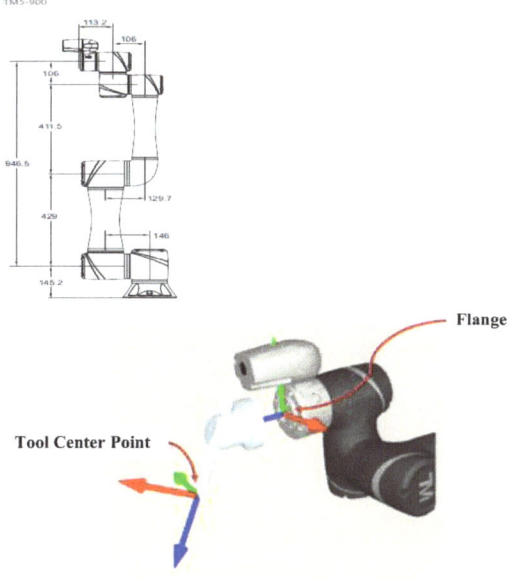
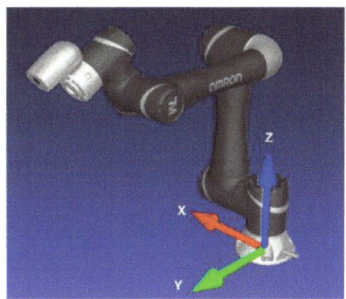

Figure 1. Collaborative robot TM5M-900.

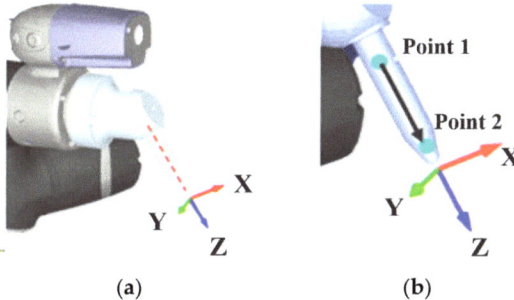

Figure 2. (**a**) Coordinate of TCP; (**b**) schematic diagram of the calculation tool axial vectors.

Secondly, we set up the laser sensor that operates by sending a Boolean value when the laser is detecting an object. The precise position of the TCP is unknown, and to ensure the TCP can be correctly detected by a laser sensor, use two laser sensors on the same plane, one in the X-axis direction and the other in the Y-axis direction, also, the calibration is executed according to this plane (Figure 3).

Figure 3. The laser sensor device.

2.2. Tool State Analysis and Error Modeling

Before deriving the various theoretical formulas, it is necessary to define the tool state before calibration. There are four types: Ideal case, offset, runout, and offset with runout (Figure 4a) [19].

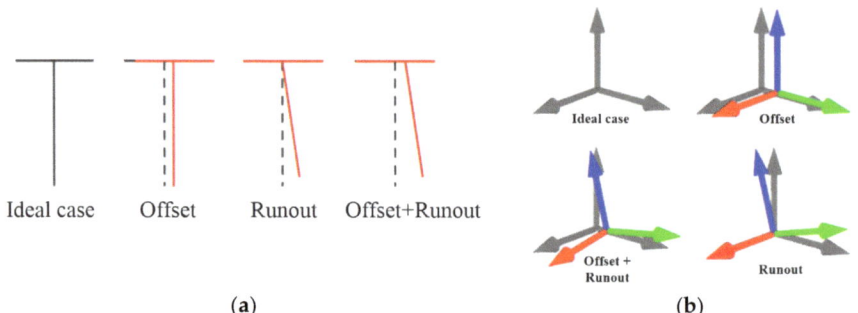

Figure 4. (a) Definition of four states of tool; (b) tool coordinate system.

In Figure 4, the horizontal line at the top of each state represents the flange surface, the solid black line below the horizontal line and the dark red solid line below represent the tools, and the black dashed line below each state is the ideal tool state for comparison. For the definition of each state, the tool offset represents the error of the tool's position relative to the ideal tool state, this means the tool has only the position displacement error concerning the ideal tool; the tool runout represents that the tool has only the rotation error concerning the ideal case. The tool offset and runout errors are the errors of tool position and rotation concerning the ideal case. However, in reality, when the tool is installed, the ideal tool coordinate system and the flange coordinate system usually have relative rotational deviation, so a more exact illustration of the tool condition is shown in Figure 4b.

After analyzing the possible states of the tool, the error modeling of the tool coordinate system can be performed using nonlinear equations [19], because in the kinematic model, the externally mounted tool can be considered as an extension of the robot arm; in addition, the orientation of the tool coordinate system in the robot arm base coordinate system can be expressed as a nonlinear function of the geometric parameters of the robot arm linkage, the geometric parameters of the tool, and the angular values of the joints, so the relationship between the ideal tool coordinate system and the ideal robot arm base coordinate system can be expressed as Equation (1),

$$P_{it} = f\left(\vec{q}, \vec{g_r}, \vec{g_t}\right) \quad (1)$$

where P_{it} represents the measured TCP posture under the ideal robot arm base coordinate system; the vector $\vec{q}$ represents the angle value of each joint of the robot arm; vector $\vec{g_r}$ represents the ideal linkage geometric parameter of the robot arm; and the vector $\vec{g_t}$ represents the ideal tool geometric parameter.

Practically, the robot model does not anticipate the exact position and orientation of the additional mounted tools. Therefore, the difference between the actual tool pose and the robot kinematic tool pose is the geometric error of the tool installation and the link geometry error of the robot, as shown in Equation (2).

$$P_{at} = f\left(\vec{q}, \vec{g_r} + \Delta\vec{g_r}, \vec{g_t} + \Delta\vec{g_t}\right) \quad (2)$$

where P_{at} represents the measured TCP posture under the actual robot arm base coordinate system; the vector $\Delta\vec{g_r}$ is the error of the geometric parameters between the ideal and the actual robot arm connecting linkage; and the vector $\Delta \vec{g_t}$ is the error of geometric parameters between the ideal tool and the actual tool.

From Equations (1) and (2), the errors of the actual robot arm and actual tool can be derived as Equations (3) and (4).

$$\Delta P_t = P_{at} - P_{it} \tag{3}$$

$$\Delta P_t = f\left(\vec{q}, \vec{g_r} + \Delta \vec{g_r}, \vec{g_t} + \Delta \vec{g_t}\right) - f\left(\vec{q}, \vec{g_r}, \vec{g_t}\right) \tag{4}$$

where ΔP_t represents the tool coordinate and the ideal tool coordinate posture error after the actual tool linked by actual robot.

The purpose of this study is to discuss TCP calibration, so it is assumed that the robotic arm has already completed its native calibration, and the main calibration error model is shown in Equation (5).

$$\Delta P_{tool} = f\left(\vec{q}, \vec{g_{ra}}, \vec{g_t} + \Delta \vec{g_t}\right) - f\left(\vec{q}, \vec{g_{ra}}, \vec{g_t}\right) \tag{5}$$

where ΔP_{tool} represents the posture error model of the actual tool coordinate system concerning the ideal tool coordinate system after the robot has been calibrated; and the vector $\vec{g_{ra}}$ represents the linkage geometric parameters after the robot arm has been calibrated.

In the next section, the calibration theory will be derived for $\vec{g_t}$ in Equation (5), i.e., the error in the geometric parameters between the actual tool and the ideal tool. In the parameter derivation section, how to obtain the actual tool attitude will be discussed, which is mainly for the calibration of runout and offset.

2.3. Runout Calibration

Runout calibration can be divided into three steps. First, the tool center offset is calculated by position method and then the projection angle is derived. Finally, the rotation matrix can be solved to express the position of an object rotating in space. The position method is used to derive the tool center offset and record the position of the tool when the laser beam is triggered. As the laser sensor is triggered, the coordinates of the TCP are recorded in real-time, and four positions of the TCP are known in each moving cycle. Figure 5a shows the actual and hypothetical trajectories and Figure 5b,c show the position of the tool when it triggers the laser sensor. There are two laser beams, one is along the X-axis and another one is along the Y-axis, I_{ideal} is the ideal into point, I_{actual} is the actual tool into point, and P1 to P4 is the position when the laser sensor is triggered.

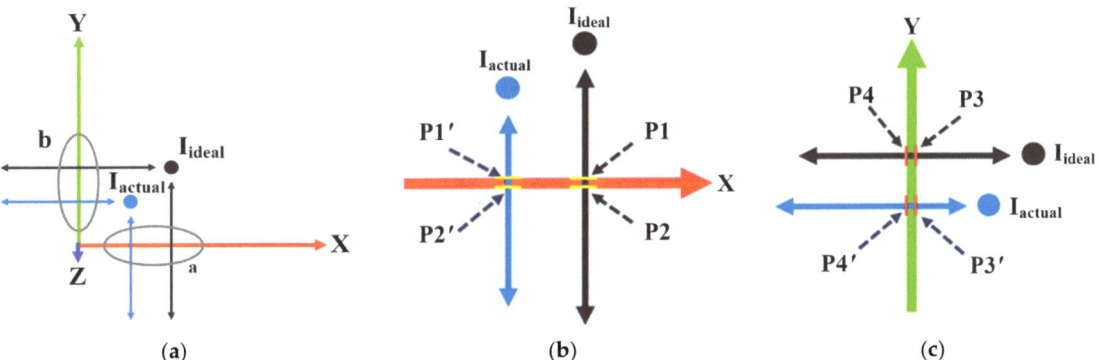

Figure 5. Using the four positions of the tool center offset in the same plane: (**a**) actual and hypothetical trajectories; (**b**) trigger point in X-axis; (**c**) trigger point in Y-axis.

After calculating the tool center offset, the runout angle can be obtained by performing a trigonometric calculation by the height difference. Figure 6a represents the ideal tool coordinate system (C_p) and the runout error of coordinate system ($C_{p'}$). When the tool has a runout error, the base point of the tool on the flange coordinate system is defined as $(\hat{X}_p, \hat{Y}_p, \hat{Z}_p)$ and the projection angle is defined as θ_y in Y-axis. As two laser sensor units are parallel to the X and Y axes of the coordinate system, then the tool contacts the X and Y laser four times during the linear motion of the laser device. The projection angle θ_y can be found by Equation (6).

$$\theta_y = \tan^{-1}\left(\frac{\Delta x2 - \Delta x1}{z2 - z1}\right) \quad (6)$$

where $\Delta x1$ and $\Delta x2$ are the tool center offset value at plane one and plane two, respectively; $z1$ and $z2$ are the value of two planes in the z-axis.

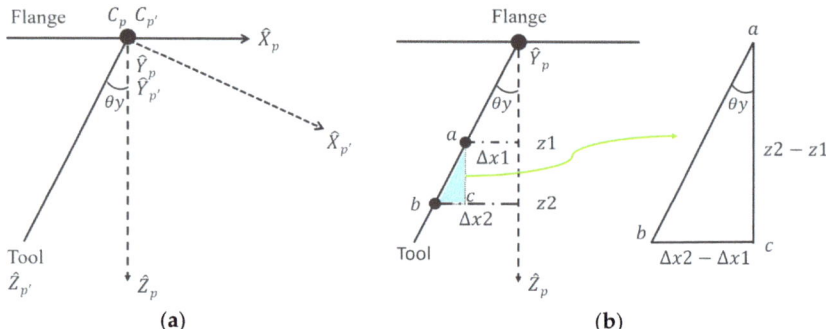

Figure 6. Tool projection angle: (**a**) with a relative displacement of Z-axis; (**b**) geometric relation.

In order to express the relationship of rotation matrix and projection angle, this study uses the Euler angle of the X-Y-Z rotation sequence. The first rotation is defined as an angle α counterclockwise around the X-axis with rotation matrix, R_A, and the second rotation is counterclockwise rotating with an angle β around the Y-axis by matrix R_B. For the rotation of the Z-axis, it does not affect the error. The final rotation matrix (R_d) for the ideal tool coordinate system is shown as Equation (7):

$$R_d = \begin{bmatrix} \cos\beta & 0 & \sin\beta \\ \sin\alpha\sin\beta & \cos\alpha & -\sin\alpha\cos\beta \\ -\cos\alpha\sin\beta & \sin\alpha & \cos\alpha\cos\beta \end{bmatrix} \quad (7)$$

Assume that there is a point P in the ideal tool coordinate system and it will become point Q after rotating to a new coordinate system. Then, position of Q can be calculated and obtained. Accordingly, the rotation matrix of the current tool (R) is derived by the initial tool rotation matrix (R_0) and the ideal tool coordinate system (R_d) shown in Equation (8).

$$R = R_0 \cdot R_d \quad (8)$$

2.4. Offset Calibration

Figure 7 shows that O, O', and O'' are the ideal tool coordinate, runout tool coordinate, and runout with offset tool coordinate system, respectively. S^c (X_s, Y_s, Z_s) is the tool installed station, and P^c (X_0, Y_0, Z_0), $P^{c'}$ (X_0', Y_0', Z_0'), and $P^{c''}$ (X_0'', Y_0'', Z_0'') are the original points of the ideal tool coordinate, runout tool coordinate, and runout with offset tool coordinate system, respectively.

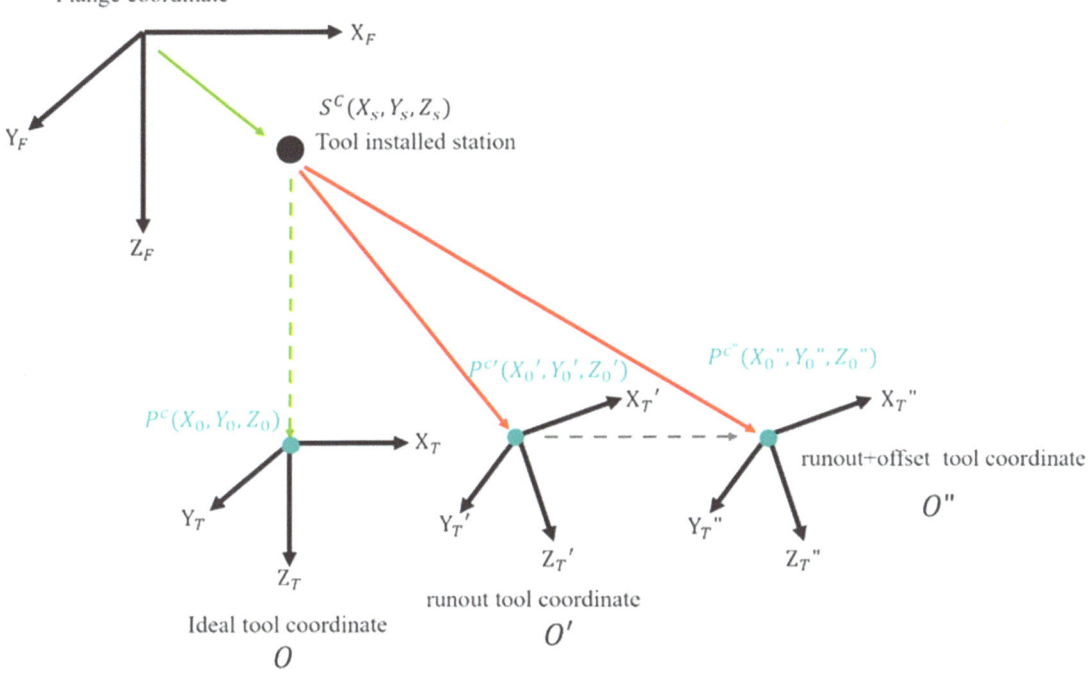

Figure 7. Coordinate space model.

According to the relationship between the coordinate systems, the relative equation can be derived as follows.

$$P^c = \begin{bmatrix} X_0 \\ Y_0 \\ Z_0 \end{bmatrix} = S^c + R_O \cdot Z^T = \begin{bmatrix} X_s \\ Y_s \\ Z_s \end{bmatrix} + R_O \cdot \begin{bmatrix} 0 \\ 0 \\ Z_h \end{bmatrix} \quad (9)$$

$$P^{c'} = \begin{bmatrix} X_0' \\ Y_0' \\ Z_0' \end{bmatrix} = S^c + R_O \cdot (R_R \cdot Z^T) = \begin{bmatrix} X_s \\ Y_s \\ Z_s \end{bmatrix} + R_O \cdot R_R \cdot \begin{bmatrix} 0 \\ 0 \\ Z_h \end{bmatrix} \quad (10)$$

$$\begin{aligned} P^{c''} &= \begin{bmatrix} X_0'' \\ Y_0'' \\ Z_0'' \end{bmatrix} = P^{c'} + R_O \cdot R_R \cdot \delta R^{T'} = \begin{bmatrix} X_0' \\ Y_0' \\ Z_0' \end{bmatrix} + R_O \cdot R_R \cdot \begin{bmatrix} \delta X \\ \delta Y \\ \delta Z \end{bmatrix} = P^c + R_O \cdot \left(R_R \cdot \delta R^{T'} - Z^T \right) \\ &= \begin{bmatrix} X_0 \\ Y_0 \\ Z_0' \end{bmatrix} + R_O \begin{bmatrix} \delta X \cos\beta + (Z_h + \delta Z) \sin\beta \\ \delta X \sin\alpha \sin\beta + \delta Y \cos\alpha - (Z_h + \delta Z) \sin\alpha \cos\beta \\ -\delta X \cos\alpha \sin\beta + \delta Y \sin\alpha + (Z_h + \delta Z) \cos\alpha \cos\beta - Z_h \end{bmatrix} \end{aligned} \quad (11)$$

where Z_h is the distance between P^c and S^c and expressed as a spatial vector Z^T.

The calibration process is shown in Figure 8. In the runout calibration, the actual tool information is obtained after two planes of motion and the tool center offset parameters can be obtained, which are then used to calculate the runout angle and rotation matrix. The first preliminary offset calibration is also performed. After completing the runout calibration, a second preliminary offset calibration is executed using the tool information obtained from the third plane motion to accurately calculate the X and Y axis offset parameters based on the tool coordinate system. The last process is the final offset calibration. The tool moves over the laser beam, triggering the laser sensor to go vertically down and find the Z-axis parameters of the TCP according to the tool coordinate system.

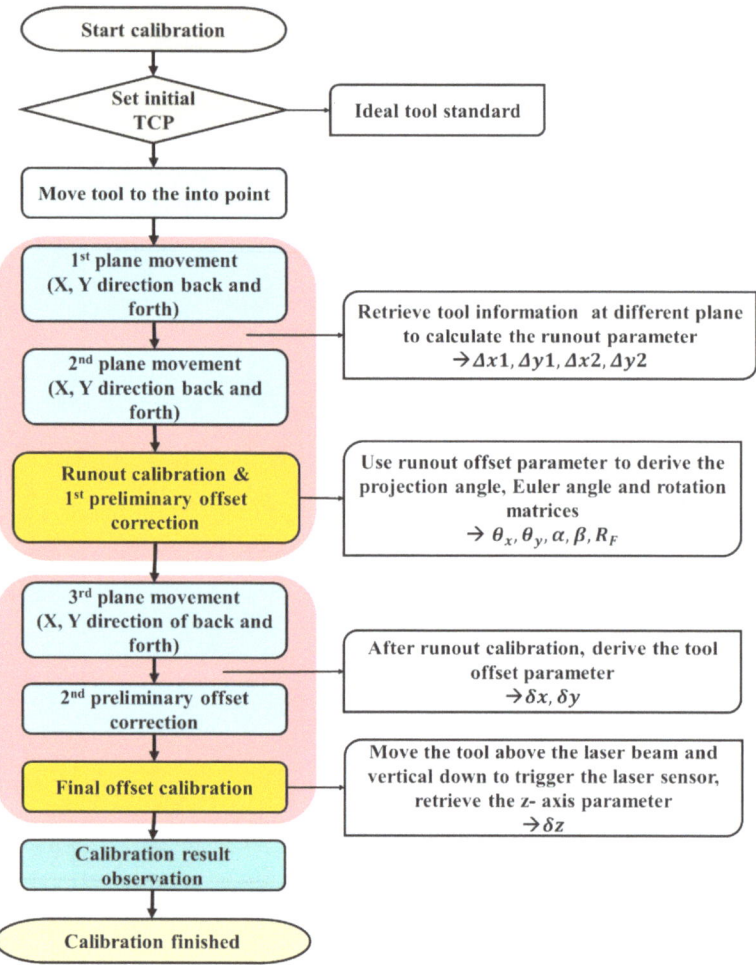

Figure 8. Flow chart of calibration process.

In Figure 8, the first preliminary offset calibration is performed after the 2nd plane movement is completed. The first preliminary offset calibration, which estimates the tool height Z_h, can be performed when the tool only has an offset. The second preliminary offset calibration is performed by deriving the tool offsets δX and δY from the tool center offset equation. Figure 9 represents how the tool height Z_h is calculated.

Looking in from the positive direction of the Y-axis of the coordinates of the tool mounting point, where O is the tool mounting station and the origin of the coordinate system, the coordinates of p^o are $(0, 0, Z_h)$ and the coordinates of p'^o are (x', y', z'). Let the first plane of calibrated motion be M1 and the second plane be M2, where $\overline{P^oO}$ intersects with plane M1 at $M1_s(0,0,z1)$ and with plane M2 at $M2_s(0,0,z2)$, $\overline{P'^oO}$ intersects the plane M1 at $M1_s(0,0,z1)$, $\overline{P'^oO}$ intersects the plane M2 at $M2_s{'}(\Delta x2, \Delta y2, z2)$, and the parameters $\Delta x1$, $\Delta x2$, $\Delta y1$, $\Delta y1$ have been obtained from Equation (6). Let the height from plane one to p^o, labeled $\overline{P^oM1_s}$ be ΔH, and the height between plane one and plane two, labeled $\overline{M1_sM2_s}$, be Δh. After defining the above information, the following Equations (12) and (13) can be obtained.

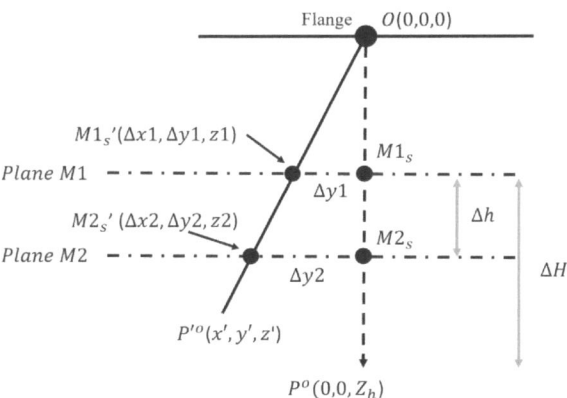

Figure 9. The tool height derivation chart of the tool runout coordinate system.

$$Z_h = Z1 + \Delta H \tag{12}$$

$$Z2 = Z1 + \Delta h \tag{13}$$

where Z_h is the tool height that needs to be derived; ΔH and Δh is the parameter set by the user. Use space linear proportion and the center offset parameter, which is $\Delta x1$, $\Delta x2$, $\Delta y1$, $\Delta y1$. Then, through relational substitution $Z1$ and $Z2$ can be solved.

To express the spatial linear scale, the coordinates of at least two points in space must be known. Herby, assume there are two points, A $(x1, y1, z1)$ and B $(x2, y2, z2)$, the line η between these two points can be expressed in the spatial linear scale as Equation (14).

$$\eta : \frac{x - x1}{x2} = \frac{y - y1}{y2} = \frac{z - z1}{z2} \tag{14}$$

Similarly, since $\Delta x1$, $\Delta x2$, $\Delta y1$, $\Delta y1$, $z1$, $z2$ can construct two points $M1_s'$ and $M2_s'$, the linear H is expressed by the spatial linear scale,

$$H : \frac{x - \Delta x1}{\Delta x2} = \frac{y - \Delta y1}{\Delta y2} = \frac{z - z1}{\Delta z2} \tag{15}$$

In this case, since the line H passes through the tool mounting point, after substituting Equation (15) and combining Equation (13) with Equation (12), then we can obtain Z_h as Equation (16).

$$Z_h = \frac{\Delta h \Delta x1}{\Delta x2 - \Delta x1} + \Delta H = \frac{\Delta h \Delta y1}{\Delta 2 - \Delta y1} + \Delta H \tag{16}$$

After the above calculation, it seems that the tool height Z_h has been solved; however, a tool with offset error will result in a denominator equal to zero, while a tool with offset and offset error will result in a line H not passing through the tool mounting point O. Therefore, for runout error or runout plus offset error, the original tool height will be used directly as the tool length Z_h.

The first preliminary offset calibration only roughly calculated the tool height, the runout plus offset error of the coordinates $\delta R^{T'}$ still can not be solved, so in the first preliminary offset calibration, use Z_h generation back to Equation (11) to obtain the tool center point that is the current tool coordinate origin, set to $P^{c''}{}_{temp1}$, and the coordinates will be returned to the robot arm controller; the first preliminary offset calibration is completed.

2.5. Calibration Environment Design

A virtual environment was used to verify the hypothesis algorithm to then be implemented in the real manipulator and laser sensor device. RoboDK software and Python were applied for the virtual environment and coding. In the real environment, the TM5-900 manipulator was used with LATC laser sensor, and the code was based on C# to complete the experiment. Figure 10 shows the actual experiment structure setting. The specification of TM5M-900 and LATC laser displacement sensor are shown in Tables 1 and 2, respectively.

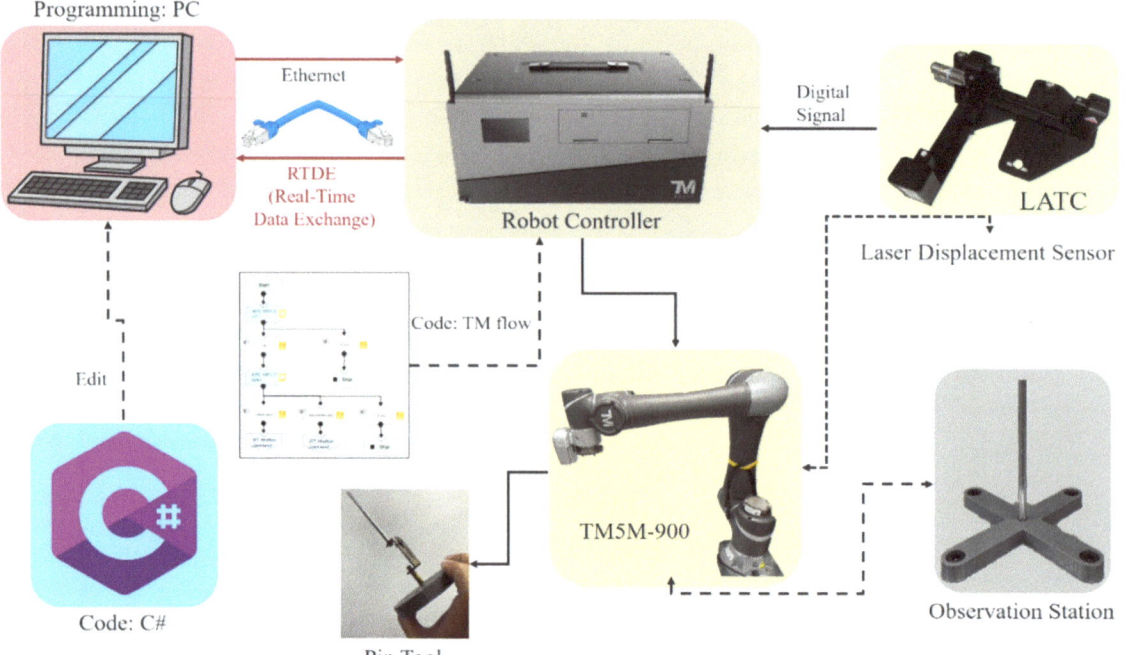

Figure 10. Experimental environment system structure.

Table 1. TM5M-900 specification.

TM5M-900 Specification			
Weight	22.6 kg	Typical speed	1.4 m/s
Max Payload	4 kg	Operating Temperature	0 to 50 °C
Reach	900 mm	Collaboration	Yes
Repeatability	±0.05 mm	DOF	6

Table 2. LATC laser displacement sensor LTC120120 specification.

LATC Laser Displacement Sensor LTC120120			
Supply Voltage	24Vdc	Tool Size	⌀ = 0.5~100 mm
Supply Current	0.2 A	Laser Type	Class 2, Red light Wavelength = 650 nm
Working Range	120 × 120 mm	Working Temperature	5 to 55 °C
Repeatability	<1 μm	Waterproof	IPX8

When the tool enters the sensing range of the sensor and triggers the sensor, it sends a digital signal to the robot controller and returns to the computer side. Figure 11a shows the lab environment setup and Figure 11b shows the standard calibration pin tool.

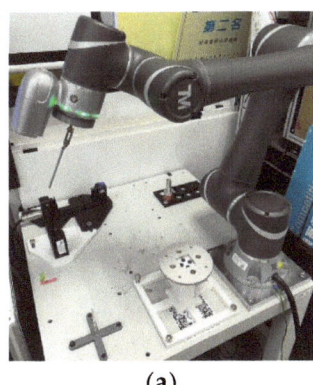
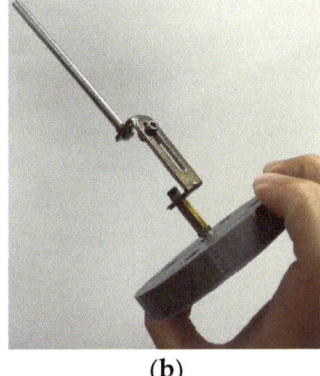

Figure 11. Experiment setup: (**a**) with a TM5-900 manipulator; and (**b**) standard calibration pin tool.

3. Results and Discussion

In each experiment, in addition to the calibrated tool center position data, the calibration error and stability of each set of calibration data are also calculated. The calculation of calibration error and stability is based on ISO 9283, the performance standard for industrial robotic arms and related test methods, which can be followed to calculate the absolute accuracy of posture and the accuracy of posture repetition. The purpose of the simulation is to analyze the feasibility of automatic TCP correction with different tools and different arm movement speeds.

3.1. Error-Free Stability Simulation with the Position Method

The first simulation uses different moving speeds, an error-free tool, and position methods to handle automatic tool center point calibration, and the whole process will be performed 30 times consecutively. The robot moving speeds are set to 20 mm/s and 50 mm/s with an 8 mm radius of pencil tool and a 20 mm radius of cylinder tool, respectively. Tables 3 and 4 show the simulation results of error-free stability by the pencil tool and cylinder tool, respectively.

Table 3. Simulation data of error-free stability using a pencil tool.

	Robot moving speed	X_d (mm)	Y_d (mm)	Z_d (mm)	θx_d (°)	θy_d (°)
Default Value	20 mm/s	0	−82.7	101.73	0	−60
	50 mm/s	0	−82.7	101.73	0	−60
	Robot moving speed	X_m (mm)	Y_m (mm)	Z_m (mm)	θx_m (°)	θy_m (°)
Mean Value	20 mm/s	0.139	−82.799	101.559	0	−60
	50 mm/s	0.218	−81.278	101.581	0.506	−60.525
	Robot moving speed	P_a (mm)		θx_a (°)		θy_a (°)
Acuracy	20 mm/s	0.241		0		0
	50 mm/s	1.841		0.506		−0.525
	Robot moving speed	P_r (mm)		θx_r (°)		θy_r (°)
Repeatability	20 mm/s	±0		±0		±0
	50 mm/s	±3.656		±4.645		±4.789

Table 4. Simulation data of error-free stability using a cylinder tool.

	Robot moving speed	X_d (mm)	Y_d (mm)	Z_d (mm)	θx_d (°)	θy_d (°)
Default Value	20 mm/s	0	0	110	0	0
	50 mm/s	0	0	110	0	0
	Robot moving speed	X_m (mm)	Y_m (mm)	Z_m (mm)	θx_m (°)	θy_m (°)
Mean Value	20 mm/s	−0.450	−0.295	110.43	0	−0.667
	50 mm/s	−0.301	0.185	110.46	−0.242	−0.32
	Robot moving speed	P_a (mm)			θx_a (°)	θy_a (°)
Acuracy	20 mm/s	0.687			0	−0.667
	50 mm/s	0.579			−0.242	−0.32
	Robot moving speed	P_r (mm)			θx_r (°)	θy_r (°)
Repeatability	20 mm/s	±0.382			±0	±0
	50 mm/s	±3.809			±5.017	±4.257

It is found that the accuracy and repeatability under 20 mm/s are better than the results of 50 mm/s. The reason is when using a high-speed tool through the laser sensor, the sensor cannot be responded immediately and the error will be increased. In addition, for the RoboDK virtual environment, the triggering of the laser sensor is interfered by the 3D model. The results of 20 mm/s and 50 mm/s are quite different and there is almost no error in the case of speed 20 mm/s or the error is fixed; however, the results for speed 50 mm/s are unstable, as shown in Figure 12.

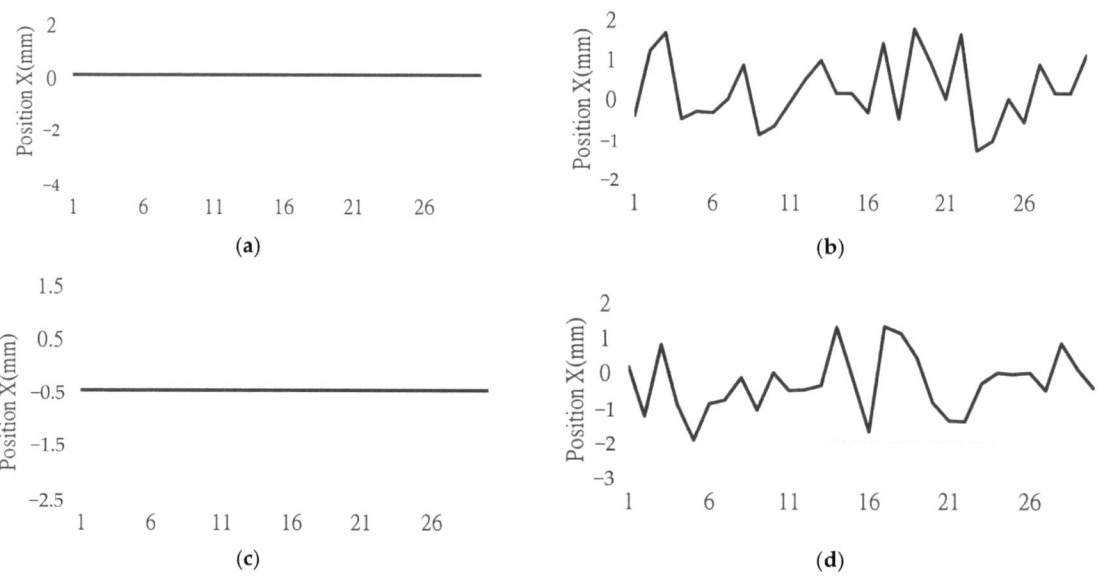

Figure 12. Calibration data of X position: (**a**) pencil tool at 20 mm/s; (**b**) pencil tool at 50 mm/s; (**c**) cylinder tool at 20 mm/s; (**d**) cylinder tool at 50 mm/s.

By using a pencil tool, the results of the proposed TCP calibration method can achieve a positioning deviation of 0.241 and 1.841 mm for the robot moving speeds of 20 and 50 mm/s, respectively. The orientation deviation (θx_a) was 0 and 0.506 degrees for the moving speed of 20 and 50 mm/s, respectively. The positioning repeatability was ±0.083 mm and the orientation repeatability (θx_a) was ±4.645 for a speed of 50 mm/s. It reveals that lower robot moving speed can obtain higher accuracy and better repeatability, as shown in Table 3.

3.2. Four-Quadrant Calibration Simulation with the Position Method

In the four-quadrant simulation section, the experiments were conducted by changing the geometric position without changing the tool coordinates. Table 5 shows the four cases of tool center point geometry position relative to the tool coordinate error setting, i.e., four quadrants.

Table 5. Parameter settings for different tools geometry position in four different quadrants.

Tool	Speed (mm/s)	Quadrant	x (mm)	y (mm)	z (mm)	R_x (°)	R_y (°)
Pencil	20 and 50	1st	−1	1	0	−5	−5
		2nd	−1	−1	0	5	−5
		3rd	1	1	0	−5	5
		4th	1	−1	0	5	5
Cylinder	20 and 50	1st	1	1	0	−5	5
		2nd	−1	1	0	−5	−5
		3rd	−1	−1	0	5	−5
		4th	1	−1	0	5	5
		1st	1	1	0	−10	10
		2nd	−1	1	0	−10	−10
		3rd	−1	−1	0	10	−10
		4th	1	−1	0	10	10

During the experiment, the calibration for one quadrant was performed 30 times in succession, each time restoring the geometric position of the tool coordinate system and the tool center point to the pre-calibration state. The calibration accuracy and repeatability of calibration are shown in Tables 6 and 7.

Table 6. Four-quadrant calibration simulation results using a 5° runout (speed at 20 mm/s).

Tool		Quadrant	1st	2nd	3rd	4th
Pencil	Accuracy	P_a (mm)	0.239	0.689	0.616	0.571
		θx_a (°)	−0.044	1.716	−1.718	−0.405
		θy_a (°)	−0.717	0.341	0.37	−1.472
	Repeatability	P_r (mm)	±0	±0	±0	±0
		θx_r (°)	±0	±0	±0	±0
		θy_r (°)	±0	±0	±0	±0
Cylinder	Accuracy	P_a (mm)	0.852	0.578	0.578	0.481
		θx_a (°)	−0.705	−0.704	0.704	0.705
		θy_a (°)	−0.067	0.164	0.164	−0.067
	Repeatability	P_s (mm)	±0	±0	±0	±0
		θx_r (°)	±0	±0	±0	±0
		θy_r (°)	±0	±0	±0	±0

Table 7. Four-quadrant calibration simulation results using a 5° runout (speed at 50 mm/s).

Tool		Quadrant	1st	2nd	3rd	4th
Pencil	Accuracy	P_a (mm)	0.239	0.689	0.616	0.571
		θx_a (°)	−0.044	1.716	−1.718	−0.405
		θy_a (°)	−0.717	0.341	0.37	−1.472
	Repeatability	P_r (mm)	±5.233	±4.710	±5.209	±5.344
		θx_r (°)	±5.869	±6.835	±6.870	±4.139
		θy_r (°)	±5.005	±4.895	±5.254	±5.335
Cylinder	Accuracy	P_a (mm)	0.786	0.568	0.509	0.803
		θx_a (°)	0.016	−0.511	0.602	0.605
		θy_a (°)	−0.652	0.276	0.166	−0.447
	Repeatability	P_s (mm)	±3.785	±3.309	±3.077	±2.901
		θx_r (°)	±5.223	±4.000	±4.391	±3.709
		θy_r (°)	±4.449	±3.036	±3.966	±3.956

The four-quadrant calibration simulation results seem to be highly inaccurate and it is obvious that some of the errors are caused by the 3D model interfering at different quadrants. The error at the second and third quadrant are higher than the first and fourth quadrants; in addition, the misalignment are almost the same in each quadrant. Due to the laser sensor delay, the results at speed 20 mm/s can reach the standard so that this calibration method still achieves the purpose of TCP calibration shown in Table 8, so it can continue to be tested in the actual experiment.

Table 8. Comparison results of four-quadrant calibration simulation.

Conditions	Accuracy			Repeatability		
Parameters	P_a (mm)	θX_a (°)	θY_a (°)	P_r (mm)	θX_r (°)	θY_r (°)
5° pencil 20 mm/s	0.52875	−0.11275	−0.3695	±0	±0	±0
5° pencil 50 mm/s	2.17225	−0.04575	0.2995	5.124	5.92825	5.12225
5° cylinder 20 mm/s	0.62225	0	0.0485	±0	±0	±0
5° cylinder 50 mm/s	0.6665	0.178	−0.16425	3.268	4.33075	3.85175
10° cylinder 20 mm/s	0.9085	0	−0.6185	±0	±0	±0
10° cylinder 50 mm/s	1.10475	0.22225	−0.4755	2.77925	3.59825	3.5135

3.3. Experiment of Error-Free Stability

In the actual error-free stability experiment, a standard of calibration tools with a processing error of ±0.05 mm are used, and the whole process is performed 30 times consecutively. After calibration, the tool is moved to the observation station to check the result. The moving speed of robot is set to be 20 mm/s and 40 mm/s, respectively.

The data of the error-free calibration experiment by the position method at different speeds are shown in Table 9. It can be seen that the misalignment at a speed of 40 mm/s is slightly larger than that at a speed of 20 mm/s. Accordingly, all results are to be considered with a positioning accuracy of ±0.05 mm for the robot arm with moving speed of 20 mm/s.

Table 9. Experimental result of error-free stability using a pin tool at different moving speeds.

Default Value			Mean Value			Accuracy			Repeatability		
Speed (mm/s)	20	40	Speed (mm/s)	20	40	Speed (mm/s)	20	40	Speed (mm/s)	20	40
X_d (mm)	0.07	0.644	X_m (mm)	0.0219	0.654	P_a (mm)	0.074	0.125	P_r (mm)	±0.083	±0.101
Y_d (mm)	468.653	468.391	Y_m (mm)	468.70	468.358	θX_a (°)	0.089	−0.184	θX_r (°)	±0.207	±0.257
Z_d (mm)	84.403	84.349	Z_m (mm)	84.434	84.228	θY_a (°)	0.182	−0.128	θY_r (°)	±0.173	±0.216
θX_d (°)	179.647	179.935	θX_m (°)	179.736	179.750						
θY_d (°)	−0.804	−0.312	θY_m (°)	−0.621	−0.44						

By using a pin tool, there was a positioning deviation of 0.074 and 0.125 mm for the robot moving speeds of 20 and 40 mm/s, respectively. The orientation deviation (θx_a) was 0.089 and −0.184 degrees for the moving speeds of 20 and 40 mm/s, respectively. The positioning repeatability was ±0.083 mm and ±0.101 mm for the moving speeds of 20 mm/s and 40 mm/s, respectively; and the orientation repeatability (θx_a) was ±0.207 and ±0.257 for the speeds of 20 mm/s and 40 mm/s, respectively. It shows that lower moving speed can achieve higher accuracy and better repeatability, as shown in Table 9.

3.4. Four-Quadrant Calibration Experiment

In the four-quadrant experiment, the tool was used in the same way as the error-free experiment. Before the experiment starts, the tool is rotated to the four-quadrant position for the experiment, and then the calibration procedure is started and calibrated 30 times continuously. After the calibration is finished, the tool is moved to the observation station and the results are checked. Due to the offset of the rotating tool, the actual TCP and tool position are difficult to measure, so the stability results of the calibration are only considered. The results of the first quadrant calibration experiment with the position method at a speed of 20 mm/s are shown in Figure 13, and the experimental results have been compiled in Table 10. It can be seen that the positioning method used in the actual experiment is much more accurate than that in the simulation environment but the error of the results in the four quadrants is slightly higher than that in the error-free experiment but the positioning repeatability can reach 0.12 mm and the positioning repeatability can reach less than 0.14°.

Table 10. Results of the calibration with the position method at 20 mm/s.

	Quadrant	P_r (mm)	θX_r (°)	θY_r (°)
	First	±0.118	±0.147	±0.129
Repeatability	Second	±0.130	±0.119	±0.123
	Third	±0.192	±0.216	±0.130
	Fourth	±0.120	±0.140	±0.108

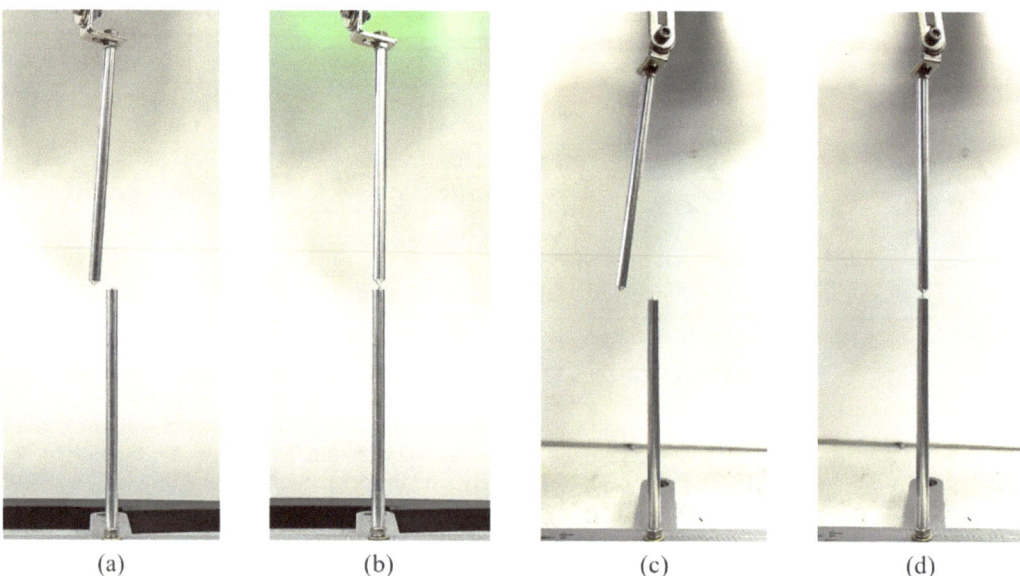

(a) (b) (c) (d)

Figure 13. Comparison of before and after calibration using the position method at 20 mm/s speed through the first quadrant calibration experiment: (**a**) x-axis before calibration; (**b**) x-axis after calibration; (**c**) y-axis before calibration; (**d**) y-axis after calibration.

3.5. Results of the with LATC

For comparing with LATC's existing product, the LATC's calibration software is used to perform the four-quadrant experiment and record the data. First, a calibration speed of 40 mm/s was setup in LATC's software compared with proposed method of 60 mm/s shown in Table 11. Second, the calibration process was almost identical but the motion flow was still a little different. For LATC's entire process, it was programmed in the TM flow system, meaning that all motion, trigger IO, matrix calculations, and position settings were in the same system. For our proposed method, the TM flow system is used to present the robot motion, sensor triggering, and data exchange to the computer, and then use C# to receive the robot information for calculation, processing algorithm, and to send the robot arm position back to the robot. Third, the whole process of using software for LATC is about 70 s, however, our proposed method was about 50 s.

Table 11. Comparison of results by the position method and LATC software.

Quadrant	P_r (mm)		θX_r		θY_r (°)	
	Position Method	LATC Software	Position Method	LATC Software	Position Method	LATC Software
1st	±0.118	±0.055	±0.147	±0.102	±0.129	±0.108
2nd	±0.045	±0.045	±0.130	±0.085	±0.119	±0.124
3rd	±0.192	±0.074	±0.216	±0.110	±0.130	±0.125
4th	±0.120	±0.088	±0.140	±0.17	±0.108	±0.128

Results of the four-quadrant calibration experiments by the positioning method achieved positioning errors below 0.12 mm and orientation errors below 0.14° but the results obtained by the LATC software were lower than our proposed experimental results, which can achieve positioning errors below 0.07 mm on average and orientation errors below 0.12° on average. To sum up, compared to LATC's software, the error could be caused by the loss of data exchange between the TM flow and C#. The calibration cycle of position method is 50 s, which is lower than 70 s of LATC.

Although our proposed method was faster than LATC's experiment, the accuracy was also affected, so in the following adjusted experiments, the speed is set to 40 mm/s, which is the same as LATC's experiment.

From the experimental results shown in Table 12, it can be found that the errors of the two calibration methods are relatively close to each other and the results before adjustment using positioning method are not as accurate as the adjusted positioning method; the reason for this is the speed of the robot's movement. The results of the position method are always inferior to LATC because there is no motion flow to the reference point in their calibration procedure. Meanwhile, when the robot moves, a positioning error is generated and the size of this error depends on the type of robot. Therefore, using LATC's software, the unadjusted experimental results can reach a high accuracy, which decreases when we add an additional motion flow to the program.

Table 12. Comparison of results by the position method and LATC software with the same calibration speed of 40 mm/s.

Quadrant	P_r (mm)		θX_r		θY_r (°)	
	Position Method	LATC Software	Position Method	LATC Software	Position Method	LATC Software
1st	±0.114	±0.126	±0.112	±0.106	±0.099	±0.142
2nd	±0.168	±0.112	±0.187	±0.118	±0.133	±0.127
3rd	±0.118	±0.065	±0.112	±0.115	±0.138	±0.133
4th	±0.151	±0.129	±0.160	±0.191	±0.074	±0.163

In summary, the comparison results are more reliable and convincing after fixing the initial conditions, reducing the robot motion speed, and using the same motion flow in both experiments for the calibration experiment.

4. Conclusions

In this study, an external laser sensor, which is relatively inexpensive, was used to implement an automatic calibration method for the tool center point of a 6-axis manipulator. After the feasibility analysis and validation of the simulation, it was confirmed that this TCP calibration method can correct the runout and offset error of the tool, moreover, this method is simple, affordable, and automatic. The highlights of this paper are listed as following:

1. The proposed method is a non-contact scheme that uses a laser displacement sensor to handle the TCP calibration procedure;
2. After feasibility analysis and laboratory verification, it is confirmed that this calibration method can calibrate runout, tool offset, and runout plus tool offset error;
3. Although the absolute positioning accuracy of the manipulator is 0.05 mm, the position calibration error is about 0.07 mm to 0.19 mm, and the calibration error of the projection angle is within 0.18°;
4. This automatic tool center point calibration method has the advantages of being simple, versatile, less time-consuming, and relatively inexpensive, and it enables an automatic workflow that maintains the flexibility of the manipulator in the work process.

In conclusion, the TCP calibration method provided in this thesis is an accurate and repeatable one that can be used to improve the pose accuracy and repeatability of industrial robots for point-to-point applications.

Author Contributions: Conceptualization, C.-J.L.; methodology, C.-J.L.; software, H.-C.W.; validation, C.-J.L., H.-C.W. and C.-C.W.; formal analysis, C.-J.L. and H.-C.W.; investigation, C.-J.L. and H.-C.W.; writing—original draft preparation, C.-J.L., H.-C.W. and C.-C.W.; writing—review and editing, C.-C.W.; visualization, C.-C.W.; supervision, C.-C.W.; project administration, C.-C.W.; funding acquisition, C.-J.L. All authors have read and agreed to the published version of the manuscript.

Funding: This research was funded by the National Science Council of the Republic of China, grant number Contract No. MOST 108-2221-E-027-112-MY3 and MOST111-2622-E-167-019. The APC was funded by Contract No. MOST 108-2221-E-027-112-MY3 and MOST111-2622-E-167-019.

Institutional Review Board Statement: Not applicable.

Data Availability Statement: Not applicable.

Conflicts of Interest: The authors declare no conflict of interest.

References

1. Yan, B.; Hao, Y.; Zhu, L.; Liu, C. Towards high milling accuracy of turbine blades: A review. *Mech. Syst. Signal Process* **2022**, *170*, 108727. [CrossRef]
2. Hao, Y.; Zhu, L.; Yan, B.; Qin, S.; Cui, D.; Lu, H. Milling chatter detection with WPD and power entropy for Ti-6Al-4V thin-walled parts based on multi-source signals fusion. *Mech. Syst. Signal Process* **2022**, *177*, 109225. [CrossRef]
3. Dun, Y.; Zhu, L.; Yan, B.; Wang, S. A chatter detection method in milling of thin-walled TC4 alloy workpiece based on auto-encoding and hybrid clustering. Mechanical Systems and Signal Processing. *Mech. Syst. Signal Process* **2021**, *158*, 107755. [CrossRef]
4. Zhu, L.; Yan, B.; Wang, Y.; Dun, Y.; Ma, J.; Li, C. Inspection of blade profile and machining deviation analysis based on sample points optimization and NURBS knot insertion. *Thin-Walled Struct.* **2021**, *162*, 107540. [CrossRef]
5. International Federation of Robotics (IFR). *World Robotics Report 2020 Press Conference Presentation*; IFR Press Conference: Frankfurt, Germany, 2020; pp. 8–18.
6. Elango, N.; Faudzi, A.A.M. A review article: Investigations on soft materials for soft robot manipulations. *Int. J. Adv. Manuf. Technol.* **2015**, *80*, 1027–1037. [CrossRef]
7. Bakhy, S.H.; Hassan, S.S.; Nacy, S.M.; Dermitzakis, K.; Arieta, A.H. Contact mechanics for soft robotic fingers: Modeling and experimentation. *Robotica* **2013**, *31*, 599–609. [CrossRef]
8. Catalano, M.G.; Grioli, G.; Farnioli, E.; Serio, A.; Piazza, C.; Bicchi, A. Adaptive synergies for the design and control of the Pisa/IIT softhand. *Int. J. Rob. Res.* **2014**, *33*, 768–782. [CrossRef]
9. Mc Graw-Hill Dictionary of Engineering Second Edition. Available online: https://www.academia.edu/8016572/Mc_Graw_Hill_Dictionary_of_Engineering_Second_Edition (accessed on 6 April 2022).
10. Bergström, G. Method for Calibrating of Off-Line Generated Robot Program. Master's Thesis, Chalmers University of Technology, Göteborg, Sweden, 2011.
11. Guo, C.; Xu, C.; Xiao, D.; Zhang, H.; Hao, J. A tool centre point calibration method of a dual-robot NDT system for semi-enclosed workpiece testing. *Ind. Robot.* **2019**, *46*, 202–210. [CrossRef]
12. Fares, F.; Souifi, H.; Bouslimani, Y.; Ghribi, M. Tool center point calibration method for an industrial robots based on spheres fitting method. In Proceedings of the 2021 IEEE International Symposium on Robotic and Sensors Environments (ROSE), Virtual, 28–29 October 2021; pp. 1–6. [CrossRef]
13. Erick, N.; Ning, X. Robot control system for multi-position alignment used to automate an industrial robot calibration approach. In Proceedings of the 2014 IEEE International Conference on Robotics and Automation (ICRA), Hong Kong, China, 31 May–7 June 2014; pp. 2126–2131.
14. Borrmann, C.; Wollnack, J. Enhanced calibration of robot tool centre point using analytical algorithm. *Int. J. Mater. Sci. Eng.* **2015**, *3*, 12–18. [CrossRef]
15. Zhang, L.; Li, C.; Fan, Y.; Zhang, X.; Zhao, J. Physician-friendly tool center point calibration method for robot-assisted puncture surgery. *Sensors* **2021**, *21*, 366–385. [CrossRef] [PubMed]
16. Liu, S.; Wu, Y.; Qiu, C.; Zou, X. Automatic calibration algorithm of robot TCP based on binocular vision. In Proceedings of the 2021 2nd International Conference on Control, Robotics and Intelligent System (CCRIS '21), Qingdao, China, 20–22 August 2021; Association for Computing Machinery: New York, NY, USA, 2021; pp. 244–249. [CrossRef]
17. Techmen Robot Specification. Available online: https://www.tm-robot.com/en/mobile-series/ (accessed on 30 April 2022).

18. Li, T.M. Research on Automatic Calibration of TCP of a Six-Axis Manipulator. Master's Thesis, National Taiwan Normal University, Taipei, Taiwan, 2021.
19. Helal, M. Robotic Tooling Calibration Based on Linear and Nonlinear Formulations. Master's Thesis, Ryerson University, Toronto, ON, Canada, 2015.

Disclaimer/Publisher's Note: The statements, opinions and data contained in all publications are solely those of the individual author(s) and contributor(s) and not of MDPI and/or the editor(s). MDPI and/or the editor(s) disclaim responsibility for any injury to people or property resulting from any ideas, methods, instructions or products referred to in the content.

Article

Optimization of the Storage Spaces and the Storing Route of the Pharmaceutical Logistics Robot

Ling Zhang [1], Shiqing Lu [2,3,*], Mulin Luo [2] and Bin Dong [4]

[1] Mechanical Engineering and Automation College, Chongqing Industry Polytechnic College, Yubei District, Chongqing 400050, China
[2] School of Mechanical Engineering, Chongqing University of Technology, Banan District, Chongqing 400045, China
[3] Robot and Intelligent Manufacturing Technology, Key Laboratory of Chongqing Education Commission of China, Chongqing 400045, China
[4] Chongqing Dile Jinchi General Machinery Co., Ltd., Chongqing 401320, China
* Correspondence: shiqing.lu@cqut.edu.cn

Abstract: Auto drug distribution systems are used popularly to replace pharmacists when drugs are distributed in pharmacies. The Cartesian robot is usually used as the recovery mechanism. Under non-dynamic storage location conditions, generally, the selected planning route of the Cartesian robot is definite, which makes it difficult to optimize. In this paper, storage spaces were distributed for different drugs, and the route of storing was broken down into multiple path optimization problems for limited pick points. The path was chosen by an improved ant colony algorithm. Experiments showed that the algorithm can plan an effective storing route in the simulation and actual operation of the robot. The time spent on the route by improved ant colony algorithm sequence (IACS) was less than the time spent of route by random sequence (RS) and the time spent of route by traditional ant colony algorithm sequence (ACS); compared with RS, the optimized rate of restoring time with iacs can improve by 22.04% in simulation and 7.35% in operation. Compared with ACS, the optimized rate of restoring time with iacs was even more than 4.70% in simulation and 1.57% in operation. To the Cartesian robot, the optimization has certain guiding significance of the application on the 3D for improving quality.

Keywords: Cartesian robot; storing route; ant colony algorithm

Citation: Zhang, L.; Lu, S.; Luo, M.; Dong, B. Optimization of the Storage Spaces and the Storing Route of the Pharmaceutical Logistics Robot. *Actuators* **2023**, *12*, 133. https://doi.org/10.3390/act12030133

Academic Editor: Chih Jer Lin

Received: 13 February 2023
Revised: 16 March 2023
Accepted: 20 March 2023
Published: 21 March 2023

Copyright: © 2023 by the authors. Licensee MDPI, Basel, Switzerland. This article is an open access article distributed under the terms and conditions of the Creative Commons Attribution (CC BY) license (https://creativecommons.org/licenses/by/4.0/).

1. Introduction

As far as the safety of patients is concerned, the safety of the drugs distributed and managed is crucial. There are errors related to dispensing in the pharmacy every day. Dispensing errors, including drug shortages, account for 20–50% of all existing medication errors, and the dispensing error rate is 0.01–0.08% of drug distribution [1–4]. These are not only caused by subjective reasons, but also by objective reasons. For example, in many Asian countries, doctors must play a dual role between examining prescription drugs and dispensing drugs. Sometimes, the number of patients exceeds the capacity of most hospitals, which means that a large number of drugs need to be allocated [5]. Many dispensing devices and robots were developed to distribute and manage drugs in hospitals and pharmacies in order to reduce human errors, achieve a high response rate, and reduce labor intensity.

In the past three decades, the development of new technologies such as control technology, Internet technology, the Internet of Things, robots, artificial intelligence, and machine manufacturing promoted the development of drug recovery, distribution, and management, such as the Pharmacy Automation System (PAS). This was studied and used in many countries, including Germany, USA, Japan, China, and the Netherlands [6]. Between 1993 and 1997, the first fast medicine dispensing device, the Automatic Pharmacy

(AP), was designed, developed, and installed by the ROWA in Germany. It can be dispensed and recovered by the operator and electronic prescriptions can be issued automatically, rather than the manual work of pharmacists. The manipulator applied for an invention patent, but it has important shortcomings in terms of recovery efficiency [7]. Later, the drug dispensing system (DDS) was designed to rapidly dispense and recover drugs to meet the needs. With the use of DDS, errors were significantly improved; is the errors were only dispensing errors, which is called a repeated error. In drug distribution [8,9], the dispensing error rate is not more than 0.03%. At the same time, a prescription takes 7 to 8 s, and DDS can save waiting time.

Under the normal operating conditions of the device, research fields such as recovering drugs, examining the names and quantities of drugs, and examining prescriptions were expanded. The difficulty and labor intensity of the operation were reduced. The recovery process is assisted by a manipulator. In this case, how to improve the working efficiency of the manipulator is actively studied.

The recovery process includes manually assisting the robot to replenish and restore drugs to the drug repository [10,11]. The recovery path is the planning path of the DDS medical logistics system, which is traditionally applied to the transportation system and warehouse storage of production and service departments [12]. Picking list is one of the main activities to be executed, which determines its operation cost and time [13,14]. Research shows that the proportion of order picking in the total operating cost of the warehouse is not less than 55–75% [15–17]. The drug warehouse is called "picking", to distinguish it from the drug warehouse when selecting to store drugs. Therefore, an effective pickup and recovery route strategy is selected in FMDS to reduce costs, manpower, and time.

The storing strategy was a major task in previous works. The storing process was defined as a traveling salesman problem (TSP) or vehicle route problem (VRP) in the pharmaceutical logistics robot [18–20]. However, the storing process cannot simply be viewed as a customer in TSP or VRP because of the characteristics of the capacity of the manipulator and actual demand. The reason that one picking may be visited more than once is because a shortage of one picking may exceed the capacity of the manipulator. A novel storing strategy was proposed to store medicine in the logistics pharmaceutical robot. All picking points are assigned in multiple and different paths. The optimization problem we are now discussing is more about the point-to-point method, which is a vector problem [21]. In many cases, we only consider the point-to-point relationship, or more importantly, the optimization of the algorithm, or the increase or decrease in the intermediate constraints [22]. However, when applied in practical engineering, we found that driving by the motor will actually affect the in optimization of this path. Research this field is relatively few; at the same time, we also found that in rectangular coordinate robots, there will be differences in path caused optimization by different driving modes, and the impact of the motor drive is often not considered. For example, during the printing process, the driving of the motor of the 3D printer's three-rectangular coordinate robot will indirectly affect the printing [23]. Therefore, different ways of the same drive will have different effects on the optimization of this path.

At present, domestic and foreign researchers made many achievements in the research of storage allocation. J. Giacomo Lanza et al. [24] proposed a mixed integer linear programming (MILP) model based on multi-commodity flow formula, which had some effective inequalities. Two relaxation methods were also proposed to estimate the quality of the model solution. Then, two mathematical methods were designed from the MILP model. Xiangbin Xu et al. [25] proposed a distributed storage location allocation (MPSSLA) strategy based on multiple sorters, which can achieve better optimization effect than the traditional centralized storage strategy in the multi-picker to parts picking system. Zhong Qiang Ma et al. [26] proposed a heuristic algorithm to select the minimum number of shelves and evaluate the optimization effect of SA-VNSSA and SCSPCC on the number of shelves transferred. At the same time, intelligent algorithms such as particle swarm optimization

(PSO), genetic algorithm (GA), and ant colony optimization (ACO) were used to optimize the storage path of the drug logistics robot [27,28]. In terms of searching optimized storage paths, PSO proposed by James Kennedy, GA [29] proposed by John Holland, and ACO proposed by Marco Dorigo can effectively reduce the number of values or manipulators of different storage paths to achieve the purpose of optimization. Because there are few adjustable parameters, the structure of PSO is simple, but more iterations are needed to obtain the best result of problem solving. For GA, the number of ants and the number of picking points 5 and 15 are close to or the same, and ACO is better than GA in searching TSP optimization solution [30].

All picking points may be assigned in multiple and different paths, and all paths form the route that is stored. The route is regarded as a multiple of TSP, and the order of drug stored is optimized in each TSP. The remainder of this paper is structured as follows. In Section 2, the structure of storage and storing manipulator are described. Storing models and optimized algorithms are shown in Section 3. Simulation, operations, and experimental analysis are conducted in Section 4. Finally, the conclusion and future work are provided in Section 5.

2. Storage Unit Structure

2.1. Description of the Storage

The structure of the medicine storage and distribution equipment proposed in this paper includes two parts: a medicine-grabbing manipulator and a slope-type medicine storage tank, as shown in Figure 1. The medicine-grabbing manipulator is mainly used to provide real-time supply of drugs, ensure sufficient inventory of drugs in the automated pharmacy, and meet the demand for dispensing drugs in the pharmacy. The slope-type drug storage tank is responsible for storing and managing a certain number of regular boxed drugs, maximizing the drug storage within the effective storage area, ensuring drug supply during peak periods of drug delivery, and facilitating connection with the dispensing and dispensing systems.

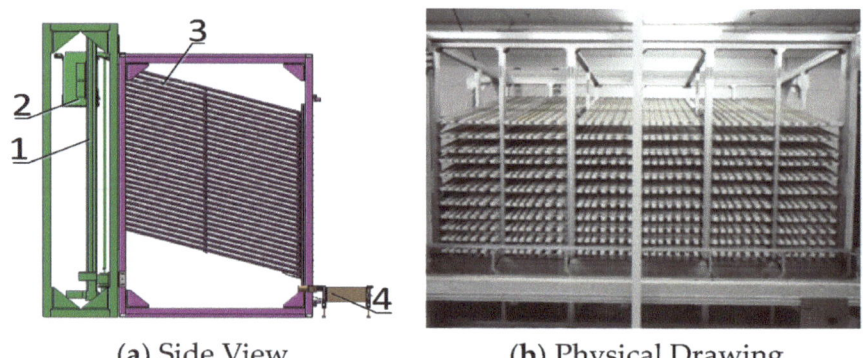

Figure 1. 3D and physical picture of the medicine storage system. 1. Cartesian coordinate robot. 2. Dispensing manipulator 3. Sloping drug storage 4. Automatic drug delivery device.

In this paper, the U-shaped storage structure of the roller and the direct-acting tilting plate type drug delivery mechanism were used. The U-shaped storage position of roller is mainly composed of a roller, roller shaft, spacer bar, frame, and beam. Generally, a nylon roller with a diameter of 10 mm is used to cover the roller shaft and evenly distributed at a spacing of 20 mm. When the roller is installed to a certain number, the spacer bar will be added. The two adjacent spacer bars form the smallest storage unit, called storage position. Several storage positions form a drug storage tank, as shown in Figure 2. Since the principle of gravity dropping is adopted, there needs to be an angle between the medicine storage tank and the horizontal plane to ensure the medicine and the size of the angle will affect the installation quantity of the medicine storage tank. Therefore, it is necessary to calculate the inclination angle of the medicine storage tank and the friction coefficient of nylon and paper $f_1 = 0.3$–0.4.

$$G \sin \alpha = \mu G \cos \alpha \qquad (1)$$

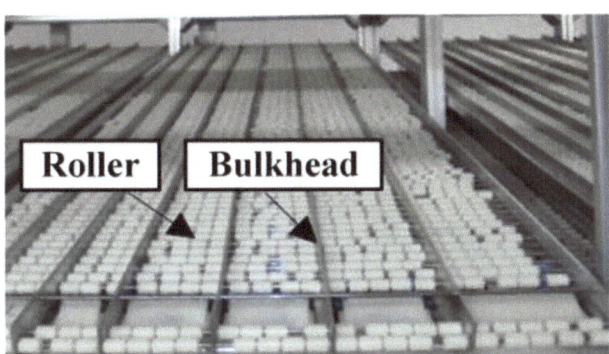

Figure 2. U-shaped storage position and medicine storage tank.

Figure 3 shows the side view of the storage tank. From this, it can obtain the α angle range, $\alpha_1 = 16.99°$, $\alpha_2 = 21.8°$. Due to air resistance, the best angle was finally set at 18°. The number of medical tanks will be reduced when the angle of the tank becomes larger, but this is beneficial for the downward movement of the medical tank. The total thickness of the frame and partition of the medicine storage tank was 32 mm, and the spacing between two adjacent drug storage tanks was 40 mm. The outer frame size of the rapid dispensing system was 3540 mm × 1440 mm × 2450 mm, and the size of its drug storage mechanism was $l_1 = 1260$ mm, $h = 2050$ mm.

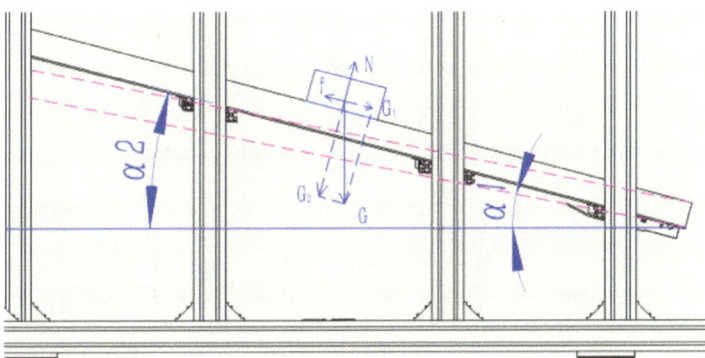

Figure 3. Side view of medicine storage tank.

According to the side diagram of the storage slot in Figure 4, the number of layers in the tank m was calculated:

$$m = \frac{h - l_1 \tan \beta}{h_2 / \cos \beta} = \frac{(h - l_1 \tan \beta) \cos \beta}{h_2} \qquad (2)$$

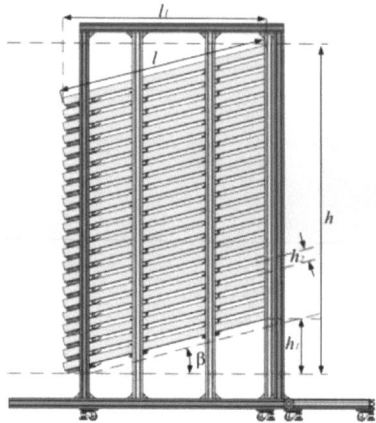

Figure 4. Side view of drug storage tank.

The results are shown in Table 1. It can be observed that the number of layers that can be stored in the storage tank varied with different angles. When the tilt angle was 18°, the space utilization rate was 10% higher than that with 21.8°. It can be observed that the use of 18° roller U-shaped storage tank was in line with the optimal distribution.

Table 1. Quantity of medicine storage tanks with different tilt angles.

β	h_1 (m)	h_2 (m)	m
18°	0.4095	1.640	20
20°	0.4586	1.591	19
21.8°	0.5040	1.546	18

The direct-acting flap type drug delivery mechanism is linked to the lifting baffle via the slider mechanism, and the movement of the lifting baffle is realized by pulling the slider through the electromagnet. The advantage of this drug delivery method is that the drug delivery mechanism will not hit the drug box during the drug delivery process. Therefore, the downward angle of the drug will not change and will not reverse. At the same time, it will also reduce the situation that a small number of drugs cannot be delivered due to the reduction in the delivery angle of the drug box.

For the direct-acting flap type drug delivery mechanism, the experimental results are shown in Tables 2 and 3, where the baffle height was 5 mm. The test equipment is shown in Figures 5 and 6.

Table 2. Delivery time of test drugs (test kit size: 130 mm × 95 mm × 12 mm).

Number of Remaining Medicine Boxes in the Medicine Storage Tank	7	6	5	4	3	2	1
Single Box Dispensing Time	0.311	0.328	0.335	0.334	0.369	0.392	0.468

Table 3. Outgoing test results of different kits.

Kit Size (mm)	Delivery Time (s)	No Drug Delivery
167 × 67 × 17	0.655	Two boxes and above
130 × 95 × 12	0.362	nothing
98 × 47 × 12	0.377	One box

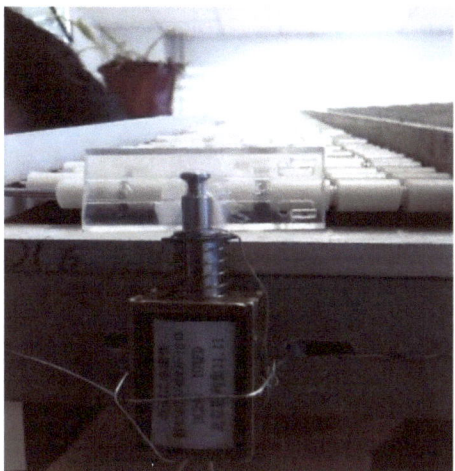

Figure 5. Simple drug delivery mechanism.

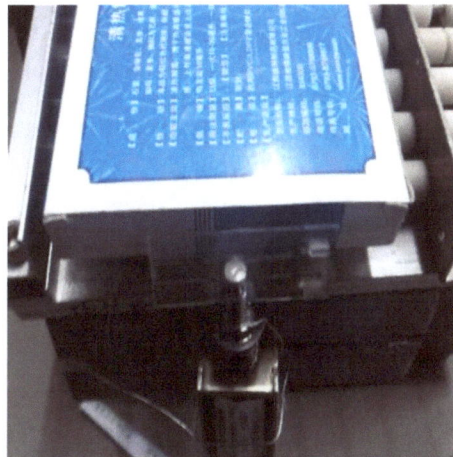

Figure 6. Experimental diagram.

It was observed in the above experiments that the control of the drug delivery time of the flap type drug delivery mechanism was not different, but when the drug was light, the gravity was not enough to overcome the friction, and the drug box would not slide. At the same time, when the medicine was heavy and the quantity in the medicine storage tank was too large, the first box of medicine was squeezed by the medicine behind, and when the electromagnetic force was small, it was not able to leave the warehouse. To sum up, the flap type drug delivery mechanism has a fast delivery speed and can meet the requirements of drug delivery in any situation.

To ensure the uniqueness of drug storage and distribution in each unit, the following settings were established:

(1) The drug storage and distribution system is a rectangular frame structure, as shown in Figure 3. The drug storage and distribution system is composed of six storage units, each of which is composed of multi-layer drug storage tanks, and each of which is composed of multiple storage locations;

(2) The length of each tank is the same, but the width and height are different. The storage height of the same layer is the same. After the equipment runs, each storage location can only store a certain drug;

(3) Set the drug storage and distribution system to be composed of M drug storage tanks. Since the width of drugs that can be stored in each unit is limited when setting, the storage unit needs to be optimized first;

(4) Set the drug r to be stored for the first time, the drug width is l_r, where the inventory of the drug storage tank is $s = 0$, and the length of the drug storage tank is L, and then, the number of replenishment required for the drug storage tank is $s_r = \lceil L/l_r \rceil$;

(5) For the storage of a certain drug, the number of storage units allocated is N. Since the length of storage units is the same, the number of this drug stored in each unit is the same;

(6) It is set that there are n rows of m layers in the reservoir rectangle, and the reservoirs in row j of layer i are recorded as (i, j), where $i = 1, 2, \ldots, m; j = 1, 2, \ldots, n$;

(7) Each chemical storage tank has a unique number, among which, the storage location number in column j of layer i is $ID_{ij} = 100 \times i + j$, and the chemical storage tank code is arranged from small to large. If a chemical storage tank has been allocated, the corresponding mark is 1, otherwise it is idle, and the mark is 0.

2.2. Description of the Sorting Mechanical Configuration and Sorting Process

The principle of the end effector is similar to that of the striker ejection, and so, it is referred to as the clip manipulator (as shown in Figure 7). Before medicine replenishment, the medicine is stored in the tray (similar to a cartridge). Because the height of each medicine box is different, the maximum number of medicine boxes stored in the tray also varies. The number of medicine boxes that can be stored in the tray is controlled and managed by the software system.

Figure 7. End actuator. 1.Chain clamping device. 2. Flexible tensioning device. 3. Detection sensor. 4. Motor.

In order to ensure the accuracy and stability of the end effector desired position during the process of medicine supplement, the form of rectangular coordinate robot is adopted in the design. The robot is composed of two groups of linear motion units. The end actuator is installed on two vertical linear motion units in the X direction and connected to the Y-axis guide rail through a slider. The end actuator and X-axis guide rail have a certain tilt angle. The movement in the Y direction is connected by the synchronous axis, and the motor drives the synchronous axis movement, which can realize the movement of the end actuator and the X-axis guide rail in the Y direction, that is, the horizontal movement of the end actuator. The movement in the X direction is directly driven by the motor to realize the movement of the end actuator on the X-axis guide rail, that is, the vertical movement of the end actuator, as shown in Figure 8.

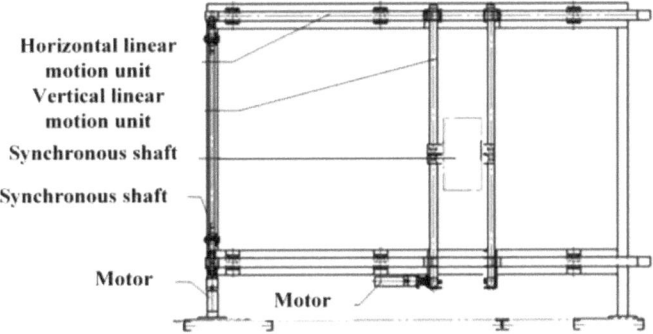

Figure 8. X-Y rectangular coordinate robot.

In the storage of certain drugs, a different number of drug storage tanks will be allocated according to their distribution frequency. Although there are many drug storage tanks, there are not many drug storage tanks for a certain drug for hundreds of drug storage types. In general, the quantity of a drug placed into the equipment shall be as much as possible to ensure its half day distribution. The resulting drugs need to be quickly replenished at any time. Generally, the pharmacy will set the times of replenishment within a day according to the drug consumption, but at the same time, if a drug is less than 1/4 of the total storage of the drug, the system will also remind the temporary replenishment.

The process of restoring (replenishment drug process) is somewhat similar to the traveling salesman problem (TSP), but it is not completely the same. Due to the limited capacity of the manipulator, the quantity of replenishment at one time is limited, and the inventory quantity of the drug storage tank is generally greater than the capacity of the manipulator. Therefore, in general, the manipulator will replenish a certain drug storage tank according to the maximum capacity. When the shortage of the drug storage tank is less than the manipulator's capacity, it will be considered to complete the replenishment together with other drug storage tanks. Therefore, for a certain drug, its replenishment process includes all the drug storage tanks storing the drug, while for the drug storage tank, its replenishment process is divided into two parts: point-to-point replenishment and mixed replenishment.

3. Optimization of Sorting Route

3.1. Amount of Storing with a Kind Drug

The storage route of drugs includes the path of point-to-point replenishment and the mixed replenishment path. Before calculating its path, it is necessary to determine the replenishment time of the point-to-point replenishment process and the points of each replenishment during the mixed replenishment path.

First of all, the point-to-point time of shortage in each drug storage tank are separated. The specific separation model is as follows:

$$\begin{cases} s_{ij}^r = \lfloor L/l_r \rfloor \\ s_{ij}^{r1} = \lfloor s_{ij}^r/Q \rfloor \\ s_{ij}^{r2} = s_{ij}^r - s_{ij}^{r1} Q \end{cases} \quad (3)$$

where Q is the maximum quantity of medicine r that can be carried by the manipulator, s_{ij}^r is the maximum quantity of medicine r stored in the ij medicine storage tank, s_{ij}^{r1} is the number of point-to-point replenishment of medicine r stored in the medicine storage tank, and s_r'' is the quantity of out-of-stock of point-to-point replenishment removed by the medicine storage tank.

The number of point-to-point supplements s_r^1 is,

$$s_r^1 = \sum_{a=1}^{N} s_{ij}^{r1} \tag{4}$$

In the Formula, the drug r is assigned N drug storage tanks, where a = 1, 2,..., N. The number of mixed replenishment s_r^2 is,

$$s_r^2 = \begin{cases} s_r^2 & \sum_{a=1}^{N} s_{ij}^{r2} \prec Q \\ s_r^2 + 1 & \sum_{a=1}^{N} s_{ij}^{r2} \leq Q, \sum_{b=a+1}^{N} s_{ij}^{r2} \succ Q \end{cases} \tag{5}$$

3.2. Design of the Ant Colony Algorithm

The ant colony algorithm (AC) is a heuristic algorithm used to simulate the behavior of real ant colony when establishing the shortest path between a food source and nest. When an ant moves, it releases a trace pheromone that can be detected by other ants. As more ants pass through the path, more pheromones are deposited. Because ants move according to the number of pheromones, the richer the pheromone tracks on the path, the greater the possibility of other ants tracking it. Therefore, ants can build the shortest path from the nest to the food source and return.

In the supplement process, whether it is point-to-point supplement or mixed supplement, its supplement path is composed of all paths from the previous supplement point (including the initial point) to the next supplement point. Set the previous supplement point of drug r as the (i, j) drug storage tank, and the next supplement point as the $(i + e, j + f)$ drug storage tank. e, f are arbitrary rational numbers. A unit of X coordinate or Y coordinate denotes that length is different in each drug storage; to simplify our modeling, the value is set to 12 cm. The corresponding supplementary distance is,

$$\begin{cases} l_{(i,\,i+e),(j,j+f)}^{rx} = 0.12e \\ l_{(i,\,i+e),(j,j+f)}^{ry} = 0.12f \end{cases} \tag{6}$$

The time it takes from (i, j) to $(i + e, j + f)$ on the X and Y axes is,

$$t_{(i,\,i+e),(j,j+f)}^{rx} = \begin{cases} 2\sqrt{\dfrac{l_{(i,\,i+e),(j,j+f)}^{rx}}{a_x}} & if \dfrac{v_x}{a_x} \succ \sqrt{\dfrac{2 l_{(i,\,i+e),(j,j+f)}^{rx}}{a_x}} \\ 2\left(\dfrac{v_x}{a_x} + \sqrt{\dfrac{l_{(i,\,i+e),(j,j+f)}^{rx} - v_x/a_x}{a_x}}\right) & other \end{cases} \tag{7}$$

$$t_{(i,\,i+e),(j,j+f)}^{ry} = \begin{cases} 2\sqrt{\dfrac{l_{(i,\,i+e),(j,j+f)}^{ry}}{a_y}} & if \dfrac{v_y}{a_y} \succ \sqrt{\dfrac{2 l_{(i,\,i+e),(j,j+f)}^{ry}}{a_y}} \\ 2\left(\dfrac{v_y}{a_y} + \sqrt{\dfrac{l_{(i,\,i+e),(j,j+f)}^{ry} - v_y/a_y}{a_y}}\right) & other \end{cases} \tag{8}$$

Pickings must be defined on a graph as $G = \{V, E\}$, where $V = \{0\} \cup N$ is the vertex set and E is the arc set.

Pheromone concentration is defined as $\tau_{ij}(t)$ at t moment on the edge between picking i and picking j. It is equal to $\Delta \tau_{ij}(t) = 0$, $t = 0$. Over time, the pheromone concentration on the path is changed because of new pheromone being applied and old pheromone evaporating. ρ is set as volatility coefficient of the pheromone and showed the speed of evaporation. When all ants completed one tour, the pheromone on each path is given as:

$$\tau_{ij}(t+1) = (1-\rho)\tau_{ij}(t) + \Delta \tau_{ij}(t) \tag{9}$$

where $\Delta \tau_{ij}(t) = \sum_{k=1}^{m} \Delta \tau_{ij}^{k}(t)$.

The pheromone concentration is defined as $\Delta\tau_{ij}(t)$, and it is released on the path from picking i to picking j by ant k. The pheromone value is determined with the ant's performance. A shorter path meant more pheromone is applied by the unit. Ant-cycle was used in this paper.

$$\Delta\tau_{ij}^k(t, t+1) = \begin{cases} Q/L_k & \text{the path from picking } i \text{ to picking } j \text{ by the ant } k \\ 0 & \text{other} \end{cases} \quad (10)$$

where Q is a constant, and L_k is the length of the tour constructed by ant k.

$$P_{ij}^k = \begin{cases} \dfrac{\tau_{ij}^\alpha \eta_{ij}^\beta}{\sum_{j \in allowed_k} \tau_{ij}^\alpha \eta_{ij}^\beta} & \text{if} \quad j \in allowed_k \\ 0 & \text{other} \end{cases} \quad (11)$$

$$\eta_{ij}(t) = 1/d_{ij} \quad (12)$$

where P_{ij}^k is the selection probability.

$allowed_k = (v_1, v_2, \ldots, v_n) - tabU_k$ represents the collection of locations that could be chosen by ant k. $tabU_k(k = 1, 2, \ldots, m)$ is the taboo list of ant k. The visited location is recorded in the taboo list, and the memory of ant k is illustrated. The inverse distance between picking i and picking j is shown as $\eta_{ij}(t)$, which is the visibility of moving from city i to city j. α is the residual degree of information on (i, j) edge. β is the heuristic degree of information. Both of them could be changed by the user.

A new d_{ij} can be obtained from Formulas (6)–(8),

$$d_{ij} = \begin{cases} l^{rx}_{(i,\ i+e),(j,j+f)} & t^{rx}_{(i,\ i+e),(j,j+f)} \geq t^{ry}_{(i,\ i+e),(j,j+f)} \\ l^{ry}_{(i,\ i+e),(j,j+f)} & \text{other} \end{cases} \quad (13)$$

Meanwhile, the L'_k is calculated by Formula (13),

$$L'_k = \sum_{i=0}^{j=N} d'_{ij} \quad (14)$$

where L'_k is the time of the tour constructed by ant k.

It can be obtained from Formula (14),

$$\Delta\tau_{ij}^k(t, t+1) = \begin{cases} Q/L'_k & \text{the path from picking } i \text{ to picking } j \text{ by the ant } k \\ 0 & \text{other} \end{cases} \quad (15)$$

Combining Equations (11), (12) and (15), the new transfer possibility, P_{ij}^k, between picking i and picking j are calculated.

3.3. The Process of the Algorithm

Based on Section 3.2, the actual pseudocode of the improved ant colony algorithm is as follows (Algorithm 1):

Algorithm 1. (improved ant colony algorithm sequence (iacs))

Initialize α, β, τ_0, ρ, Q, n, Nc_max, S_x, S_y, V_x, V_y, a_x, a_y
While($nc <= Nc\_max$)
    Initalize *refilling time* T_k, $tabU_k$, and $allowed_k$
    For ($k=1$; $k < m$; $k++$)
        While ($allowed_k$!= null)
        Build $tabU_k$ by applying n-1 times the following step
        $tabU_k = tabU_k + (i, j)$ and $allowed_k = allowed_k - (i, j)$
        Choose the next node j probability, calculate P_{ij}^k according to the Formula (11):

$$P_{ij}^k = \begin{cases} \dfrac{\tau_{ij}^\alpha \eta_{ij}^\beta}{\sum_{j \in allowed_k} \tau_{ij}^\alpha \eta_{ij}^\beta} & if \quad j \in allowed_k \\ 0 & other \end{cases}$$

        Calculate the optimal tour L_k of ant k
        End while
        For every edge (i, j)
            Update the pheromone according to the Formulas (9) and (12):
$$\tau_{ij}(t+1) = (1 - \rho)\tau_{ij}(t) + \Delta\tau_{ij}(t)$$
        $\Delta\tau_{ij}(t)$ by applying the rule:
$$\Delta\tau_{ij}(t) = \sum_{k=1}^{m} \Delta\tau_{ij}^k(t)$$
        where $\Delta\tau_{ij}^k$ is the same as Equation (15):
$$\Delta\tau_{ij}^k(t) = \frac{Q}{L_k'}$$
        End for
    End for
    Update the historical optimal storing path d_{ij}
        where d_{ij} is updated as Equation (13)
    For every edge (i, j) do
    $\tau_{ij} = (nc+1) = \tau_{ij} (nc)$
    End For
End while

4. Experiments and Analysis

4.1. Test Samples

Some experiments were conducted to test the quality of searching restoring route to different selection methods by the random sequence (rs), the traditional ant colony algorithm sequence (acs), and the improved ant colony algorithm sequence (iacs). Test samples, two kinds of medicine, are described as follows. The width of an ordinary medicine box was in the range of 35–110 mm and the height was in the range of 10–60 mm. The test samples are shown in Table 4.

Table 4. Medicine information of test samples.

Number	Medicine Name	Pharmaceutical Manufacturers	H (mm)	Max
1	Ritodrine Hydrochloride Tablets	Biotech, Ltd. TAIWAN	10.85	30
2	Polyferose Capsules	Qingdao Guofeng Pharmaceutical Co., Ltd.	15.90	20

The test category included 5, 10, 15, and 20 selections. As shown in Table 5, the test categories of two test samples are listed. SPP indicates a picking shortage, and ORT indicates picking order according to storage time. For example, 101 in the table represents the first of the first layer in the drug storage equipment. In the first test sample, the number of SPPs was balanced, so the robot had five pick points for each drug recovery. In the second test sample, the number of SPPs was different, and so, the picking points of the robot for each drug for recovery were changed from 8 to 2.

Table 5. Shortage and order of test samples.

Number	Picking	101	103	109	202	208	211	212	303	305	307
	ORT	11	5	19	17	6	3	18	10	14	20
	SPP	6	3	4	5	7	5	2	7	7	10
1	Picking	408	409	502	507	512	601	608	707	709	711
	ORT	16	15	8	4	12	7	13	2	9	1
	SPP	4	6	3	8	4	4	5	8	7	3
Number	Picking	103	104	107	201	207	202	210	302	305	301
	ORT	19	1	4	16	20	14	3	8	2	13
	SPP	5	4	2	6	5	1	2	1	4	6
2	Picking	409	403	404	509	108	602	606	611	706	707
	ORT	10	7	12	18	17	6	11	15	5	9
	SPP	4	3	1	10	3	3	2	2	1	3

4.2. Results and Analysis of Test Samples

The simulation experimental data of the test samples are expressed in Table 6.

Table 6. Simulation results of different storing models.

Number	Storing Model	Route	Storing Route Value (m)	Route Rate	Storing Time (s)	Time Rate
a1-1	rs	0-103-507-211-707-711-0	3.7261	0	11.3895	0
b1-1	acs	0-103-211-711-707-507-0	3.0827	17.26%	10.3353	9.26%
c1-1	iacs	0-103-507-707-211-711-0	3.5261	5.37%	10.1490	10.89%
a1-2	rs	0-208-601-502-709-303-0	7.3718	0	22.8189	0
b1-2	acs	0-303-208-709-502-601-0	6.0045	18.55%	19.7200	13.58%
c1-2	iacs	0-208-709-601-502-303-0	6.4561	12.42%	19.2517	15.63%
a1-3	rs	0-512-608-305-409-101-0	10.9175	0	33.9041	0
b1-3	acs	0-305-409-512-608-101-0	8.7301	20.03%	29.6905	12.43%
c1-3	iacs	0-512-409-608-305-101-0	9.3088	14.74%	28.3970	16.24%
a1-4	rs	0-408-202-212-109-307-0	14.8050	0	45.9859	0
b1-4	acs	0-109-212-408-307-202-0	11.3332	23.45%	39.0516	15.08%
c1-4	iacs	0-212-109-307-408-202-0	12.1811	17.72%	37.3981	18.67%
a2-1	rs	0-104-305-210-107-706-602-403-302-0	3.2082	0	12.0624	0
b2-1	acs	0-104-305-107-210-706-602-403-302-0	3.0132	6.08%	11.1822	7.30%
c2-1	iacs	0-302-104-305-107-210-706-403-602-0	3.3867	-5.56%	10.8914	9.71%
a2-2	rs	0-611-202-301-404-707-409-606-0	7.8976	0	25.3772	0
b2-2	acs	0-202-309-611-707-606-404-301-0	5.7759	26.87%	20.9768	17.34%
c2-2	iacs	0-611-409-707-606-404-202-301-0	6.4479	18.36%	19.7844	22.04%
a2-3	rs	0-108-509-207-0	10.2047	0	32.5624	0
b2-3	acs	0-108-207-509-0	8.1137	20.49%	27.8004	14.62%
c2-3	iacs	0-108-207-509-0	8.7857	13.91%	26.6080	18.29%
a2-4	rs	0-103-204-0	11.1096	0	36.4167	0
b2-4	acs	0-103-204-0	9.0186	18.82%	31.6547	13.08%
c2-4	iacs	0-103-204-0	9.6906	12.77%	30.4623	16.35%

The comparison can be summarized as follows.

(1) In the first test sample, compare the recovery paths between the three models. They were completely different from other models. The recovery path includes all recovery paths in the storage process of a drug. The recovery path lengths of rs, acs, and iacs were 14.8050, 11.3332, and 12.1811, respectively. Based on this, we can see that the recovery path of acs was the shortest. Compared with rs, the best recovery path rate

of acs and iacs was more than 23.45% and 17.72%, respectively. It can be seen that the optimized rate between the acs and the iacs based on the rs was more than 5.37%. Moreover, restoring time of the acs and the iacs were the difference. Both of them required shorter time than the rs on restoring route. The optimized rate of restoring time of both the acs and the iacs were shorter than 15.08% and 18.67% when compared with that of the rs. It can be seen that the acs can search shorter sorting routes than the rs and the iacs. In terms of storing time, the iacs was better than the rs and the acs. Accordingly, the objective function of the iacs was to achieve the shortest sorting time, and the objective function of the acs was to search the shortest storing route.

(2) In the second test sample, comparing restoring paths among the three models, the rs were different from the acs and the iacs. Only one storing route was same as the acs and the iacs in four storing paths. Because it included two picking points which only has kind of storing route. it could not be optimized. The lengths of storing route of the rs, the acs, and the iacs were 11.1096, 9.0186, and 9.6906, respectively. Therefore, the acs is the best storing model to search the shortest storing route. The results showed that when the storage path of rs was properly arranged, the storage path of iacs was longer than that of rs. However, this rarely happens. However, iacs was better than rs and ac in terms of storage time. Compared with rs and acs, the optimal storage time of iacs was 16.35% and 3.27% shorter, respectively. It can be seen that iacs was the best storage model for searching storage paths and needed the shortest time.

(3) In Table 6, as the number of pickings increased, the optimized rate on the storing route value and storing time also increased. The iacs was obviously shorter than the acs on storing time.

The operation experimental data of the test samples are expressed in Table 7.

Table 7. Operation test results of different storing models.

Number	Storing Model	Operation Time (s)	Storing Route Time (s)	Operation Rate	Storing Rate
a1-1	rs	46.490	11.390	0	0
b1-1	acs	45.435	10.335	2.32%	9.26%
c1-1	iacs	45.249	10.149	2.67%	10.89%
a1-2	rs	94.319	22.819	0	0
b1-2	acs	91.220	19.720	3.29%	13.58%
c1-2	iacs	90.752	19.252	3.78%	15.63%
a1-3	rs	141.804	33.904	0	0
b1-3	acs	137.591	29.691	2.97%	12.43%
c1-3	iacs	136.297	28.397	3.88%	16.24%
a1-4	rs	186.386	45.986	0	0
b1-4	acs	179.452	39.052	3.72%	15.08%
c1-4	iacs	177.798	37.398	4.61%	18.67%
a2-1	rs	38.062	12.062	0	0
b2-1	acs	37.182	11.182	2.31%	7.30%
c2-1	iacs	36.891	10.891	3.08%	9.71%
a2-2	rs	76.077	25.377	0	0
b2-2	acs	71.677	20.977	5.78%	17.34%
c2-2	iacs	70.484	19.784	7.35%	22.04%
a2-3	rs	107.962	32.562	0	0
b2-3	acs	103.200	27.800	4.41%	14.62%
c2-3	iacs	102.008	26.608	5.51%	18.29%
a2-4	rs	124.817	36.417	0	0
b2-4	acs	120.055	31.655	3.82%	13.08%
c2-4	iacs	118.862	30.462	4.77%	16.35%

The comparison can be summarized as follows.

(1) In Table 7, comparing with the operation time or storing route time of the first test sample, the iacs was better than the rs and the acs. The optimized rate of operation time was more than 4.61% when comparing the iacs with the rs. Additionally, the optimized rate of operation time was more than 18.67% when comparing the iacs with the rs. The reason for this is that the storage time included in the operation time was not optimized. Optimization of the acs and the iacs were negative to the storing route. It can be seen that the iacs was better than the acs. The reason for this is that acceleration and deceleration of motor were set as "S" curve in actual operation and were set as oblique straight line in the simulation. The storing route of the iacs was longer than the acs's in the first test sample. However, the iacs required less time to operate through this storing path than the acs. Additionally, the time used in operation, which was spent on storing path, was less than that used in the simulation when the storing path was short.

(2) In Table 7, comparing with the operation time and storing route time of the second test sample, the iacs was also better than the rs and the acs. The optimized rate of operation time was more than 4.77% when comparing the iacs with the rs. The optimized rate of operation time was more than 16.35% when comparing the iacs with the rs. It can be seen that the optimized effect was similar to the first test sample. The optimization of the iacs was better than the acs. Compared with the rs, optimized rate of operation storing time with the iacs and the acs were both reducing instead. The reason for this is that with the increase in selection, more rs storage time was required, which was set as the denominator and had a greater impact.

(3) In conclusion, the iacs was better than the rs on storing time. As picking increased, the iacs was obviously better than the acs. Based on the storing time of the rs, compared with the acs, the optimized rate of storing time with the iacs was more than 0.35%, even up to 1.57%.

5. Conclusions

This paper first increased the maximum storage capacity of the device by optimizing the storage space of the drug storage device, and then, optimized the main path of the device. It not only improved the effective use of the space of the device, but also enhanced the efficiency of drug delivery by the device.

Through the research on the inclination of the storage tank, 18° was selected to meet the minimum angle used. Therefore, under the condition of ensuring the smooth delivery of the drug delivery equipment and the largest number of storage, the storage space of the equipment was optimized, and the optimization result was 10% higher than the space of the most conventional angle (21.8°).

Based on this space, a multiple local TSP model based on improved ant colony optimization was proposed to increase storing efficiency and consider the acceleration and deceleration of the motor during actual storing process. Compared with the rs, both of the acs and the iacs were better on searching storing route and the storing time of the storing route was shorter. Compared with the rs, the optimized rate of storing route with the acs was more than 3.72% in actual operation. Compared with the rs, the optimized rate of storing route with the iacs was more than 4.61% in actual operation. Additionally, therefore, the iacs is a suitable storing model in the pharmaceutical logistics robot. Meanwhile, it was observed that storing time of the iacs was shorter than that of the acs. Moreover, it was more evident when the pickings increased. Based on the storing time of the rs, compared with the acs, the optimized rate of storing time with the iacs was more than 0.35%, even up to 1.57%. So, acceleration and deceleration were considered in the model, which is necessary.

Author Contributions: Conceptualization, L.Z. and M.L.; methodology, M.L. and B.D.; validation, L.Z., S.L. and M.L.; resources, L.Z. and S.L.; experiment, M.L.; writing, L.Z., M.L. and B.D.; funding acquisition, L.Z. All authors have read and agreed to the published version of the manuscript.

Funding: This research was funded by the Science and Technology Research Program of Chongqing Education Commission of China (KJQN202003202, KJQN202001127).

Data Availability Statement: The research simulation and experimental data of our paper are reflected in the pictures and tables in the manuscript. Therefore, no new dataset link was established.

Acknowledgments: We thank Hu for his suggestions on revising the English style of this paper.

Conflicts of Interest: The authors declare that there are no conflict of interest regarding the publication of this paper.

References

1. Tan, L.; Chen, W.; He, B.; Zhu, J.; Cen, X.; Feng, H. A survey of prescription errors in paediatric outpatients in multi-primary care settings: The implementation of an electronic pre-prescription system. *Front. Pediatr.* **2022**, *10*. [CrossRef] [PubMed]
2. Anjalee, J.A.L.; Rutter, V.; Samaranayake, N.R. Application of failure mode and effects analysis (FMEA) to improve medication safety in the dispensing process—A study at a teaching hospital, Sri Lanka. *BMC Public Health* **2021**, *21*, 1430. [CrossRef]
3. Nermeen, A.; Safa, A.; Sayed, S.; Mojeba, H. Studying the medication prescribing errors in the egyptian community pharmacies. *Asian J. Pharm.* **2018**, *12*, 25–30.
4. Hesse, M.; Thylstrup, B.; Seid, A.K.; Tjagvad, C.; Clausen, T. A retrospective cohort study of medication dispensing at pharmacies: Administration matters! *Drug Alcohol Depend.* **2021**, *225*, 108792. [CrossRef] [PubMed]
5. Jacobson, M.G.; Chang, T.Y.; Earle, C.C.; Newhouse, J.P. Physician agency and patient survival. *J. Econ. Behav. Organ.* **2017**, *134*, 27–47. [CrossRef]
6. Chen, F.Y. Current situation and new progress of automated pharmacy. *Mingyi Dr.* **2020**, *4*, 277.
7. Cai, Y.X.; Zhang, M.L. Study on dispensing optimization of integrated traditional Chinese medicine dispensing system. *Light Ind. Mach.* **2019**, *37*, 80–88.
8. Jin, H.; Yun, C.; Wang, W.; Li, D.J. Application and research of the Clip Type Manipulator. In *Mechanisms and Machine Science*; Springer: Cham, Switzerland, 2016; pp. 841–851. [CrossRef]
9. Lin, Y.; Cai, Z.; Huang, M.; Gao, X.; Yu, G. Evaluation of development status and application effect of outpatient pharmacy automatic dispensing system in mainland China. *Chin. J. Mod. Appl. Pharm.* **2020**, *37*, 1131–1138.
10. Ozden, S.G.; Smith, A.E.; Gue, K.R. A computational software system to design order picking warehouses. *Comput. Oper. Res.* **2021**, *132*, 105311. [CrossRef]
11. Mulac, A.; Mathiesen, L.; Taxis, K.; Granås, A.G. Barcode medication administration technology use in hospital practice: A mixed-methods observational study of policy deviations. *BMJ Qual. Saf.* **2021**, *30*, 1021–1030. [CrossRef]
12. Jin, H.; He, Q.; He, M.; Lu, S.; Hu, F.; Hao, D. Optimization for medical logistics robot based on model of traveling salesman problems and vehicle routing problems. *Int. J. Adv. Robot. Syst.* **2021**, *18*, 17298814211022539. [CrossRef]
13. Sng, Y.L.; Ong, C.K.; Lai, Y.F. Approaches to outpatient pharmacy automation: A systematic review. *Eur. J. Hosp. Pharm.* **2019**, *26*, 157–162. [CrossRef]
14. Ahtiainen, H.K.; Kallio, M.M.; Airaksinen, M.; Holmström, A.R. Safety, time and cost evaluation of automated and semi-aut-omated drug distribution systems in hospitals:a systematic review. *Eur. J. Hosp. Pharm.* **2020**, *27*, 253–262. [CrossRef] [PubMed]
15. Zhu, S.; Wang, H.; Zhang, X.; He, X.; Tan, Z. A decision model on human-robot collaborative routing for automatic logis-tics. *Adv. Eng. Inform.* **2022**, *53*, 101681. [CrossRef]
16. Keung, K.; Lee, C.; Ji, P. Industrial internet of things-driven storage location assignment and order picking in a resource synchronization and sharing-based robotic mobile fulfillment system. *Adv. Eng. Inform.* **2022**, *52*, 101540. [CrossRef]
17. Boysen, N.; de Koster, R.; Füßler, D. The forgotten sons: Warehousing systems for brick-and-mortar retail chains. *Eur. J. Oper. Res.* **2020**, *288*, 361–381. [CrossRef]
18. Jin, H.; Yun, C.; Gao, X. Application and research of the refilling process with Clip Type Manipulator. In Proceedings of the 2015 IEEE International Conference on Robotics and Biomimetics (ROBIO), Zhuhai, China, 6–9 December 2015; pp. 775–780. [CrossRef]
19. Jin, H.; He, Q.; He, M.; Hu, F.; Lu, S. New method of path optimization for medical logistics robots. *J. Robot. Mechatronics* **2021**, *33*, 944–954. [CrossRef]
20. Yang, D.; Wu, Y.; Ma, W. Optimization of storage location assignment in automated warehouse. *Microprocess. Microsyst.* **2020**, *80*, 103356. [CrossRef]
21. Wu, L.; Huang, X.; Cui, J.; Liu, C.; Xiao, W. Modified adaptive ant colony optimization algorithm and its application for solving path planning of mobile robot. *Expert Syst. Appl.* **2023**, *215*, 1–22. [CrossRef]
22. Li, S.; Zhang, M.; Wang, N.; Cao, R.; Zhang, Z.; Ji, Y.; Li, H.; Wang, H. Intelligent scheduling method for multi-machine cooperative operation based on NSGA-III and improved ant colony algorithm. *Comput. Electron. Agric.* **2023**, *204*, 107532. [CrossRef]
23. Shi, E.; Lou, L.; Warburton, L.; Rubinsky, B. 3D Printing in Combined Cartesian and Curvilinear Coordinates. *J. Med. Devices* **2022**, *16*, 044502. [CrossRef]

24. Lanza, G.; Passacantando, M.; Scutellà, M.G. Assigning and sequencing storage locations under a two level storage policy: Optimization model and matheuristic approaches. *Omega* **2021**, *108*, 102565. [CrossRef]
25. Xu, X.; Ren, C. A novel storage location assignment in multi-pickers picker-to-parts systems integrating scattered storage, demand correlation, and routing adjustment. *Comput. Ind. Eng.* **2022**, *172*, 1–15. [CrossRef]
26. Ma, Z.; Wu, G.; Ji, B.; Wang, L.; Luo, Q.; Chen, X. A novel scattered storage policy considering commodity classification and correlation in robotic mobile fulfillment systems. *IEEE Trans. Autom. Sci. Eng.* **2022**, 1–14. [CrossRef]
27. Zuñiga, J.B.; Martínez, J.A.S.; Fierro, T.E.S.; Saucedo, J.A.M. Optimization of the storage location assignment and the picker-routing problem by using mathematical programming. *Appl. Sci.* **2020**, *10*, 534. [CrossRef]
28. Lu, F.; Feng, W.; Gao, M.; Bi, H.; Wang, S. The fourth-party logistics routing problem using ant colony system-improved grey wolf optimization. *J. Adv. Transp.* **2020**, *2020*, 8831746. [CrossRef]
29. Zhang, Z.; Xu, Z.; Luan, S.; Li, X.; Sun, Y. Opposition-Based Ant Colony Optimization Algorithm for the Traveling Salesman Problem. *Mathematics* **2020**, *8*, 1650. [CrossRef]
30. Jin, H.; Wang, W.; Cai, M.; Wang, G.; Yun, C. Ant colony optimization model with characterization-based speed and multi driver for the refilling system in hospital. *Adv. Mech. Eng.* **2017**, *9*, 1687814017713700. [CrossRef]

Disclaimer/Publisher's Note: The statements, opinions and data contained in all publications are solely those of the individual author(s) and contributor(s) and not of MDPI and/or the editor(s). MDPI and/or the editor(s) disclaim responsibility for any injury to people or property resulting from any ideas, methods, instructions or products referred to in the content.

Article

High Precision Hybrid Torque Control for 4-DOF Redundant Parallel Robots under Variable Load

Shengqiao Hu [1,2], Houcai Liu [1], Huimin Kang [1,*], Puren Ouyang [3], Zhicheng Liu [1] and Zhengjie Cui [1]

[1] Department of Mechanical Engineering, Hunan University of Science and Technology, Xiangtan 411201, China; hushengqiao@knu.ac.kr (S.H.); houcailiu@163.com (H.L.)
[2] Department of Mechanical Engineering, Kyungpook National University, Daegu 41566, Republic of Korea
[3] Department of Aerospace Engineering, Toronto Metropolitan University, Toronto, ON M5B2K3, Canada; pouyang@torontomu.ca
* Correspondence: xykanghm@163.com

Abstract: As regards the impact and chattering of 4-DOF redundant parallel robots that occur under high-speed variable load operating conditions, this study proposed a novel control algorithm based on torque feedforward and fuzzy computational torque feedback hybrid control, which considered both the joint friction torque and the disturbance torque caused by the variable load. First of all, a modified dynamic model under variable load was established as follows: converting terminal load change to terminal centroid coordinate change, then mapping to the calculation of terminal energy, and lastly, establishing a dynamic model for each branch chain under variable load based on the Lagrange equation. Subsequently, torque feedforward was used to compensate for the friction torque and the disturbance torque caused by the variable load. Feedforward torques include friction torque and nonlinear disturbance torque under variable load. The friction torque is obtained by parameter identification based on the Stribeck friction model, while the nonlinear disturbance torque is obtained by real-time calculation based on the modified dynamic model under variable load. Finally, dynamic control of the robot under variable load was realized in combination with the fuzzy computational torque feedback control. The experimental and simulation results show that the motion accuracy of the fuzzy calculation torque feedback and torque feedforward control of the three drive joints of the robot under variable loads is 49.87%, 70.48%, and 50.37% lower than that of the fuzzy calculation torque feedback. Compared with pure torque feedback control, the speed stability of the three driving joints under fuzzy calculation torque feedback and torque feedforward control is 23.35%, 17.66%, and 25.04% higher, respectively.

Keywords: redundant parallel robot; joint space; variable load; torque feedback; torque feedforward

Citation: Hu, S.; Liu, H.; Kang, H.; Ouyang, P.; Liu, Z.; Cui, Z. High Precision Hybrid Torque Control for 4-DOF Redundant Parallel Robots under Variable Load. *Actuators* **2023**, *12*, 232. https://doi.org/10.3390/act12060232

Academic Editor: Chih Jer Lin

Received: 13 March 2023
Revised: 18 April 2023
Accepted: 23 April 2023
Published: 5 June 2023

Copyright: © 2023 by the authors. Licensee MDPI, Basel, Switzerland. This article is an open access article distributed under the terms and conditions of the Creative Commons Attribution (CC BY) license (https://creativecommons.org/licenses/by/4.0/).

1. Instruction

Parallel robots are widely used in industrial production lines due to their high stiffness, speed, motion accuracy, and compact structure, making them particularly suitable for handling, sorting, and packing light objects at high speeds. However, when the parallel robot grasps and carries objects of different shapes and masses, the load mass becomes unknown and time-varying. This not only causes significant variation in the force and inertia matrix of each component of the robot arm but also results in chattering and impact when the robot runs at high speeds. These issues can trigger dynamic coupling between the robot mechanisms and affect the motion control accuracy of the robot.

As parallel robots are often used to handle objects of varying shapes and masses, researchers have conducted a series of studies on the motion control of robots under variable loads. For instance, Ref. [1] proposed an improved PSO algorithm for the parameter identification of SCARA robots, taking into account the influence of loads below 1.2 kg on the inertial matrix, while ignoring friction. The approach effectively improves the identification accuracy. Ref. [2] proposed a control scheme for a 3-DOF parallel robot to deal with uncertain disturbances while ignoring friction torque. For real-time control of redundant parallel robots facing unknown loads, Ref. [3] adopted a control algorithm with a high gain observer to automatically adjust the robot's load parameters and achieve motion control under variable load conditions. Ref. [4] proposed an improved scheme combining a variable structure compensator and calculated torque control for Delta robots with an uncertain load. Although the independent control scheme based on the joint friction torque and the variable load can reduce the harm to control stability, it is difficult to identify dynamic load parameters in practical applications, and the change of load will lead to the synchronous change of friction torque of each joint, making it difficult to meet high precision control requirements using independent control methods that rely solely on individual parameters.

Therefore, some research focuses on the design of force control algorithms, such as adaptive control, model reference adaptive control, fuzzy control, etc., to achieve the stable movement of robot end-effectors under external load changes. Ref. [5] presents our research on the adaptive finite-time neural network control scheme for redundant parallel manipulators. The proposed controller is based on a fully-tuned radial basis function neural network (RBFN), non-singular fast terminal sliding mode control (NFTSMC), and nonlinearity in the output feedback. The RBFN, with fully online updating of output weights and Gaussian function center and variance, is used to estimate system uncertainties and disturbances. The proposed method has several advantages over other existing methods, such as robustness, fast response, no singularity, higher accuracy, finite-time convergence, and better tracking control performance. Finally, the stability of the parallel manipulator is guaranteed by the Lyapunov theory. Ref. [6] proposes a controller design method based on fuzzy sliding mode control. The controller uses adaptive algorithms to estimate the uncertainty of the mechanism's parameters and uses fuzzy logic to control the motion trajectory of the mechanism. At the same time, sliding mode control is used to suppress external disturbances and uncertainties in the system. It is important to find suitable sliding surfaces and sliding modes, as the sliding mode controller is prone to chattering and its parameter tuning can be complex. The control effect is also affected by parameter changes.

Some researchers have focused on neural network control for robot motion control under unknown and varying loads. In [7], an adaptive control method based on neural dynamic surface control was proposed to address this problem. This method learns the unknown load dynamics model using a neural network to achieve the adaptive adjustment of robot motion. The method has the advantages of being real-time and having strong adaptability, making it suitable for various robot systems. Another approach proposed in [8] is a robot motion control method based on robust adaptive neural network control, which can handle unknown loads and disturbances at the end effector of the robot. This method has good robustness and adaptivity and can improve the control accuracy of the robot under varying loads. However, achieving effective robot motion control under varying loads is a complex and important problem that requires comprehensive consideration of the dynamic characteristics of the robot system, the design of control algorithms, and real-time feedback control factors. Currently, research in related fields is still ongoing and developing.

This paper proposes a hybrid torque control approach that combines feedforward compensation and feedback control to achieve improved control accuracy. By using feedforward compensation, dynamic response time is reduced, while fuzzy control helps handle uncertainty. Additionally, the approach incorporates calculating torque based on the dynamic model to better describe the motion characteristics of the robot. It takes 4-DOF redundant parallel robots as the research object, takes 0–5 kg varying load as the excitation, and proposes a novel hybrid control algorithm that considers both the joint friction torque and the disturbance force feedforward and fuzzy computational torque feedback under the action of the variable load. Firstly, the improved dynamic model was constructed under variable loads. Then, according to the pose and velocity of the system, the joint torques, friction torques, and disturbance torques were obtained by using the Lagrange equation and the Stribeck model. Finally, the fuzzy calculated torque was used to adjust the torque feedback, and the time-varying characteristics of friction torque and disturbance torque of each joint were converted into the current change control of each joint drive motor by the torque feedforward control algorithm, to realize the stability control of the drive motor control system.

2. Construction of Time-Varying Dynamics Model of 4-DOF Redundant Parallel Robots under Variable Load

Figure 1 shows the structural model of the 4-DOF redundant parallel robot and the coordinate relationship of each joint at any time.

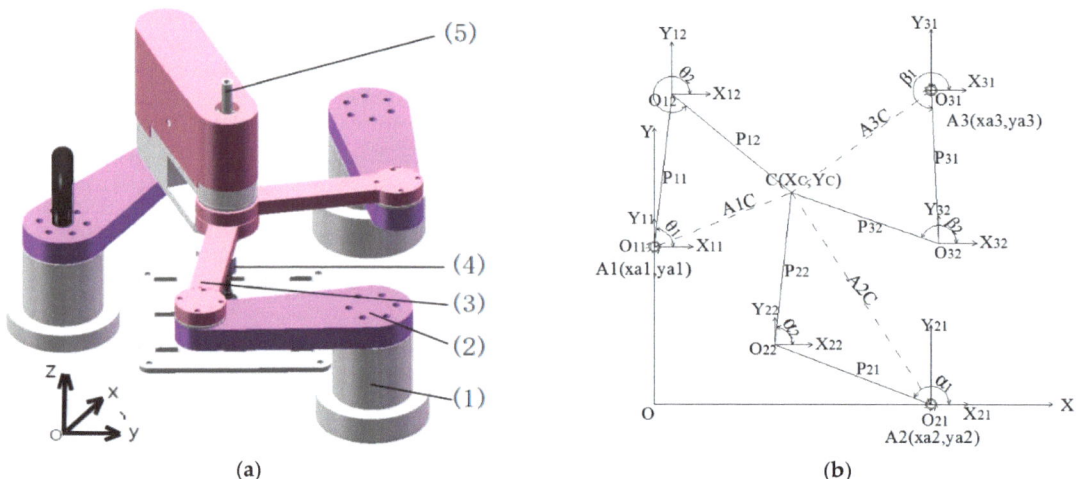

Figure 1. Overall structure diagram of a 4-DOF redundant parallel robot. (**a**) Structure diagram of a 4-DOF redundant parallel robot: (1) Pedestal; (2) Driving shaft; (3) Driven shaft; (4) Air gripper; (5) End effector. (**b**) Construction of coordinate system for 4-DOF redundant Parallel Robot.

To facilitate the analysis of the forces on each joint, any branched chain in Figure 1a was separated from the redundant mechanism that generated over-constraint, and the separated structure was shown in Figure 2.

As shown in Figure 2, in the process of the robot extracting goods, the change of load will be reflected in its mass and volume, thus leading to the change of the position of the system's centroid. When applying the Lagrange equation to calculate the torque of each joint, the position and posture of each joint and the coordinate position of the system's centroid should be determined first. Suppose the three branch chains of the 4-DOF redundant parallel robot subscripts i are 1, 2, and 3, respectively. For a single chain, its base center at point $A\left(x_{a_i}, y_{a_i}, h\right)$, driving shaft AB and driven shaft BC joint points for $B\left(x_{b_i}, y_{b_i}, h\right)$, driven shaft, BC, and end-effector HE joint points for $C\left(x_{c_i}, y_{c_i}, h\right)$. The AB rod length, centroid, centroid distance AQ, and mass are l_1, $Q\left(x_{qi}, y_{qi}, z_{qi}\right)$, P_{i1}, and m_{i1}, respectively. The BC rod length, centroid, centroid distance BG, and mass are l_2, $G\left(x_{gi}, y_{gi}, z_{gi}\right)$, and m_{i2}, respectively. The AB axis and BC axis are in the same plane of xoy. End-effector HE is a member of the yoz plane with rod length l_3, centroid $F(x_{fi}, y_{fi}, z_{fi})$, centroid distance HF P_F, and mass $\frac{1}{3}m_F$, which can move up and down in the yoz plane. $D\left(x_{di}, y_{di}, z_{di}\right)$ is the load loaded by end-effector HE. Its distance between the load and base plane is Z_C, and the mass is $\frac{1}{3}m_D$. h_x is from the base plane to D. The relevant model parameter information can be found in reference Ref. [9].

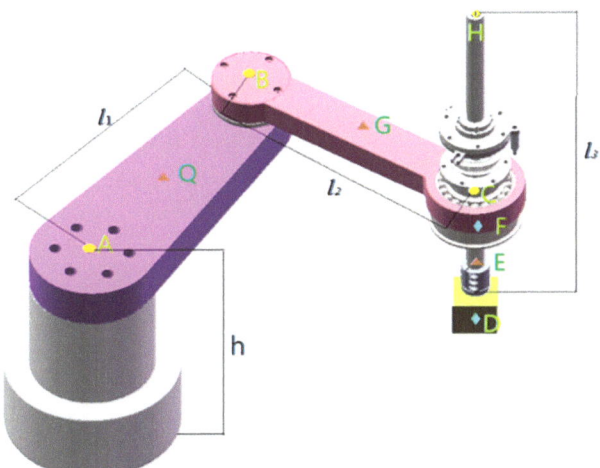

Figure 2. Simplified schematic diagram of the single-chain mechanism.

According to the above assumptions, the Lagrange multiplier is applied to obtain the joint torque of the driving joints, as shown in Equation (1) [10].

$$\tau = \frac{d}{dt}\frac{\partial L}{\partial \dot{q}} - \frac{\partial L}{\partial q} + \Phi_q^T \lambda \tag{1}$$

L is the sum of the system's kinetic energy and potential energy in Equation (1), where $q_{(6\times1)} = [\theta_1, \theta_2, \alpha_1, \alpha_2, \beta_1, \beta_2]^T$ is the vector of the system coordinates, $\tau_{(6\times1)}$ is the vector of the corresponding external force/torque, and $\lambda_{(6\times1)}$ is the vector of the Lagrangian multipliers associated with the constraint Torques $\Phi_q^T \lambda$.

The complete constraint equations [10] are derived by expression (with i = 1, 2, 3) as follows:

$$\begin{cases} f(q) = [f_1^T, f_2^T, f_3^T]^T \\ \Phi_q \dot{q} = 0 \text{ with } \Phi_q \dot{q} = \frac{\partial f(q)}{\partial q} \end{cases} \tag{2}$$

where:

$$\Phi_q^T = \begin{bmatrix} \Phi_{q1}^T, \Phi_{q2}^T, \Phi_{q3}^T \end{bmatrix}^T \tag{3}$$

Assume that the centroid of end-effector HE in Figure 2 is point F and the load centroid is point D. Then the end-effector HE and the load are regarded as a whole and expressed as FD, and the equivalent centroid point of the two is E(x_{ei}, y_{ei}, z_{ei}). When the load is a variable, the coordinates of the centroid E (x_{ei}, y_{ei}, z_{ei}) of the end-effector FD are as follows:

$$\begin{cases} x_{ei} = x_{ai} + l_1 \cos(q_{i1}) + l_2 \cos(q_{i2}) \\ y_{ei} = y_{ai} + l_1 \sin(q_{i1}) + l_2 \sin(q_{i2}) \\ z_{ei} = h_x + L_{CD2} \end{cases} \tag{4}$$

In this equation, q_{i1} is the driving angular displacement. q_{i2} is the driven angular displacement. The driven angle is not a variable here, but is only for the convenience of representing and calculating the energy and force/moment of each branch chain. The calculation formula of the driven angle is as follows:

$$\begin{cases} X_C = x_{a1} + l_1 \cos(q_{11}) + l \cos(q_{12}) = x_{a2} + l_1 \cos(q_{21}) + l_2 \cos(q_{22}) \\ \qquad = x_{a3} + l_1 \cos(q_{31}) + l_2 \cos(q_{32}) \\ Y_C = y_{a1} + l_1 \sin(q_{11}) + l_2 \sin(q_{12}) = y_{a2} + l_1 \sin(q_{21}) + l_2 \sin(q_{22}) \\ \qquad = y_{a3} + l_1 \sin(q_{31}) + l_2 \sin(q_{32}) \end{cases} \tag{5}$$

The driven angle q_{i2} is expressed as:

$$q_{i2} = \arccos\left(\frac{X_C - x_{ai} - l_1 \cos q_{i1}}{l_2}\right) \qquad (6)$$

At the end effector, L_{CD2} is satisfied with the following equation:

$$\begin{cases} m_F L_{CD1} = m_D L_{CD2} \\ L_{CD1} + L_{CD2} = L_3 - P_C \end{cases} \qquad (7)$$

In this equation, L_{CD1} and L_{CD2} are the lengths of EF and DE respectively, and both are variables.

Because the end effector and the load are jointly supported by three branches, the weight of the end effector and the end load is distributed to the three branches. Thus, it can be concluded that the equivalent mass m_{FD} of the single-chain end-effector HE and the load equivalent volume FD is:

$$m_{FD} = \frac{1}{3}(m_D + m_F) \qquad (8)$$

In this equation, m_F and m_D are the mass of the end-effector and load, respectively.

In combination with Equation (1), the Lagrange multiplier of equivalent volume FD is assumed to be L_3, and its calculation equation is as follows:

$$L_3 = E_{K3} - E_{P3} \qquad (9)$$

In this equation, the potential energy of equivalent volume FD is $E_{P3} = m_{FD} g h_x$; I kinetic energy E_{K3} is:

$$E_{K3} = \frac{1}{2} m_{FD} \left(\dot{x}_{ei}^2 + \dot{y}_{ei}^2 + \dot{z}_{ei}^2\right) \qquad (10)$$

Similarly, the Lagrange multiplier L_1 of the AB manipulator is:

$$L_1 = E_{K1} - E_{P1} \qquad (11)$$

$$E_{K1} = \frac{1}{2} m_{i1} \left(\dot{x}_Q^2 + \dot{y}_Q^2 + \dot{z}_Q^2\right) \qquad (12)$$

In this equation, $\dot{x}_Q$, $\dot{y}_Q$, and $\dot{z}_Q$ can be obtained according to the centroid coordinates of the AB axis, namely:

$$\begin{cases} x_Q = x_{a_i} + P_{i1} \cos q_{i1} \\ y_Q = y_{a_i} + P_{i1} \sin q_{i1} \\ z_Q = h \end{cases} \qquad (13)$$

Similarly, the Lagrange multiplier L_2 of the BC manipulator is:

$$L_2 = E_{K2} - E_{P2} \qquad (14)$$

$$E_{K2} = \frac{1}{2} m_{i2} \left(\dot{x}_G^2 + \dot{y}_G^2 + \dot{z}_G^2\right) \qquad (15)$$

The centroid coordinate equation of the BC axis is:

$$\begin{cases} x_G = x_{a_i} + l_1 \cos q_{i1} + P_{i2} \cos q_{i2} \\ y_G = y_{a_i} + l_1 \sin q_{i1} + P_{i2} \sin q_{i2} \\ z_G = h \end{cases} \qquad (16)$$

To sum up, the Lagrange multiplier L of a single-chain system can be expressed as follows:

$$L = \frac{1}{2} m_{i1} \left(\dot{x}_Q^2 + \dot{y}_Q^2 + \dot{z}_Q^2\right) + \frac{1}{2} m_{i2} \left(\dot{x}_G^2 + \dot{y}_G^2 + \dot{z}_G^2\right) + \frac{1}{2} m_{FD} \left(\dot{x}_{ei}^2 + \dot{y}_{ei}^2 + \dot{z}_{ei}^2\right) + m_{FD} g h_x \qquad (17)$$

Combining Equations (1) and (17), the solution of joint torques of every single chain is as follows:

$$\tau_{di} = \begin{pmatrix} \tau_{d1} \\ \tau_{d2} \end{pmatrix} = \begin{pmatrix} D_{11} & D_{12} \\ D_{21} & D_{22} \end{pmatrix}\begin{pmatrix} \ddot{q}_{i1} \\ \ddot{q}_{i2} \end{pmatrix} + \begin{pmatrix} E_{11} & E_{12} \\ E_{21} & E_{22} \end{pmatrix}\begin{pmatrix} \dot{q}_{i1}^2 \\ \dot{q}_{i2}^2 \end{pmatrix} + \begin{pmatrix} F_{11} & F_{12} \\ F_{21} & F_{22} \end{pmatrix}\begin{pmatrix} \dot{q}_{i1}\dot{q}_{i2} \\ \dot{q}_{i2}\dot{q}_{i1} \end{pmatrix} + G(q)\begin{pmatrix} q_{i1} \\ q_{i2} \end{pmatrix} + \Phi_{qi}^T\lambda \quad (18)$$

In the equation of (18):

$$D_{11} = m_{i1}p_{i1}^2 + m_{i2}l_1^2 + m_{i2}p_{i2}^2 + m_{i2}l_1p_{i2}\cos(q_{i1} - q_{i2}) + m_{FD}l_1^2 + m_{FD}l_2^2 + m_{FD}l_1l_2\cos(q_{i1} - q_{i2})$$
$$D_{12} = m_{i2}p_{i2}^2 + m_{i2}l_1p_{i2}\cos(q_{i1} - q_{i2}) + m_{FD}l_2^2 + m_{FD}l_1l_2\cos(q_{i1} - q_{i2})$$
$$D_{21} = m_{i2}l_1l_2 + m_{i2}l_1p_{2i}\cos(q_{i1} - q_{i2}) + m_{FD}l_1l_2 + m_{FD}l_1l_2\cos(q_{i1} - q_{i2})$$
$$D_{22} = m_{i2}p_{i2}^2$$
$$E_{11} = E_{22} = 0$$
$$E_{12} = E_{21} = m_{i2}l_1p_{i2}\sin(q_{i1} - q_{i2}) + m_{FD}l_1l_2\sin(q_{i1} - q_{i2})$$
$$F_{12} = F_{22} = 0$$
$$F_{11} = F_{21} = -m_{i2}l_1p_{i2}\sin(q_{i1} - q_{i2}) - m_{FD}l_1l_2\sin(q_{i1} - q_{i2})$$

Compared with the state without load, $m_{FD}l_1^2$, $m_{FD}l_2^2$, $m_{FD}l_1l_2$, $m_{FD}l_1l_2\cos(q_{i1} - q_{i2})$, and $m_{FD}l_1l_2\sin(q_{i1} - q_{i2})$ all show nonlinear time-varying disturbance characteristics, which can be further sorted out as:

$$\tau_{di} = D_i'q\ddot{q} + H_i'(q,\dot{q}) + G_i'(q) + \Phi_{qi}^T\lambda \quad (19)$$

In the equation of (19):

$$H_i'(q,\dot{q}) = \begin{pmatrix} E_{11} & E_{12} \\ E_{21} & E_{22} \end{pmatrix}\begin{pmatrix} \dot{q}_{i1}^2 \\ \dot{q}_{i2}^2 \end{pmatrix} + \begin{pmatrix} F_{11} & F_{12} \\ F_{21} & F_{22} \end{pmatrix}\begin{pmatrix} \dot{q}_{i1}\dot{q}_{i2} \\ \dot{q}_{i2}\dot{q}_{i1} \end{pmatrix}$$

In this equation, D_i' is the inertial matrix, H_i' is the Coriolis and centrifugal matrix, and G_i' is the gravity matrix.

$\begin{pmatrix} E_{11} & E_{12} \\ E_{21} & E_{22} \end{pmatrix}$ and $\begin{pmatrix} F_{11} & F_{12} \\ F_{21} & F_{22} \end{pmatrix}$ are the Positive Definite Symmetric Matrix. The energy corresponding to the inertia matrix under a single branch is discussed separately. For a single branched chain, the second type of Lyapunov is used to determine its stability, $E = \frac{1}{2}\dot{q}^T D\dot{q}$; thus, the derivative of E is as follows:

$$\dot{E} = \frac{1}{2}\dot{q}^T D\dot{q} + \dot{q}^T D\ddot{q} = -\frac{1}{2}\dot{q}^T\dot{q}(3m_{i2}l_1p_{i2} + 3m_{FD}l_1l_2)\sin(q_{i1} - q_{i2})(\dot{q}_{i1} - q_{i2} + q_{i1} - \dot{q}_{i2}))$$

Thus, the result of $\dot{E} < 0$ shows this system is stable.

To sum it up, the entire parallel robot system's dynamic model [11] is as follows:

$$\tau_d = D'q\ddot{q} + H'(q,\dot{q}) + G'(q) + \Phi_q^T\lambda \quad (20)$$

where $D' = \text{dig}(D_1', D_2', D_3')$, $H' = [H_1'^T, H_2'^T, H_3'^T]^T$, $\tau = [\tau_1^T, \tau_2^T, \tau_3^T]^T$, $\Phi_q^T\lambda = [\Phi_{q1}^T\lambda, \Phi_{q2}^T\lambda, \Phi_{q3}^T\lambda]^T$.

To eliminate the assumed Lagrangian multipliers of ideal constraint torque from Equation (1), a matrix $R_{(6\times3)}$ is assumed, which determines the null space of the matrix $\Phi_q(\Phi_q R = 0)$ refer to Refs. [12,13].

$$R^T\tau_d = R^T[D'q\ddot{q} + H'(q,\dot{q}) + G'(q)] \quad (21)$$

Φ_{qi} and R are shown as follows:

$$\Phi_{qi} = \begin{bmatrix} -l_{i1}S_{\theta i} - l_{i2}S_{\gamma i} - l_{i2}S_{\gamma i} \\ -l_{i1}C_{\theta i} + l_{i2}C_{\gamma i} - l_{i2}C_{\gamma i} \end{bmatrix} \quad (22)$$

$$R = \begin{bmatrix} \frac{C_{\gamma 1}}{l_{11}} & \frac{S_{\gamma 1}}{l_{11}} & \frac{-C_{\gamma 1}}{l_{11}} \\ \frac{-l_{11}C_{\theta 1}-l_{12}C_{\gamma 1}}{l_{11}l_{12}} & \frac{-l_{11}S_{\theta 1}-l_{12}S_{\gamma 1}}{l_{11}l_{12}} & \frac{-l_{11}C_{\theta 1}+l_{12}C_{\gamma 1}}{l_{11}l_{12}} \\ \frac{C_{\gamma 2}}{l_{21}} & \frac{S_{\gamma 2}}{l_{21}} & 0 \\ \frac{-l_{21}C_{\theta 2}-l_{22}C_{\gamma 2}}{l_{21}l_{22}} & \frac{-l_{21}S_{\theta 2}-l_{22}S_{\gamma 2}}{l_{21}l_{22}} & 0 \\ \frac{C_{\gamma 3}}{l_{31}} & \frac{S_{\gamma 3}}{l_{31}} & 0 \\ \frac{-l_{31}C_{\theta 3}-l_{32}C_{\gamma 3}}{l_{31}l_{32}} & \frac{-l_{31}S_{\theta 3}-l_{32}S_{\gamma 3}}{l_{31}l_{32}} & 0 \end{bmatrix} \quad (23)$$

where:

$$\begin{cases} \gamma i = q_{i2} \\ \theta i = q_{i1} \end{cases}$$

3. Establishment of Hybrid Torque Control Model

Under the condition of variable load, the load has the characteristics of time-varying and nonlinear strong coupling. In this case, the dual-torque feedforward decoupling control method is proposed for the measurable but uncontrollable friction torque and the disturbance torque caused by the operation of the variable load. Meanwhile, as the feedforward control has difficulty resisting other unknown disturbances, the fuzzy computational torque feedback control method is adopted to improve the stability and motion accuracy of the system. The control block diagram of the system after the combination of the two is shown in Figure 3.

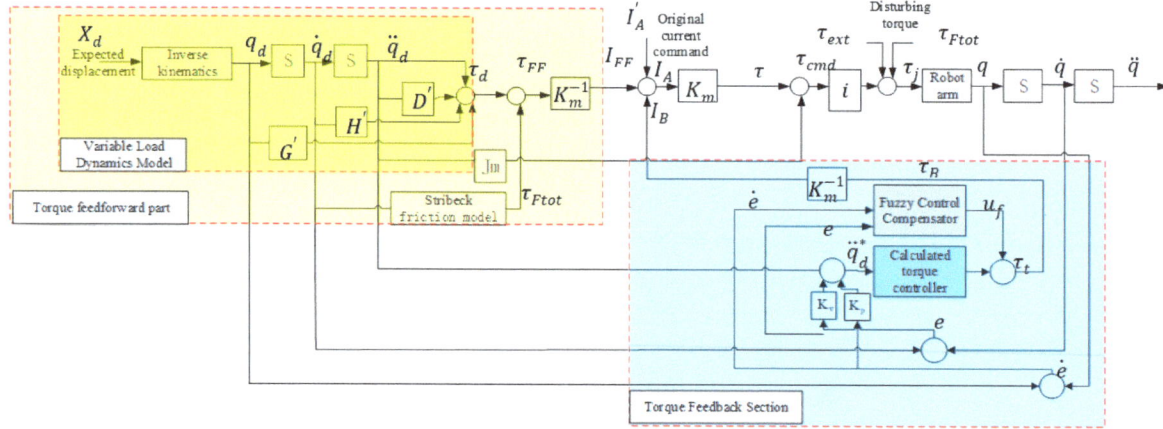

Figure 3. Torque control structure diagram.

3.1. Feedforward Control of Variable Load Disturbance and Friction Torque

The control system based on error feedback has the characteristics of delay and slow response. Therefore, it is particularly important to estimate the load and friction disturbances in advance and decouple the robot joints with torque feedforward. In feedforward control, the end-execution trajectory was mapped to the joint space using inverse kinematics, and the velocity and acceleration expressions of the joint space at any time of the optimized trajectory were obtained under variable load. Based on the dynamic theory of the robot under variable load, the nonlinear time-varying disturbance τ_d of the robot joint under variable load was predicted by the Lagrange operator. The predicted disturbance torque τ_{d1} decoupled the torque loop system by feedforward compensation. The Stribeck friction model and parameter identification technology are used to predict the friction loss torque τ_f of robot joints, and the predicted friction torque is decoupled by feedforward compensation. The structure diagram of the torque feedforward compensation part is shown in Figure 4.

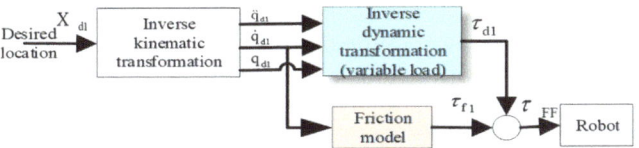

Figure 4. Feedforward controller structure diagram.

The feedforward torque τ_{FF} is expressed as:

$$\tau_{FF} = \tau_{d1} + \tau_f \tag{24}$$

In this equation, the friction torque τ_f is obtained by identification, and the predicted disturbance torque τ_{d1} (for robot arm movement under variable load disturbance) is obtained by Equation (19), which is calculated from the improved dynamics model in Section 2.

3.2. Calculation and Fuzzy Torque Feedback Control

Feedforward control alone has difficulty resisting other unknown disturbances and the system stability is poor. Therefore, a fuzzy calculation torque feedback control is proposed to compensate for the motion error. The structure diagram of torque feedback control is shown in Figure 5.

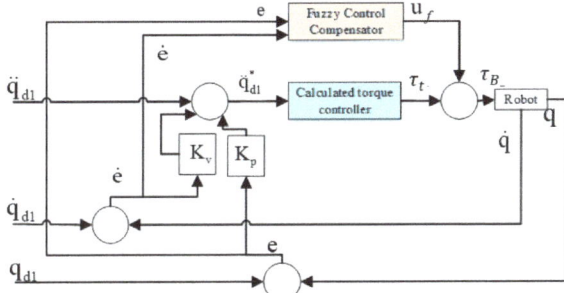

Figure 5. Feedback controller structure diagram using a combination of computational torque control and fuzzy control to feedback the robot's traveling torque.

$$\tau_B = \tau_t + u_f \tag{25}$$

The output of the calculated torque controller is τ_t, the output torque of the fuzzy controller is u_f, and the feedback control torque is τ_i.

3.2.1. Calculation Torque Controller Design

The torque control method of each branch chain is as follows:

$$\begin{cases} \tau_t = \hat{M}(q)\ddot{q}^* + \hat{C}\dot{q}^2 + \hat{G}\hat{q} + \tau_F \\ \ddot{q}^* = \ddot{q}_d + K_v \dot{e} + K_p e \end{cases} \tag{26}$$

In this equation, $\hat{M}$, $\hat{C}$, and $\hat{G}$ are the inertia matrix, centrifugal and Coriolis matrix, and gravity matrix estimated by Lagrange equation, respectively; $\ddot{q}^*$ is the control variable, and the angular displacement error and angular velocity error are e and $\dot{e}$, respectively, as follows:

$$\begin{cases} e = q_d - q \\ \dot{e} = \dot{q}_d - \dot{q} \end{cases} \tag{27}$$

3.2.2. Fuzzy Controller Design

Firstly, the language variables of the fuzzy [14] logic controller are determined. Both the single joint angular displacement deviation e and angular displacement deviation variation rate $\dot{e}$ of the robot are selected as input variables, and the fuzzy logic compensation moment u_j^* is used as an output variable to design the fuzzy controller [14]. Firstly, the input variable of the robot single joint is defined: angular displacement deviation e and angular displacement deviation variation rate $\dot{e}$ is $[-2, -1, 0, 1, 2]$. Fuzzy subset definition:

$\{$NB(Negative Big), NM(Negative Middle), ZO(Zero), PM(Positive Middle), PB(Positive Big) $\}$

Secondly, the membership function is determined as shown in Figure 6, and fuzzy rules are established as shown in Table 1.

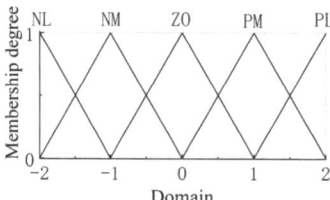

Figure 6. Membership function of fuzzy controller.

Table 1. Fuzzy rules table.

e	u_f	$\dot{e}$				
		NB	NM	ZO	PM	PB
NB		NB	NB	NB	NM	NM
NM		NM	NM	NM	NM	PM
ZO		NB	NM	PM	PM	PB
PM		NM	PM	PM	PM	PB
PB		PM	PM	PB	PB	PB

Finally, fuzzy logic reasoning and defuzzification are used. In MATLAB, the fuzzy control toolbox is used to write input and output membership functions and fuzzy control rule table. According to the regular control statement, the Mamdani method is used to deduce the corresponding relationship between the angular displacement deviation, angular displacement deviation variation rate, and the output u_f. Finally, the center of gravity method is used to denazify the output, so that the output u_f can partially compensate for the torque.

Using the second Lyapunov theory analysis, the stability of calculating torque control, the analysis process is as follows:

$$\tau_t = M(q)\ddot{q} + C(q,\dot{q}) + G(q) = K_p e - K_v \dot{q} + G(q)$$

$$M(q)\ddot{q} + C(q,\dot{q}) + K_v \dot{q} + K_p q = K_p q_d \tag{28}$$

Its energy equation: $E = \frac{1}{2}\dot{q}^T M(q)\dot{q} + \frac{1}{2}e^T K_p e$, where the M and K_p are greater than zero.

Then, the derivative of the energy equation is:

$$\dot{E} = \dot{q}^T M(q)\ddot{q} + \frac{1}{2}\dot{q}^T \dot{M}(q)\dot{q} - e^T K_p \dot{q} = \frac{1}{2}\dot{q}^T \dot{M}(q)\dot{q} - \dot{q}^T K_v \dot{q} + \dot{q}C(q,\dot{q}) = -\dot{q}^T K_v \dot{q} \tag{29}$$

Because the K_v is always positive, and the $\dot{E}$ is always non-positive, the system is stable.

4. Simulation and Experiment

4.1. Simulation Results

Parameter Identification Results of Stribeck Friction Model

The parameter identification process of the basic Stribeck friction model is based on multiple off-line measurements of the robot's single joint at different constant velocities (when the robot moves at constant velocities, the inertia matrix, centrifugal, and Coriolis moment are zero; because when the 4-DOF redundant robot moves in xoy plane, the heavy torque is zero, namely: $\tau = \tau_{Ftot}$), the relationship between friction torque and rotational speed can be obtained referring to Refs. [15–21]. The Stribeck friction model function of the joint is shown as:

$$\tau_f = f_{Fv}\omega + [\tau_{Fc} + (\tau_s - \tau_{Fc})e^{(-\frac{\omega}{\omega_s})\delta}]sgn(\omega) \qquad (30)$$

The four parameters to be identified were calculated at four points. Finally, the L-M (Levenberg-Marquardt) algorithm was used to fit the model. The final parameter identification results are shown in Table 2:

Table 2. Parameter identification results of the Stribeck model.

τ_{Fc}	τ_{Fs}	ω_s	f_{Fv}
0.11	0.14	11.077	4.0216×10^{-4}

Converting the joint friction torque to the torque-producing motor needs to be multiplied by the reduction ratio, so the comparison between the measured friction torque and parameter identification results of the robot in normal operation is shown in Figure 7.

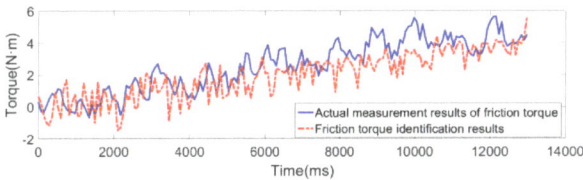

Figure 7. Comparison of friction torque measurement and identification results.

It can be seen from Figure 8 that the parameter identification result of the friction torque is close to the actual measured value, which is evaluated by the goodness of fit R^2 evaluation model in Ref. [15]. The goodness of fit is 0.9355, close to 1, which is good. Therefore, the parameter identification result of this model is relatively reliable.

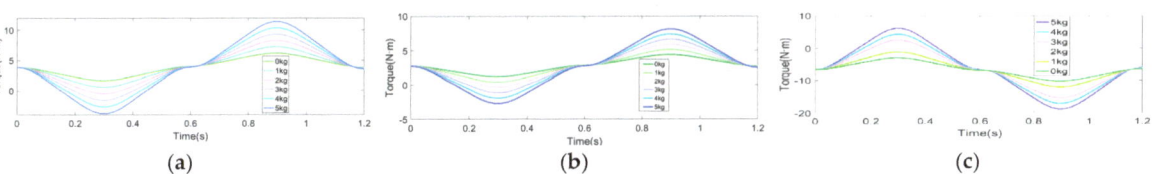

Figure 8. Feedforward compensation torque diagram of each drive joint under variable load: (**a**) Joint 1; (**b**) Joint 3; (**c**) Joint 5.

According to the friction torque τ_f, and the calculation and identification of the torque τ_d of the robot arm under the disturbance of the variable load in the dynamic model calculation of the variable load in 2.1, the simulation results of the feedforward compensation torque of the driving joint under different loads are shown in Figure 8.

4.2. Experiment

4.2.1. Experimental Design

The GPM-II 4-DOF redundant parallel robot was connected to the control PC through a serial port, the motion mode was changed to torque mode in the driver debugging software "Servo Studio", and the forward and inverse dynamics module under variable loads were established under the Gtrbox toolbox developed in MATLAB.

In "Torque Mode", a Control strategy combining feedforward and feedback with friction torque and variable load dynamics model is used to control the joint torque of a 4-DOF redundant parallel robot in "Control".

In the feedforward experiment of a 4-DOF redundant parallel robot under variable load, the loading weight is determined to be in the range of 0–5 kg, according to the rated load capacity of the driving motor. Therefore, loads of 0 kg, 1 kg, 2 kg, 3 kg, 4 kg, and 5 kg were applied to the end-effector, respectively. The field equipment and experimental figure are shown in Figure 9.

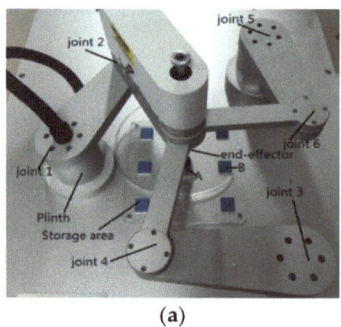
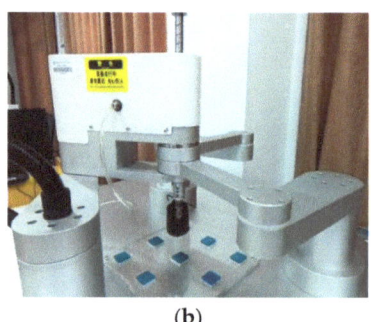

(a) (b)

Figure 9. Loading experiment diagram. (**a**) 4-DOF redundant parallel robot structure diagram. (**b**) Loading external load experiment diagram.

The single driving joint is operated according to the planned trajectory, and the trajectory tracking experiment is performed in Torque Mode based on the calculated torque and fuzzy control combined with feedback Refs. [11,22–27], supplemented by variable load disturbance and friction torque feedforward. The encoder was used to collect the angular displacement variation data under various load conditions, and the angular velocity, angular acceleration, and angular acceleration change rate were obtained through differential and filtering processing. According to the current data recorded by the driving software, the real-time situation of the joint torques in motion under various loads was calculated.

The trajectory in the operating space is the linear motion from A to B. In the joint space, the 12-phase sinusoidal shock curve is used as the motion trajectory; refer to Ref. [9]. The structure and dynamic parameters of the robot are shown in Table 3.

Table 3. The structure and dynamic parameters of the 4-DOF parallel robot.

Parameter	Quality (Kg)	Length (m)	Distance from Center of Mass to Joint (m)	Moment of Inertia $Kg \times m^2$
1	2.1	0.2440	0.1096	0.0252
2	8.5	0.2440	0.0957	0.0778
3	2.1	0.2440	0.1096	0.0252
4	0.4	0.2440	0.1260	0.0064
5	2.1	0.2440	0.1096	0.0252
6	0.4	0.2440	0.1260	0.0064

4.2.2. Experimental Results and Analysis

For robot joints within 0–5 kg, the load and the stability of the angular displacement and angular velocity remain the same, and the response speed of the torque control and control stability are improved, respectively, through the simple torque feedback and torque feedback and feedforward to control the robot joints, by comparison with the experimental running characteristics of two kinds of control mode (trajectory tracking error, response time, velocity stability, and control moment), showing the advantages of the combined model of variable load disturbance and friction torque feedforward and fuzzy computational torque feedback control in the aspects of motion accuracy, operation stability, response speed, and control stability.

(A) Trajectory tracking error comparison

The feedback control and feedback amp are obtained through experiments under different external loads. Figure 10 shows the comparison of angular displacement under feedforward compensation control.

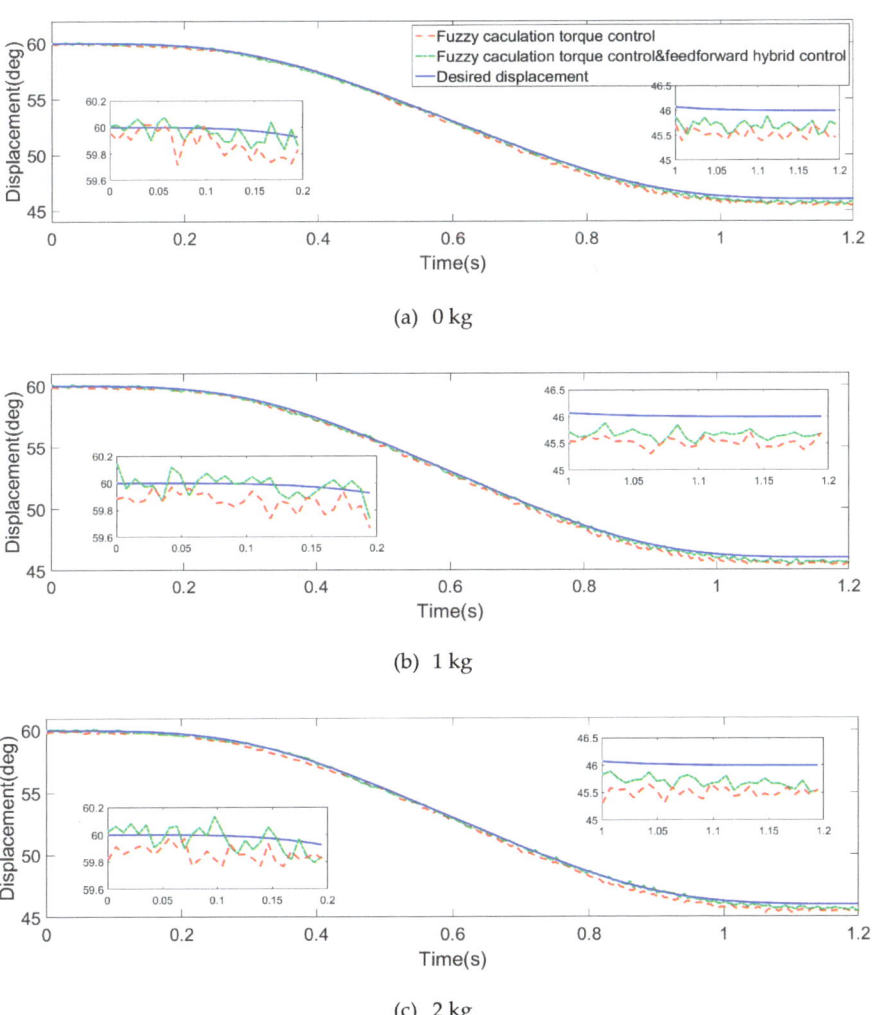

(a) 0 kg

(b) 1 kg

(c) 2 kg

Figure 10. *Cont.*

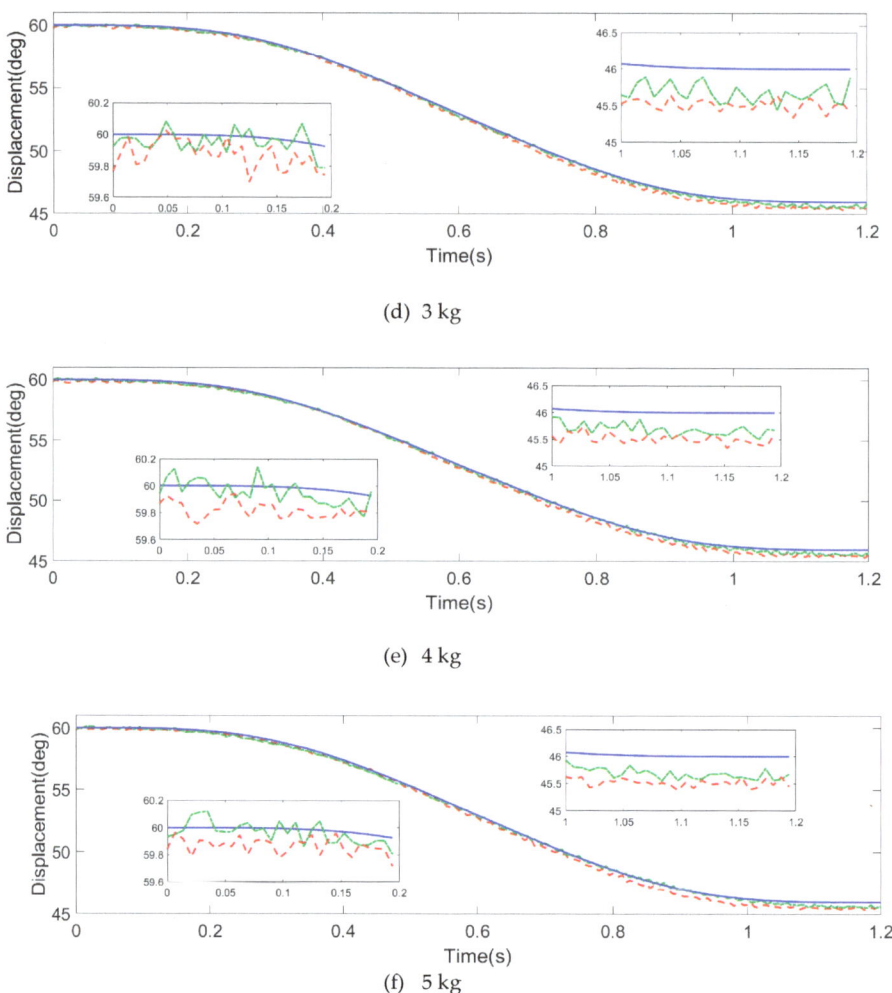

(d) 3 kg

(e) 4 kg

(f) 5 kg

Figure 10. Comparison of angular displacements of joint 1 under different external loads with and without torque feedforward compensation control.

By Figures 10–12, at different loads within 0–5 kg, three driving joints of 4-DOF parallel robotic angular displacements of the experiment are close to the planned trajectory, both in fuzzy computing torque feedback control, and fuzzy calculation and feedforward torque hybrid control. Two kinds of control modes of the angular displacement track are bigger than the planned value because of the influence of accumulated error, and the deviation increases with the increase in movement time. Additionally, the overall motion deviation of the robot under the hybrid torque control is less than that of the fuzzy computational torque feedback control.

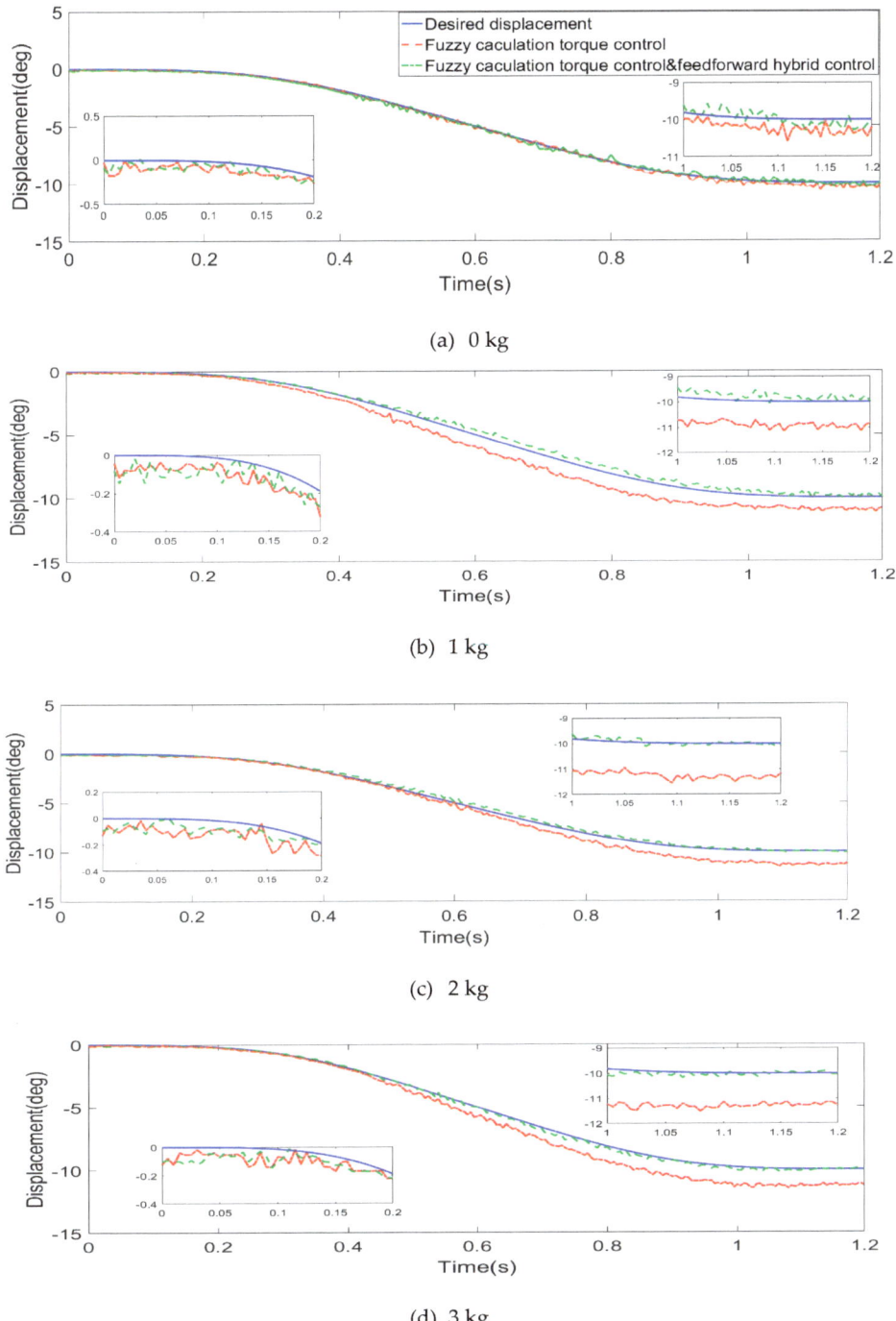

(a) 0 kg

(b) 1 kg

(c) 2 kg

(d) 3 kg

Figure 11. *Cont.*

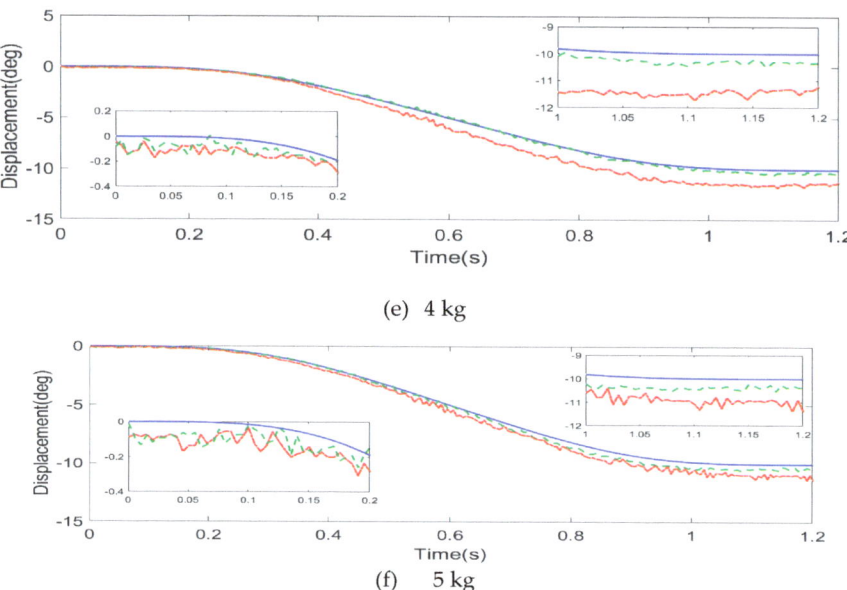

(e) 4 kg

(f) 5 kg

Figure 11. Comparison of angular displacements of joint 2 under different external loads with and without torque feedforward compensation control.

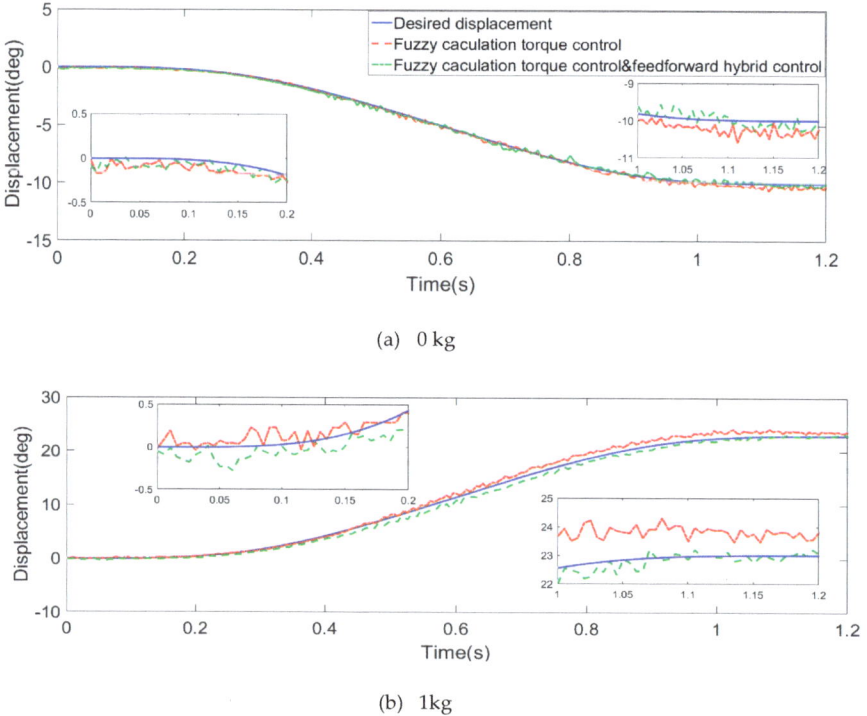

(a) 0 kg

(b) 1 kg

Figure 12. *Cont.*

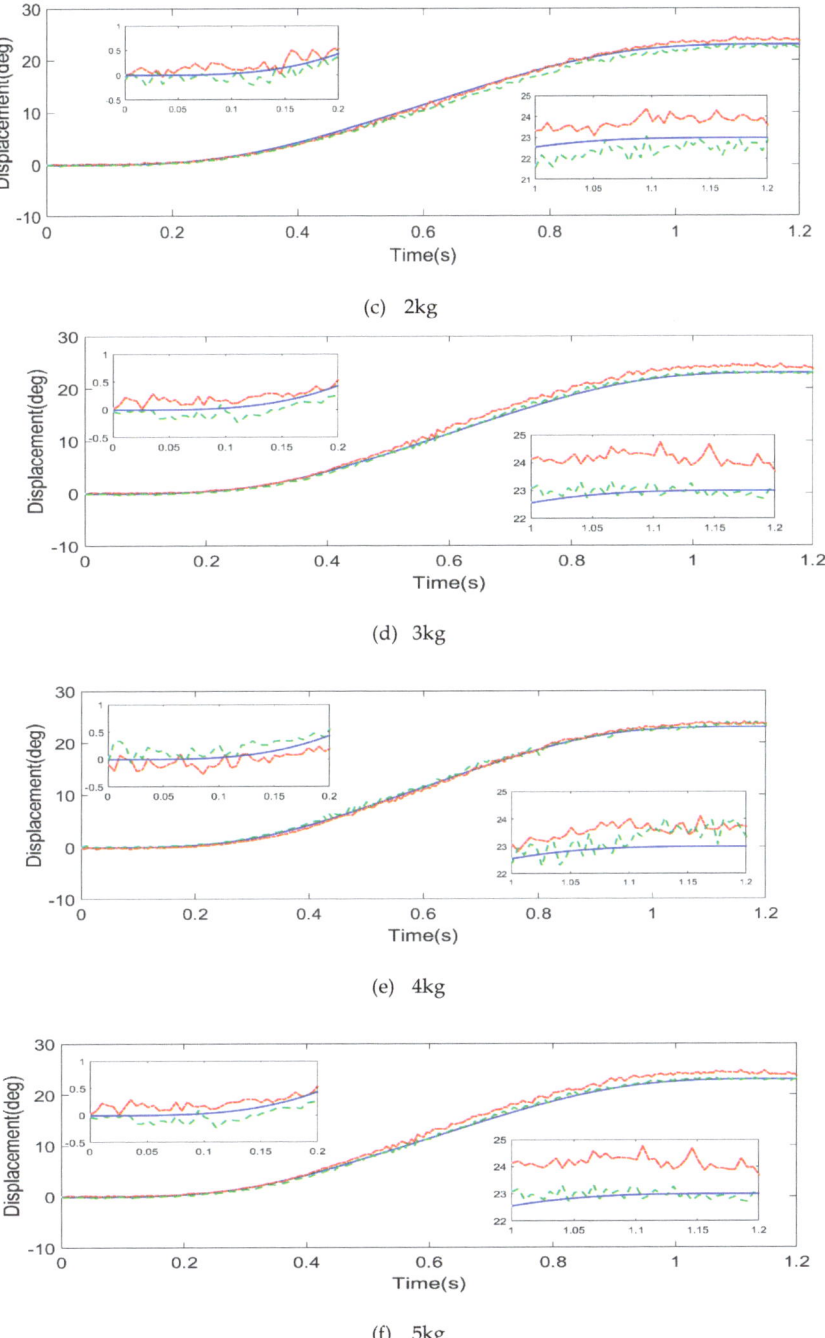

Figure 12. Comparison of angular displacements of joint 3 under different external loads with and without torque feedforward compensation control.

The specific numerical analysis of RMS error is shown in Figures 13–15.

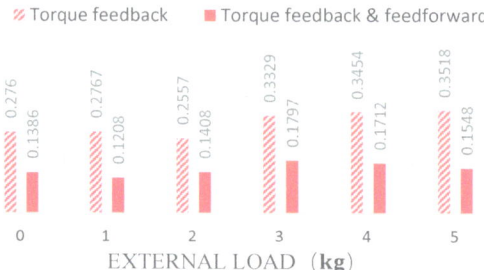

Figure 13. The comparison of the root of tracking error of joint 1 under variable load between torque feedback and torque feedforward and feedforward hybrid control.

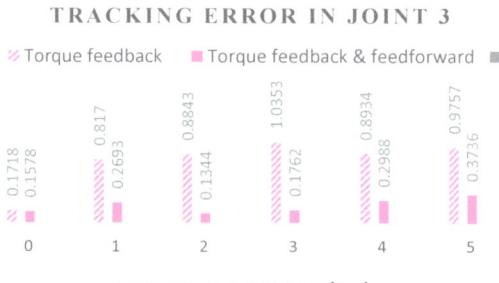

Figure 14. The comparison of the root of tracking error of joint 3 under variable load between torque feedback and torque feedforward and feedforward hybrid control.

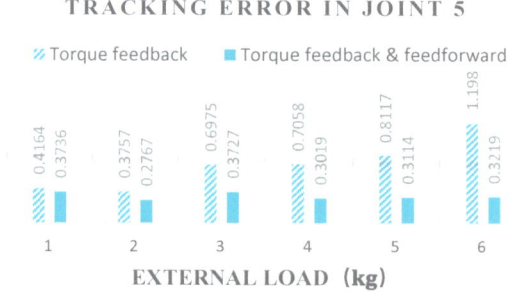

Figure 15. The comparison of the root of tracking error of joint 5 under variable load between torque feedback and torque feedforward and feedforward hybrid control.

(B) Comparison of velocity stability

By differentiating angular displacement and filtering, feedback control and feedback and feedforward hybrid control are obtained under 0–5 kg load and different external loads. Figures 16–18 show the comparison of the angular velocity under feedforward compensation control.

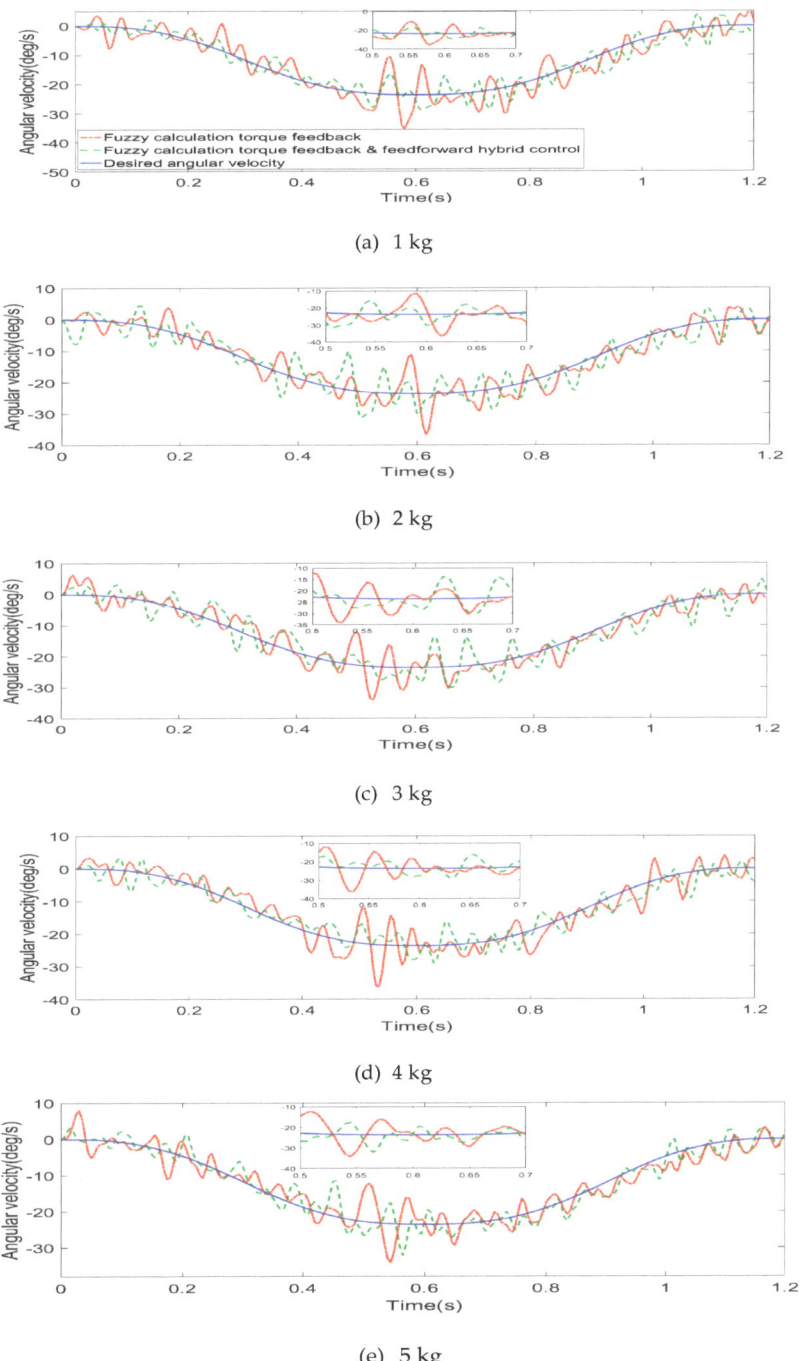

Figure 16. Comparison of the angular velocity of joint 1 under different external loads with and without torque feedforward compensation control.

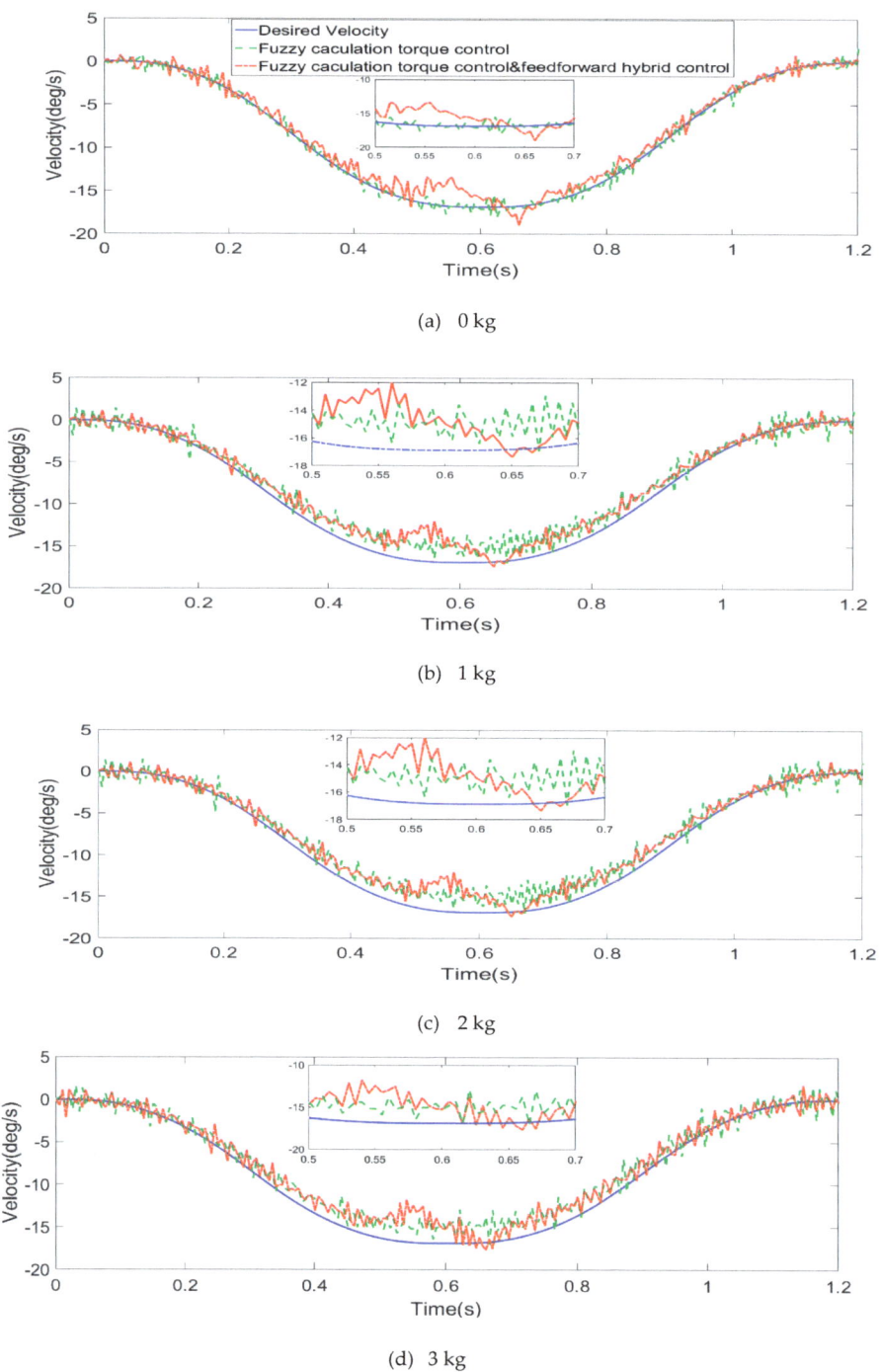

(a) 0 kg

(b) 1 kg

(c) 2 kg

(d) 3 kg

Figure 17. *Cont.*

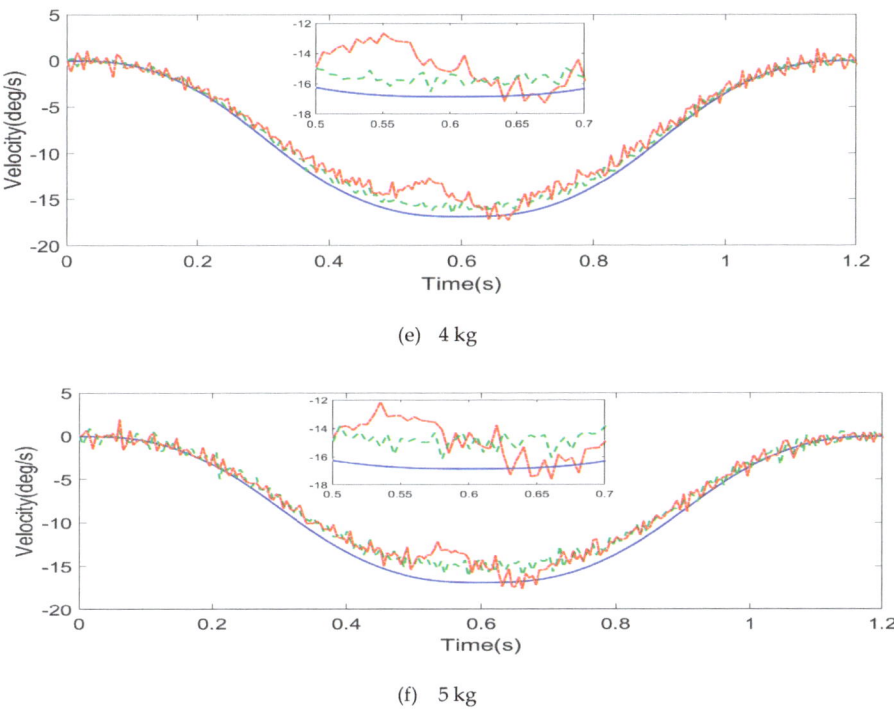

Figure 17. Comparison of the angular velocity of joint 2 under different external loads with and without torque feedforward compensation control.

According to Figures 16–18, the velocities of the robot joints under the two torque control methods in different loads of 0–5 kg fluctuate around the planned velocities, and both of them are close to the expected velocities. However, the velocity deviations under the feedback and feedforward hybrid torque control are lower than those under the single torque feedback control. The analysis of the velocity stability of the robot joint analyzes the RMS error value between the motion speed and the expected speed. By Figures 13–15, the internal angular velocity fluctuation of 0–5 kg load in the feedback and feedforward hybrid toque compensation control is compared with torque feedback control, and its velocity stability is obtained through analysis and calculation, as shown in Figures 19–21, respectively.

It can be seen from Figures 19–21 that when there is no load, the RMS values of the velocity error of joints 1, 3, and 5 in the torque hybrid control are slightly lower than those of the torque feedback. With the increase in load, the RMS value of velocity error increases in fluctuation, and the overall trend is upward. Through calculation, the average RMS values of the three driving joints' speed error under torque feedback control are 4.3879, 1.3709, and 1.2684, respectively; the average RMS values of speed error under torque feedback and feedforward control are 3.3632, 1.1288, and 0.9508, respectively; and the speed stabilities of torque hybrid control are relatively higher by 23.35%, 17.66%, and 25.04%, respectively.

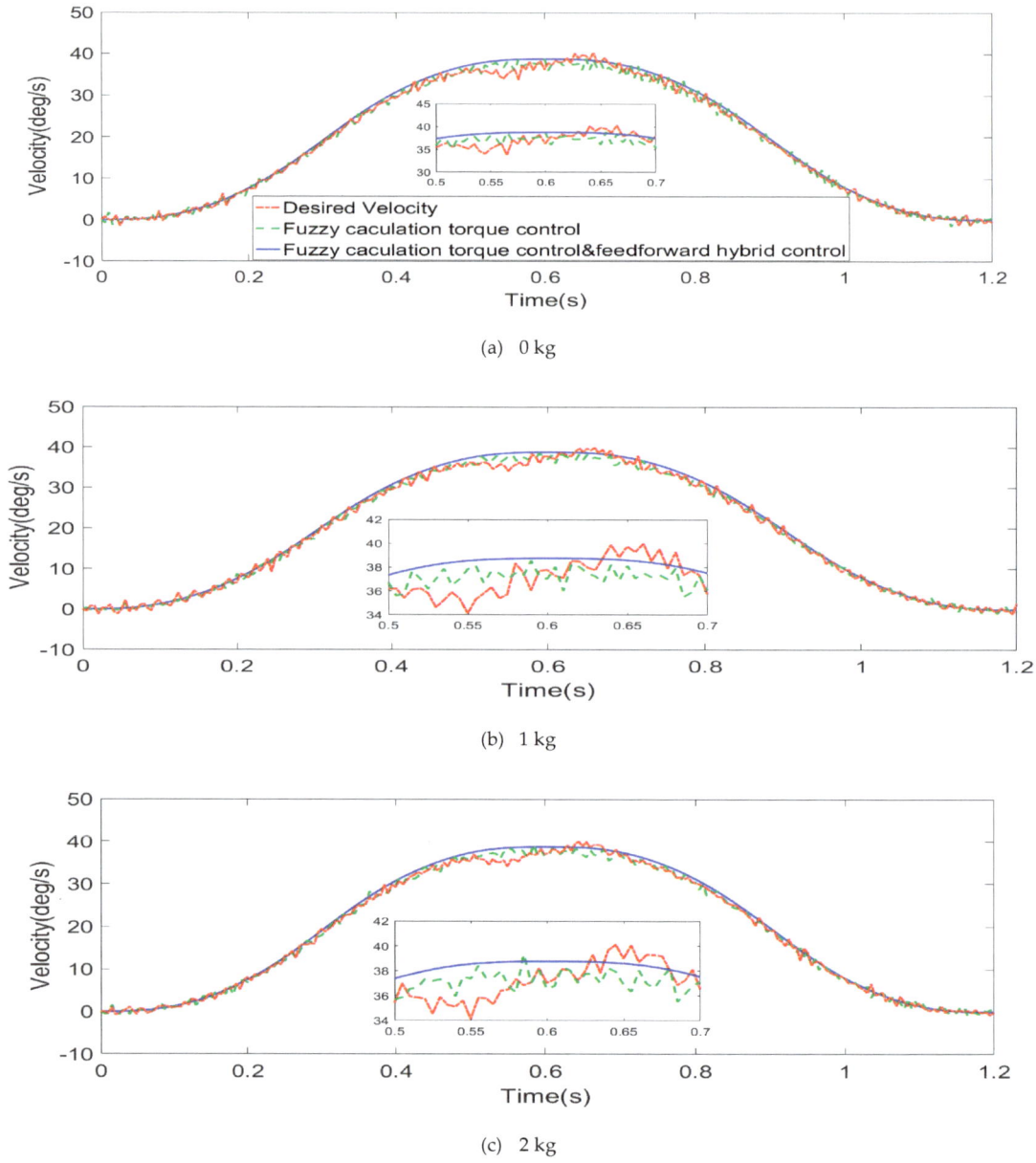

Figure 18. *Cont.*

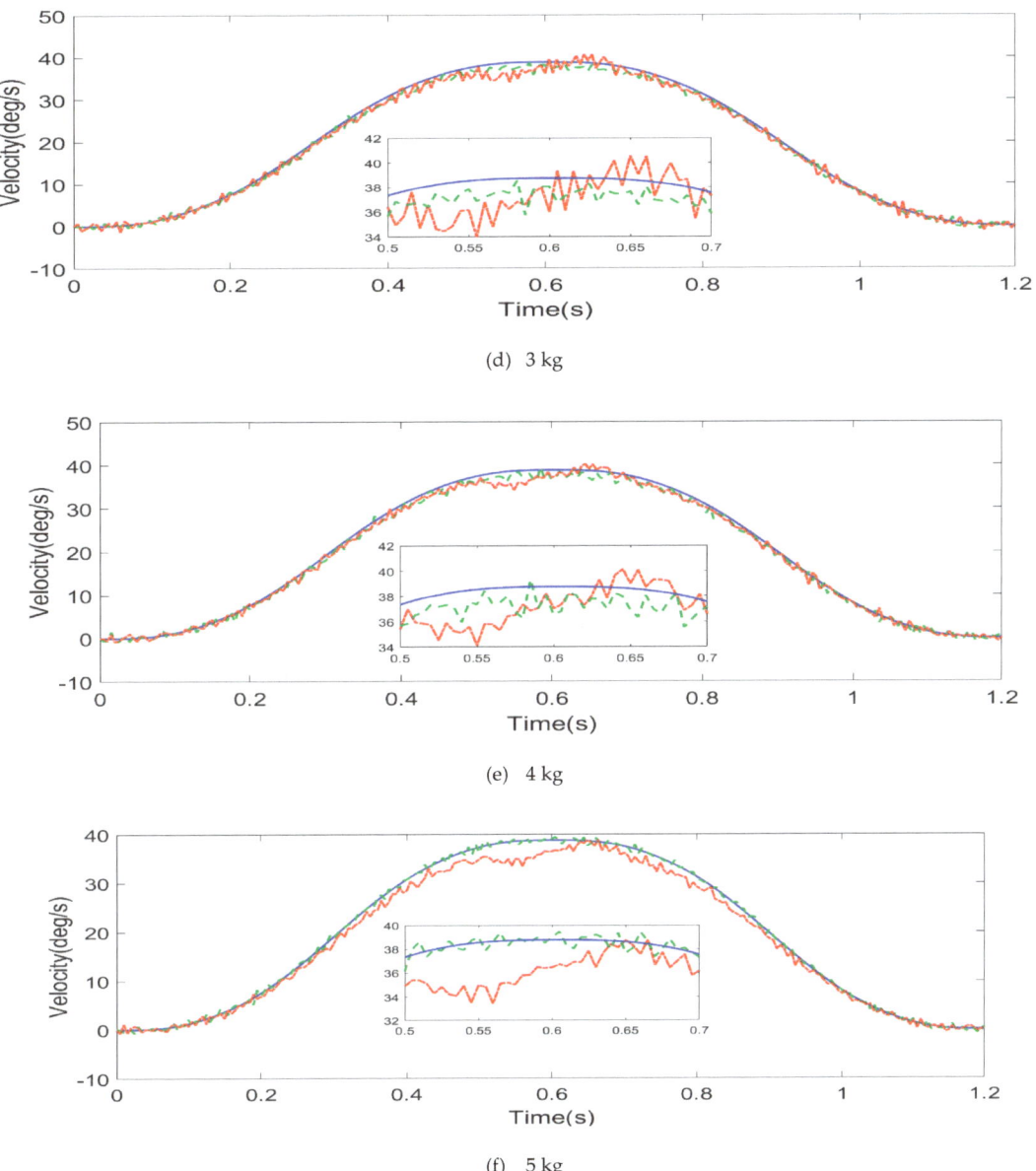

(d) 3 kg

(e) 4 kg

(f) 5 kg

Figure 18. Comparison of the angular velocity of joint 3 under different external loads with and without torque feedforward compensation control.

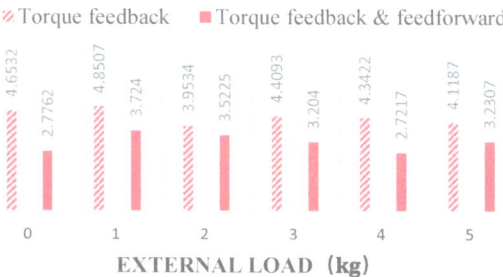

Figure 19. Speed stability analysis of torque feedforward and feedback and feedback under the variable load of joint 1.

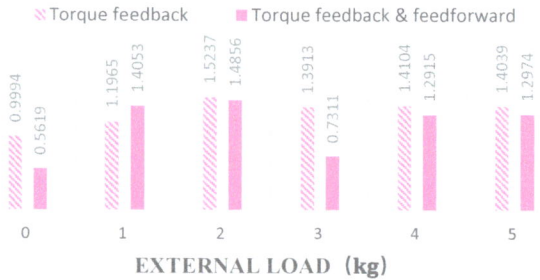

Figure 20. Speed stability analysis of torque feedforward and feedback and feedback under the variable load of joint 3.

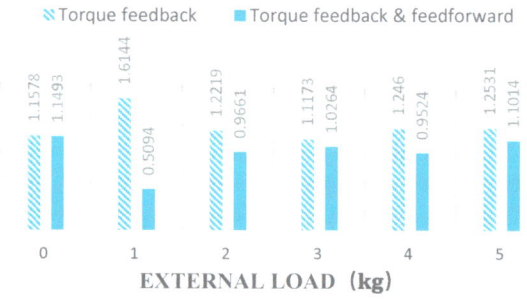

Figure 21. Speed stability analysis of torque feedforward and feedback and feedback under the variable load of joint 5.

5. Conclusions

Based on the modified dynamic model and the Stribeck friction model of joints for a 4-DOF redundant parallel robot under variable loads, this study used the torque feedforward for compensation control and combined the fuzzy computational torque feedback for hybrid control. Through relevant simulation and experiment, a comparison of the key characteristic parameters between fuzzy computational torque feedback and fuzzy computational torque feedback & torque feedforward hybrid control was performed. The conclusions of this study are as follows:

1. When the robot's joints move under variable load, compared with the fuzzy computational torque feedback, the fuzzy computational torque feedback and torque feedforward hybrid control decreased the RMS values of tracking errors by 49.87%, 70.48%, and 50.37%, respectively, and increased the kinematic precision at the same time.
2. Compared with simple torque feedback control, the hybrid torque control increased the velocity stability by 23.35%, 17.66%, and 25.04%, respectively; that is, the velocity stability of the hybrid torque control method was better than only the feedback torque control method.

Author Contributions: Conceptualization, S.H.; methodology, S.H.; software, S.H.; validation, S.H.; formal analysis, S.H.; investigation, S.H.; resources, S.H.; data curation, S.H.; writing—original draft preparation, S.H.; writing—review and editing, S.H., H.L., H.K., P.O., Z.L. and Z.C. visualization, S.H.; supervision, H.L., H.K. and P.O.; project administration, H.L. and H.K.; funding acquisition, H.K. All authors have read and agreed to the published version of the manuscript.

Funding: National Natural Science Foundation of China (51875198) and the Xiangtan City Joint Fund Project (2021JJ50118).

Data Availability Statement: Not applicable.

Conflicts of Interest: The authors declare that they have no known competing financial interest or personal relationships that could have appeared to influence the work reported in this paper.

References

1. Wang, B.; Qi, Z.; Yan, R.; Liu, H.Q. Dynamic Parameter Identification of SCARA Robot Based on Stochastic Weighted Particle Swarm Optimization. *J. Xi'an Jiaotong Univ.* **2021**, *55*, 20–27.
2. Cazalilla, J.; Vallés, M.; Mata, V.; Díaz-Rodríguez, M.; Valera, A. Adaptive control of a 3-DOF parallel manipulator considering payload handling and relevant parameter models. *Robot. Comput. Integr. Manuf.* **2014**, *30*, 468–477. [CrossRef]
3. Lee, K.W.; Khalil, H.K. Adaptive output feedback control of robot manipulators using high-gain observer. *Int. J. Control* **1997**, *67*, 869–886. [CrossRef]
4. Binbin, D.; Ling, Q.; Wang, S. Improvement of the conventional computed-torque control scheme with a variable structure compensator for delta robots with uncertain load. In Proceedings of the 2014 International Conference on Mechatronics and Control (ICMC), Jinzhou, China, 3–5 July 2014.
5. Nguyen, V.T.; Su, S.F.; Wang, N.; Sun, W. Adaptive finite-time neural network control for redundant parallel manipulators. *Asian J. Control* **2020**, *22*, 2534–2542. [CrossRef]
6. Zhang, H.; Fang, H.; Zhang, D.; Luo, X.; Zou, Q. Fuzzy sliding mode control for a 3-DOF parallel manipulator with parameters uncertainties. *Complexity* **2020**, *2020*, 2565316. [CrossRef]
7. Zhang, Y.; Li, Y.; Zhang, S. Adaptive neural dynamic surface control of robot manipulators with unknown loads. *IEEE Trans. Ind. Electron.* **2019**, *67*, 2332–2342.
8. Lin, J.; Tan, J.; Xue, J.; Liu, G. Robust adaptive neural network control of robot manipulators with uncertain loads. *IEEE Trans. Neural Netw. Learn. Syst.* **2018**, *29*, 1004–1014.
9. Hu, S.; Kang, H.; Tang, H.; Cui, Z.; Liu, Z.; Ouyang, P. Trajectory Optimization Algorithm for a 4-DOF Redundant Parallel Robot Based on 12-Phase Sine Jerk Motion Profile. *Actuators* **2021**, *10*, 80. [CrossRef]
10. Do Thanh, T.; Kotlarski, J.; Heimann, B.; Ortmaier, T. Dynamics identification of kinematically redundant parallel robots using the direct search method. *Mech. Mach. Theory* **2012**, *52*, 277–295. [CrossRef]
11. Gao, L.; Yuan, J.; Han, Z.; Wang, S.; Wang, N. A friction model with velocity, temperature and load torque effects for collaborative industrial robot joints. In Proceedings of the 2017 IEEE/RSJ International Conference on Intelligent Robots and Systems (IROS), Vancouver, BC, Canada, 24–28 September 2017.

12. Simoni, L.; Villagrossi, E.; Beschi, M.; Marini, A.; Pedrocchi, N.; Tosatti, L.M.; Visioli, A. On the Use of a Temperature Based Friction Model for a Virtual Force Sensor in Industrial Robot Manipulators. In Proceedings of the 2017 22nd IEEE International Conference on Emerging Technologies and Factory Automation (ETFA), Limassol, Cyprus, 12–15 September 2017.
13. Jalon, D.; Garcia, J.; Bayo, E. *Kinematic and Dynamic Simulation of Multibody Systems: The Real-Time Challenge*; Springer: Berlin/Heidelberg, Germany, 2012.
14. Abbasimoshaei, A.; Mohammadimoghaddam, M.; Kern, T.A. Adaptive fuzzy sliding mode controller design for a new hand rehabilitation robot. In Proceedings of the Haptics: Science, Technology, Applications: 12th International Conference, EuroHaptics 2020, Leiden, The Netherlands, 6–9 September 2020; Springer International Publishing: Cham, Switzerland, 2020.
15. Dong, L.; Tao, T.; Xuesong, M.; Dongsheng, Z. Decoupling identification method for linear characteristics and nonlinear friction of servo system. *Chin. J. Sci. Instrum.* **2010**, *31*, 782–788.
16. Wu, X.; Liu, D.; He, M.; Gao, F.; Shao, G. Research on Modeling and Compensation of robot joint Friction. *J. Instrum.* **2018**, *39*, 44–50. [CrossRef]
17. Kim, T.J.; Ahn, K.H.; Song, J.B. Friction Model of a Robot Joint Considering Torque and Moment Loads. *Trans. Korean Soc. Mech. Eng. A* **2021**, *45*, 27–34. [CrossRef]
18. Wen, S. *Tribology Principle*; Tsinghua University Press: Beijing, China, 1990; pp. 7–11.
19. Choi, J.H.; Sehoon, O.; Jinung, A. Sensorless Force Control with Observer for Multi-functional Upper Limb Rehabilitation Robot. *J. Korea Robot. Soc.* **2017**, *12*, 356–364. [CrossRef]
20. Roveda, L.; Pallucca, G.; Pedrocchi, N.; Braghin, F.; Tosatti, L.M. Iterative learning procedure with reinforcement for high-accuracy force tracking in robotized tasks. *IEEE Trans. Ind. Inform.* **2017**, *14*, 1753–1763. [CrossRef]
21. Hajnayeb, A.; Ghasemloonia, A. Nonparametric Identification of the Surgeon's Hand Vibration in Haptic Devices. In Proceedings of the 2018 23rd International Conference on Methods & Models in Automation & Robotics (MMAR), Miedzyzdroje, Poland, 27–30 August 2018.
22. Zubizarreta, A.; Marcos, M.; Cabanes, I.; Pinto, C.; Portillo, E. Redundant sensor based control of the 3RRR parallel robot. *Mech. Mach. Theory* **2012**, *54*, 1–17. [CrossRef]
23. Serdar, K. Energy minimization for 3-RRR fully planar parallel manipulator using particle swarm optimization. *Mech. Mach. Theory* **2013**, *62*, 129–149.
24. Harandi MR, J.; Khalilpour, S.A.; Taghirad, H.D.; Romero, J.G. Adaptive control of parallel robots with uncertain kinematics and dynamics. *Mech. Syst. Signal Process.* **2021**, *157*, 107693. [CrossRef]
25. Moshaii, A.A.; Moghaddam, M.M.; Niestanak, V.D. Fuzzy sliding mode control of a wearable rehabilitation robot for wrist and finger. *Ind. Robot. Int. J. Robot. Res. Appl.* **2019**, *46*, 839–850. [CrossRef]
26. Abbasi Moshaei, A.R.; Mohammadi Moghaddam, M.; Dehghan Neistanak, V. Analytical model of hand phalanges desired trajectory for rehabilitation and design a sliding mode controller based on this model. *Modares Mech. Eng.* **2020**, *20*, 129–137.
27. Wenxiang, W.; Shiqiang, Z.; Xuanyin, W.; Huashan, L. Robot low speed control based on friction fuzzy modeling and compensation. *Electr. Mach. Control.* **2013**, *17*, 70–77.

Disclaimer/Publisher's Note: The statements, opinions and data contained in all publications are solely those of the individual author(s) and contributor(s) and not of MDPI and/or the editor(s). MDPI and/or the editor(s) disclaim responsibility for any injury to people or property resulting from any ideas, methods, instructions or products referred to in the content.

Article

Hybrid Visual Servo Control of a Robotic Manipulator for Cherry Tomato Harvesting

Yi-Rong Li, Wei-Yuan Lien, Zhi-Hong Huang and Chun-Ta Chen *

Department of Mechatronic Engineering, National Taiwan Normal University, 162, Section 1, He-Ping East Road, Taipei 106, Taiwan; abc714855@gmail.com (Y.-R.L.); 80773004h@ntnu.edu.tw (W.-Y.L.); dniel41211598888@gmail.com (Z.-H.H.)
* Correspondence: chenct@ntnu.edu.tw; Tel.: +886-2-77493528; Fax: +886-2-23583074

Abstract: This paper aims to develop a visual servo control of a robotic manipulator for cherry tomato harvesting. In the robotic manipulator, an RGB-depth camera was mounted to the end effector to acquire the poses of the target cherry tomatoes in space. The eye-in-hand-based visual servo controller guides the end effector to implement eye–hand coordination to harvest the target cherry tomatoes, in which a hybrid visual servo control method (HVSC) with the fuzzy dynamic control parameters was proposed by combining position-based visual servo (PBVS) control and image-based visual servo (IBVS) control for the tradeoff of both performances. In addition, a novel cutting and clipping integrated mechanism was designed to pick the target cherry tomatoes. The proposed tomato-harvesting robotic manipulator with HVSC was validated and evaluated in a laboratory testbed based on harvesting implementation. The results show that the developed robotic manipulator using HVSC has an average harvesting time of 9.40 s/per and an average harvesting success rate of 96.25% in picking cherry tomatoes.

Keywords: hybrid visual servo control; robotic manipulator; cherry tomato; harvesting

1. Introduction

With the elderly population increasing gradually, insufficient available labor has arisen everywhere. Especially in agriculture, a serious lack of manpower may threaten crop production in the world. Therefore, research in smart agriculture offers an advantage to reduce the labor required. Among the attempts made, crop or fruit harvesting using an agricultural robot is an important priority [1–3].

An available agricultural robot can successfully pick crops and fruits that are grown in a complex, unknown, and unstructured environment. Hence, the agricultural robot must have the ability to detect targets. In this regard, vision is required for the agricultural robot to identify the positions and postures of targets. Moreover, fruits and crops have different shapes, colors, sizes, and types; therefore, harvesting algorithms must be developed for robots to perform successful picking. Currently, the key technique for overall performance of a harvesting robot lies in the performance of vision-based feedback control [4].

Vision-based control aims to detect and recognize the target crops and fruits via camera; their position and pose in space are acquired so that the coordinates and orientations are then used to control the motion of the robotic manipulator. In the detection and recognition of target fruit, many approaches rely on deep learning algorithms. Ji et al. [5] proposed the Shufflenetv2-YOLOX-based apple object detection to enable the picking robot to detect and locate apples in the orchard's natural environment. This method provides an effective solution for the vision system of the apple picking robot. Xu et al. [6] used an improved YOLOv5 for apple grading. The experiments indicated that this method has a high grading speed and accuracy for apples. Sa et al. [7] presented deep convolutional neural networks for fruit detection. The proposed detector can handle approximately 50% of scaled-down object detection. However, control by visual servo is also essential for the successful

Citation: Li, Y.-R.; Lien, W.-Y.; Huang, Z.-H.; Chen, C.-T. Hybrid Visual Servo Control of a Robotic Manipulator for Cherry Tomato Harvesting. *Actuators* **2023**, *12*, 253. https://doi.org/10.3390/act12060253

Academic Editor: Zhuming Bi

Received: 18 May 2023
Revised: 10 June 2023
Accepted: 13 June 2023
Published: 16 June 2023

Copyright: © 2023 by the authors. Licensee MDPI, Basel, Switzerland. This article is an open access article distributed under the terms and conditions of the Creative Commons Attribution (CC BY) license (https://creativecommons.org/licenses/by/4.0/).

operation of the robotic harvesting system. Based upon error signals, the visual servo controls are generally classified as PBVS and IBVS [8,9].

In the PBVS algorithm, a 3D model of target objects and camera parameters are required. The relevant 3D parameters are computed through the pose of the camera within a reference frame. The absolute or relative positions of the harvesting robot with respect to target objects can thus be determined using the visual 3D parameter information [10]. The controllers are then designed based on the position errors so that the robotic manipulator can move to an operation position to execute a picking action. For the application of PBVS to agricultural harvesting, Jun et al. [11] proposed a harvesting robot that combines robotic arm manipulation, object 3D perception, and an end cutting mechanism. For software integration, the Robot Operating System (ROS) was used as a framework to integrate the robotic arm, gripper, and related sense tester. Edan et al. [12] described the intelligent sensing, planning, and control of a robotic melon harvester. Image processing for PBVS is used to detect and locate the melons. Planning algorithms with the integration of task, motion, and trajectory were presented. Zhao et al. [13] developed an apple-harvesting robot that is composed of a manipulator, an end effector, and an image-based vision servo control system. The apple was detected using a support vector machine-based fruit recognition algorithm. The apple harvesting success rate was evaluated through PBVS. Lehnert et al. [14] presented a robotic harvester that can autonomously pick sweet pepper. A PBVS algorithm acquires 3D localization to determine the cutting pose and then to grasp the target with an end effector. Field trials demonstrated the efficacy of this approach. However, for PBVS, exact knowledge of the intrinsic parameters of the camera is required for control performance. Even very small errors in the camera calibration may greatly affect the control accuracy of robots [15].

IBVS directly uses image features that are converted from pixel-expressed images by the camera system to design the controllers. Visual features are first extracted from the image space. The errors are computed from points or vectors by the visual features [16]. Mehta et al. [17] developed a vision-based harvesting system for robotic citrus fruit picking. The cooperative visual servo controller was presented to servo the end effector to the target fruit location using a pursuit-guidance-based hybrid translation controller. The visual servo control experiment was performed and analyzed. Li et al. [18] investigated an image-based uncalibrated visual servoing control for harvesting robots and tried to resolve the overlapping effects of the target motion and the uncalibrated parameter estimation. The effectiveness of the proposed control was demonstrated by the comparative experiments. Barth et al. [19] reported the agricultural robotics in dense vegetation with software framework design for eye-in-hand sensing and motion control. An image-based visual servo control was designed to correct the motion of the robot so that the geometrical feature error was minimized. Qualitative tests were performed in the laboratory using an artificial dense vegetation sweet pepper crop. Li et al. [20] proposed an IBVS controller that mixes proportional differential control and sliding mode control. However, the visual servo controller is not completely designed to be perfect 100%, and there are unexpected interference phenomena in different environments or different hardware devices. Although the IBVS schemes are robust against the calibration errors in the camera, large calibration errors may cause the closed-loop system to be unstable [21–23]. As a result, an advanced control design is required for stability. Moreover, an IBVS using a fixed camera on a robotic manipulator is limited to a field of view. That is, the target may always move out of the field of view as the manipulator turns, so that the IBVS controller will fail to control the manipulator.

In this paper extending from our previous study [24], a robotic manipulator for cherry tomato harvesting was investigated in greater detail. The main contributions are highlighted as follows. A novel cutting and clipping integrated mechanism was designed for cherry tomato harvesting. The position of the cherry tomato in space was determined by the proposed feature geometry algorithm. To accurately and efficiently pick the target cherry tomato, an HVSC that improves PBVS and IBVS without camera calibration or a target model was proposed for visual feedback control. HVSC combines the Cartesian and image measurements for error functions. The rotation and the scaled translation of the camera between the current and desired views of an object were thus estimated as the displacement of the camera, and thus the harvesting system may perform with better stability.

2. Robotic Manipulator System for Harvesting

Harvesting robotic manipulators aim to perform effective picking on fruits and vegetables. Designs for harvesting robotic manipulators must take into account the machine perception of crops, and thus a machine vision system is required to recognize the status and postures of the target crops. Based on the identified crops, the robotic manipulator moves to a position where it is appropriate to harvest the detected crops in an uncertain, unstructured, and varying environment. The manipulation is always performed by visual servo control to make an end effector reach to the planned location and orientation. End effectors for harvesting are developed according to different harvesting methods, crops, and separated points from the stems. As a consequence, the proposed robotic manipulator in the paper for cherry tomato harvesting will be developed and designed according to these concepts.

2.1. Architecture Design and Software Setup

The architecture setup of the robotic manipulator for cherry tomato harvesting is presented in Figure 1, in which the hardware is composed of a 6-DOF UR5 manipulator, a harvesting mechanism, and an RGB-D camera (Intel Realsense D435i). The RGB-D camera is mounted to the end effector of the manipulator in an eye-in-hand setup to transmit the data of the detected tomato to the embedded board. The images taken by the camera are used for visual recognition and visual servo feedback control such that the harvesting mechanism can be driven precisely and robustly by the manipulator to perform picking.

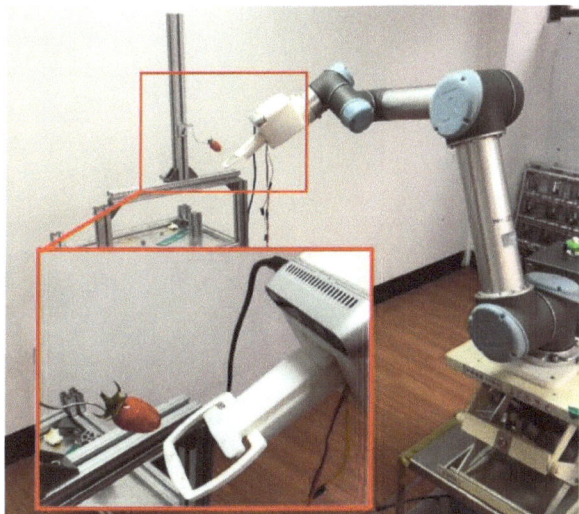

Figure 1. Architecture setup of the robotic manipulator system for cherry tomato harvesting.

The software system of the harvesting robot manipulator is defined in the Robot Operating System (ROS) environment. Each subsystem can be represented as a node. The ROS supports Python and C++ programming languages, and the software is running on Ubuntu 18.04. Image data and depth data are processed by Python. The visual servo control is developed using C++ for tomato harvesting. Various open software libraries are linked for function implementation. The robotic manipulator moves by enabling the motion controller via software ROS packages.

2.2. Harvesting Mechanism

Many harvesting mechanisms have been designed to pick cherry tomatoes. Traditionally, a scissor type of cutting method must rely on the detection of the fruit stem by vision. However, it is not easy to identify the fruit stems because fruit stems are often occluded by leaves and fruits or easily misidentified as twigs. As a result, it is preferred to detect the target fruits directly but try to cut them from the fruit stems.

In this paper, a novel cutting and clipping integrated mechanism was proposed to pick cherry tomatoes, as presented in Figure 2. Two blades are, respectively, mounted at the front and back of the rectangle sleeve. The rectangle sleeve can stretch out to pick cherry tomatoes and then return to its initial position. When the rectangle sleeve captures the target cherry tomato, the back blade moves forward to cut the fruit stem and clip the fruit.

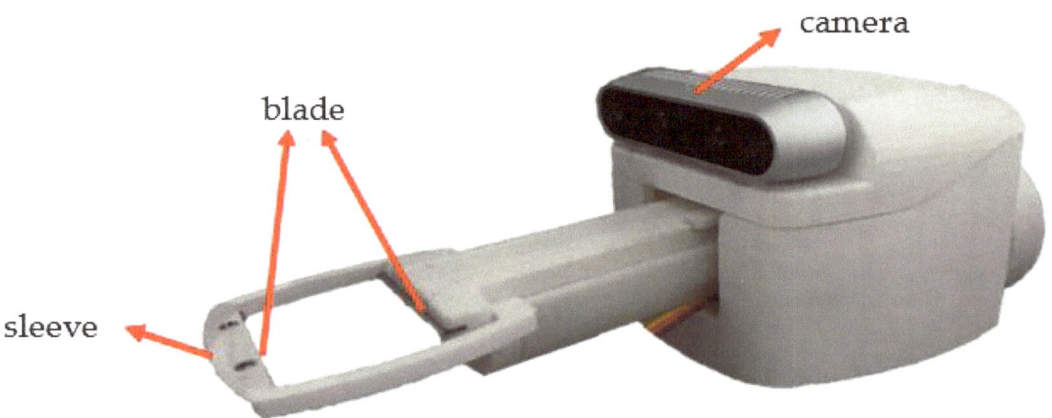

Figure 2. Harvesting module with RGB-D camera.

2.3. Determination of Feature Points

As the basis of our architecture setup of the robotic manipulator system for cherry tomato harvesting, the orthogonal frames, as shown in Figure 3, F_B, F_e, F_c, F_{C^*}, and F_T are defined and, respectively, attached to the base of the robotic manipulator, the end effector, the camera, the initial operable position, and the cherry tomato center. For simplicity, the eye-in-hand camera is installed so that the camera frame {c} and end effector frame {e} are purely translational, and there is a rotational matrix $R_c^e = I$. Because the interrelationships between these assigned coordinate frames affect the success rate of reaching target fruits, the coordinate transformation relationship is essential. And the coordinate transformation is characterized by a rigid transformation including rotations and translations. The homogeneous transformation matrices H_C^T, $H_C^{C^*}$, and $H_T^{C^*}$, respectively, represent the transformations from the camera coordinate frame to the tomato coordinate frame and from the camera coordinate frame to the initial operable position. Accordingly, the operation position needed to cut the fruit stem can be estimated using the relationships of the homogeneous transformation matrices, which enables the robotic manipulator to reach the harvesting position to pick cherry tomatoes.

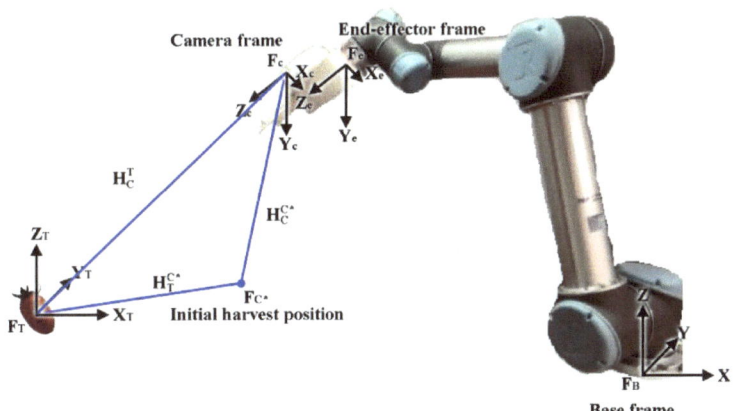

Figure 3. Coordinate frames for the robotic manipulator, tomato centroid, camera, and end effector.

To pick fruits effectively, target detection and the determination of positions and orientations are required functions for the proposed harvesting robotic manipulator. The recognition and localization process rely on reliable recognition algorithms in a visual system. Most recognition algorithms adopt multiple-feature fusion approaches to extract the desired information of the target fruits. Among them, color, geometry, and texture are popular extracting features for target fruits [25]. Color can be used to facilitate the segregation of target fruit from a complex environmental background. In general, the RGB images first captured by the camera are transformed to the YCrCb color space. Since a mature cherry tomato always appears red in color, only the Cr images that indicate the concentration offset of a red color are taken into account for mature cherry tomatoes. The color threshold values in OpenCV were applied to the filtered images [26], in which a color value range is specified. The pixels in the image that satisfy the specified range will be registered; otherwise, the pixels out of the range are labeled as different colors or values. This method allows for the extraction or segmentation of specific color regions in the image, and thus the locations of tomatoes can be distinguished and determined.

The shape of the cherry tomato in space may be regarded as an ellipsoid, and the corresponding image is a 2D ellipse as projected onto the image plane. Due to its efficiency, this shape in the image plane is first recognized using the contour method [27]. For the contour determination, a boundary point in the image must be determined as the starting point. This point will serve as the starting point to search the contour. All adjacent boundary points are traversed from this initial point along a closed boundary path. For each boundary point, the connectivity to its neighboring points must be examined to determine whether it is a branch point or a cross point. If there exist branch points or cross points, the topological structure features need to be updated. These features may contain a number of holes or connected regions. Finally, the shape in the image is thus determined after finishing the contour-following process until returning to the initial point.

The proposed image processing permits us to further find geometric feature points to recognize the status and orientations of the target cherry tomatoes. To identify the orientations of cherry tomatoes, the centroid of the shape is first determined by an image moment approach [28]. Shape and distribution can be obtained by calculating the moments of an image. Furthermore, based on moment invariants, features remain unchanged under transformations such as rotation, scaling, and translation. As a result, the center point of the image is inferred by the central moments as shown in Figure 4a for the centroid of the cherry tomato. The point $P_1(u_1, v_1)$ on the contour of the ellipse with the maximum distance from the centroid is detected and defined as one of the endpoints of the major axis of the ellipse. Taking the equal length to P_1C to obtain the point Q, the point Q must be located outside of the ellipse, as shown in Figure 4b. And hence it may not be the other endpoint of the major axis. By searching the points along the contour of the ellipse, the closest point P₂(u₂, v₂) to Q will become the other modified endpoint of the major axis of the ellipse, as shown in Figure 4c. These feature points are extracted to further recognize a tomato's posture for reliable harvesting.

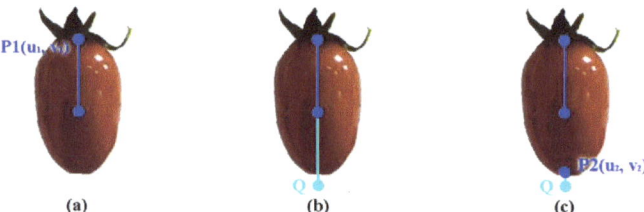

Figure 4. Feature point calibration with (**a**) the farthest point from the centroid, (**b**) the opposite point of P1, and (**c**) Q, the closest point to the tomato P2.

2.4. Pose of the Cherry Tomato

The control and motion guidance of a robotic manipulator for target cherry tomato harvesting are influenced by the targets' poses in space. In general, the orientation of a fruit can be suitably expressed in spherical coordinates with respect to the image plane. The parameters describing the status of a fruit are the length l of the major axis and two angles, φ and θ, respectively, referred to as the polar and azimuthal angles. As shown in Figure 5, the polar angle φ is the angle between the x axis and the projection of the major axis on the image plane and can be determined using the extracted feature points P_1 and P_2 as

$$\varphi = tan^{-1}[(u_1 - u_2)/(v_1 - v_2)] \quad (1)$$

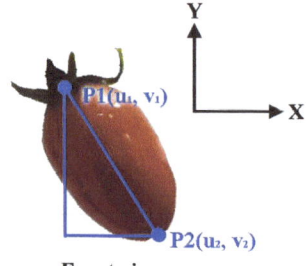

Front view

Figure 5. Polar angle of a cherry tomato in the XY plane of the tool frame.

The azimuthal angle is defined as the angle between the actual major axis and the y axis. As shown in Figure 6, the azimuthal angle can be determined by the projected length l of the actual major axis onto the image plane and the depth difference d_e of both feature points P_1 and P_2 in the z direction such that

$$\theta = tan^{-1}(d_e/l) \quad (2)$$

in which the depth difference $d_e = z_1 - z_2$, with z_1, z_2 being acquired by the depth camera of the visual system. The projected length $l = v_1 - v_2$ is the difference of the y coordinates of the two feature points in the image frame.

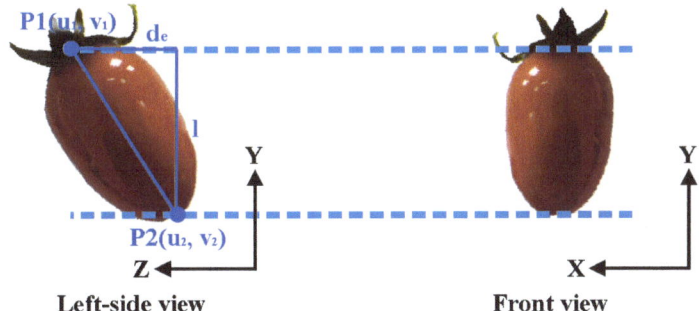

Figure 6. Azimuthal angle of the tomato in the YZ plane of the tool frame.

3. Visual Servo Controller for the Robotic Manipulator

A harvesting robotic manipulator must be capable of searching for a target and then driving to the desired position for the ensuing actions. Therefore, machine vision must be installed for visual servo control to realize the point-to-point localization. So, in this section, the visual servo control design will be presented for fruit picking.

3.1. PBVS for Cherry Tomato Harvesting

A PBVS is usually referred to as a 3D feedback control in the inertial frame. Features are extracted from the image to estimate the pose of the target tomato with respect to the camera. In this way, the error between the current and the desired pose of the target in the task space can be used to synthesize the control input to the robotic manipulator.

In the PBVS control method, the target is identified by the color depth camera with respect to the base frame. The image-expressed information is first processed and then converted to the position with respect to the camera frame according to the ideal pinhole camera model and further transformed to the coordinates with respect to the base frame using the relationship between the object frame and the camera frame. As such, the transformation from the coordinates of the object point (X, Y, Z) expressed in the base frame to the corresponding image point (u, v) is written as

$$z\begin{bmatrix} u & v & 1 \end{bmatrix}^T = AB\begin{bmatrix} X & Y & Z & 1 \end{bmatrix}^T \tag{3}$$

in which A is the camera intrinsic matrix, with $A = \begin{bmatrix} f_x & \gamma & m_x \\ 0 & f_y & m_y \\ 0 & 0 & 1 \end{bmatrix}$ representing the relationship between the camera frame and the image frame. It can be obtained through measurement or calculation using the given FOV; f_x and f_y are the effective focal length in pixels of the camera along the x_c and y_c axes; γ is the camera skew factor, and (m_x, m_y) indicate the difference between the camera center and the image center. In addition, the extrinsic coordinate transformation matrix $B = \begin{bmatrix} R_T^C & t \end{bmatrix}$ expresses the relationship between the object frame and the camera frame with R_T^C being defined as the rotational matrix and t as the translational displacement from the camera to the object. The rotational matrix R_T^C can be determined from the equivalent angle-axis representation that is constructed by the polar and azimuthal angles, as discussed in Section 2.4.

To harvest cherry tomatoes with camera alignment control, PBVS first serves as a coarse alignment and is then followed by IBVS for image-based fine alignment control. The coarse alignment control will enable the manipulator to move to a desired operation position ready to cut. The desired operation position is assigned as (u_c, v_c) near the principal point of the image plane. The corresponding desired position with respect to the base frame is determined as noted above. Since the rotation of the tomato around its central axis is considered invariant, only these two angles between the tomato central axis and the x and z axes are taken into account.

Utilizing the pixel error values, the depth values obtained from the depth camera, and the external and internal parameter matrices, the translational displacement is thus calculated. The PBVS control for cherry tomato harvesting is shown in Figure 7.

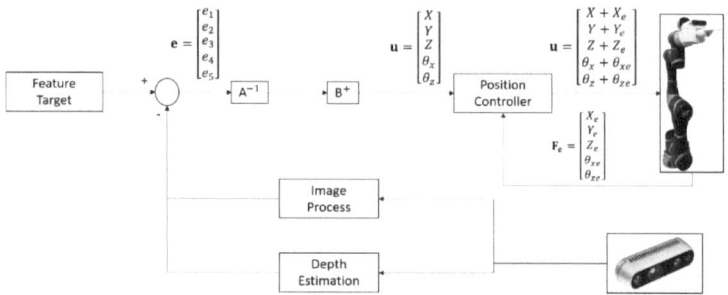

Figure 7. Control structure of the PBVS.

3.2. IBVS for Cherry Tomato Harvesting

IBVS calculates the control input to the manipulator directly using image feature errors to reduce computational delay and thus is less sensitive to calibration. The control design of IBVS and the selections of the associated control gains need to be examined in an image Jacobian matrix that relates the feature velocity to the camera velocity in an image coordinate. Let $v_c = \begin{bmatrix} v_x & v_y & v_z \end{bmatrix}^T$ and $\omega_c = \begin{bmatrix} \omega_x & \omega_y & \omega_z \end{bmatrix}^T$ be the linear velocity and the angular velocity of the camera expressed with respect to the camera frame. The image Jacobian matrix L of a point $P(X, Y, Z)$ in the camera frame with the corresponding projected coordinate in image space $P(u, v)$ can be written as [29]

$$\begin{bmatrix} \dot{u} \\ \dot{v} \end{bmatrix} = \begin{bmatrix} -\frac{f}{Z} & 0 & \frac{u}{Z} & \frac{uv}{f} & \frac{-f^2-u^2}{f} & v \\ 0 & -\frac{f}{Z} & \frac{v}{Z} & \frac{f^2+v^2}{f} & -\frac{uv}{f} & -u \end{bmatrix} \begin{bmatrix} v_c \\ \omega_c \end{bmatrix} \qquad (4)$$
$$= L v_c$$

For feedback control by IBVS for the robotic manipulator, the errors in the image frame are required. If the desired image position is defined as $(u_d, v_d) = (u_0, v_0)$, the desired depth distance of the centroid z_d and (φ_d, θ_d) are referred to as the desired polar and azimuthal angles. Conventionally, six control errors should be defined in the image space for feedback control. However, the amount of rotation about the principal axis does not affect the picking motion due to our harvesting mechanism design. So, one may define the five errors of feedback control of the robotic manipulator for harvesting as follows:

$$(e_1, e_2) = (u - u_0, v - v_0). \qquad (5)$$

$$e_3 = z_d - z_C. \qquad (6)$$

$$e_4 = \theta_d - \theta = \theta_d - \tan^{-1}\left(\frac{z_1 - z_2}{z_C |v_1 - v_2|/f_y}\right). \qquad (7)$$

$$e_5 = \varphi_d - \varphi = \varphi_d - \tan^{-1}[(u_1 - u_2)/(v_1 - v_2)]. \qquad (8)$$

These five errors that encompass three main feature points, i.e., the two end points P_1, P_2 and the centroid point P_C in the pixel plane, are used to compensate for the alignment positioning and orientation errors during the reaching and harvesting phase. The basic visual controller design for a conventional IBVS almost employs proportional control to generate the control signal. However, this method cannot have a faster control convergence and a smaller error. In this paper, a PD control with fuzzy gains is adopted to improve the visual feedback quality.

The proposed PD control scheme in the alignment of the tomato centroid to the center position of the image plane is described as [30]

$$\begin{bmatrix} v_x & v_y \end{bmatrix}^T = \begin{bmatrix} k_{p1}e_1 + k_{d1}\dot{e}_1 & k_{p2}e_2 + k_{d2}\dot{e}_2 \end{bmatrix}^T, \tag{9}$$

in which (v_x, v_y) is the translation velocity relative to the current camera frame; $k_{pi}, k_{di}, i = 1, 2$, are positive gains. Taking the derivative of Equation (5) and from the image Jacobian matrix, Equation (4), along with the controller, Equation (9), the error dynamics are obtained as

$$\begin{bmatrix} \dot{e}_1 + \left(\frac{fz_C^{-1}k_{p1}}{1+fz_C^{-1}k_{d1}}\right)e_1 & \dot{e}_2 + \left(\frac{fz_C^{-1}k_{p2}}{1+fz_C^{-1}k_{d2}}\right)e_2 \end{bmatrix}^T = 0 \tag{10}$$

It is seen that the controller in Equation (9) drives the errors to zero.

Moreover, to reach the desired depth z_d for the centroid of the cherry tomato and to rotate the end effector for the harvesting, the PD control law is used when $e_1 = e_2 = 0$

$$v_z = k_{p3}e_3 + k_{d3}\dot{e}_3. \tag{11}$$

$$\omega_x = k_{p4}e_4 + k_{d4}\dot{e}_4. \tag{12}$$

$$\omega_z = k_{p5}e_5 + k_{d5}\dot{e}_5. \tag{13}$$

Following the above procedures, the error dynamics for the depth, polar, and azimuthal angle are, respectively, derived as

$$\dot{e}_3 = -\left(\frac{k_{p3}}{1-k_{d3}}\right)e_3. \tag{14}$$

$$\dot{e}_4 = -\left(\frac{k_{p4}}{1+\theta^2+k_{d4}}\right)e_4. \tag{15}$$

$$\dot{e}_5 = -\left(\frac{k_{p5}}{1+\varphi^2+k_{d5}}\right)e_5. \tag{16}$$

The stability is examined by formulating a Lyapunov function as $V = \frac{1}{2}(e_3^2 + e_4^2 + e_5^2)$, and then taking a derivative of the function, one leads to

$$\dot{V} = -\left(\frac{k_{p3}}{1-k_{d3}}\right)e_3^2 - \left(\frac{k_{p4}}{1+\theta^2+k_{d4}}\right)e_4^2 - \left(\frac{k_{p5}}{1+\varphi^2+k_{d5}}\right)e_5^2. \tag{17}$$

If the gains $k_{p3}, k_{p4}, k_{p5}, k_{d4}, k_{p5}$, are chosen larger than zero, and $0 < k_{d3} < 1$, the asymptotic stability is guaranteed. Thus, the steady state errors (e_3, e_4, e_5) are driven to zero.

3.3. Adaptive Fuzzy Gains for IBVS

In the PD type of IBVS, the control gains $k_{pi}, k_{di}, i = 1, \ldots, 5$ are constants that are determined from the Lyapunov stability theorem. However, the control gains can be further determined dynamically to improve the visual feedback performance of the robotic harvesting manipulator. In this regard, a fuzzy inference system based on the Mamdani fuzzy theory [31] is proposed for the design of the gains. Seven fuzzy partitions for the two error inputs $e_i, \dot{e}_i$ and outputs k_{pi}, k_{di} are, respectively, denoted to perform fuzzy reasoning according to the rules in the fuzzy rule base. From the stability proof and many trials, the corresponding membership functions of input and output linguistic variables are presented, respectively, in Figure 8 for the control gains k_{pi}, k_{di}. In addition, the triangular membership functions were adopted because of their simplicity and computational efficiency. The input–

output relationships in the fuzzy inference system are determined as shown in Table 1 based on the fuzzy logic IF–THEN rule base. The centroid defuzzification- based correlation-minimum inference is used for the fuzzy implications, and thus the corresponding control gains can be adjusted adaptively according to the tracking errors and the corresponding rate errors. The whole IBVS control structure is shown in Figure 9.

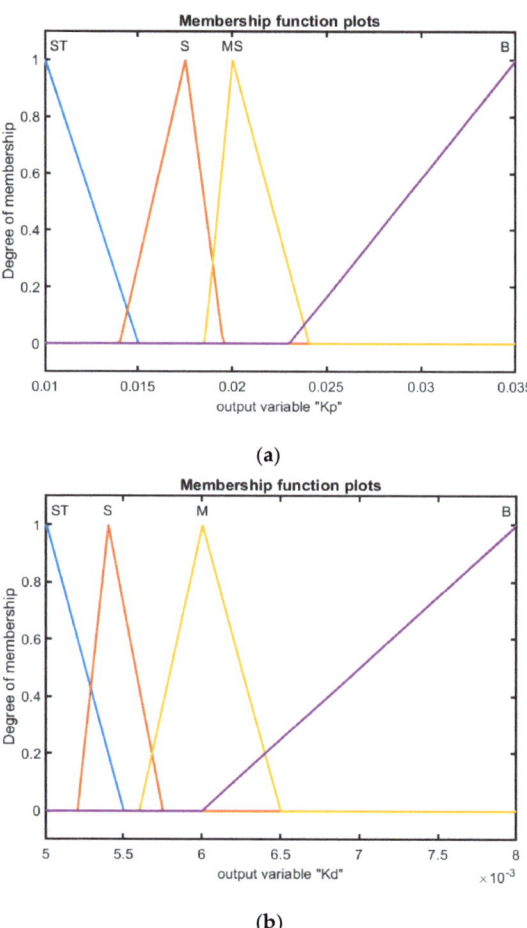

Figure 8. Membership functions of input and output linguistic variables for control gains (**a**) k_p, (**b**) k_d.

Table 1. Fuzzy rules in the fuzzy inference system.

e_2 \ e_1	NB	NM	NS	ZE	PS	PM	PB
NB	B	B	M	M	M	B	B
NM	B	M	M	S	M	M	B
NS	B	M	M	ST	M	M	B
ZE	B	M	S	ST	S	M	B
PS	B	M	M	ST	M	M	B

Table 1. *Cont.*

e_2 \ e_1	NB	NM	NS	ZE	PS	PM	PB
PM	B	M	M	S	M	M	B
PB	B	B	M	M	M	B	B

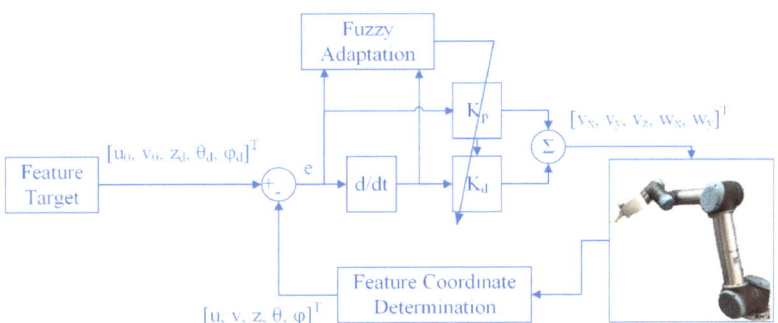

Figure 9. Control structure of the IBVS.

3.4. HVSC Algorithm

As mentioned in the preceding, PBVS makes use of a depth stereo camera to identify the target, and then the associated position is calculated by converting the desired point in the image frame to spatial coordinates. However, the conversion may result in an uncertain error because of the intrinsic and external camera parameters. Also, in the process of traveling, the position errors of the end effector will cause a serious localization deviation due to unexpected external disturbances. The errors of position are even accumulated more and more with the traveling distance. IBVS takes advantage of pixel coordinates in the image plane for feedback control without conversion to spatial coordinates, and thus the required calculation loading is comparatively lessened. Moreover, the target information is constantly returned for feedback control while traveling, so it has a higher localization accuracy than PBVS under the identical disturbances. However, the pixel-based control may cause the robotic manipulator to generate a larger response in space. The main drawback of IBVS using a fixed camera is the limited field of view. When the robotic manipulator rotates, the target may be out of the field of view, and the IBVS will fail to control the manipulator. Therefore, an HVSC integrating PBVS and IBVS was proposed for the tradeoff.

As HVSC is applied to cherry tomato harvesting, the PBVS is first executed for the point-to-point coarse localization of the end effector for efficiency. Afterwards, IBVS will be implemented to continue the ensuing movement to reach the desired operation position. Then, the remaining cutting task is performed by the PBVS again. The switching mechanism between PBVS and IBVS is under the following conditions:

(1) PBVS is first executed for the point-to-point localization until the prescribed condition $e_u \leq 5, e_v \leq 5, e_d \leq 0.2$.
(2) The mechanism switches to the fuzzy-based IBVS to continue a fine alignment to the desired operation position.
(3) When the target cherry tomato is aligned, the mechanism switches to PBVS to execute cutting off the fruit stem.

4. Experimental Results and Discussions

As shown in Figure 10, the proposed visual servo control algorithms for cherry tomato harvesting were demonstrated by the robotic manipulator. The laboratory-based experimental field as shown in Figure 1 was set for the implementation of harvesting, in which an artificial cherry tomato is installed on stainless steel wires with supposed different growth angles.

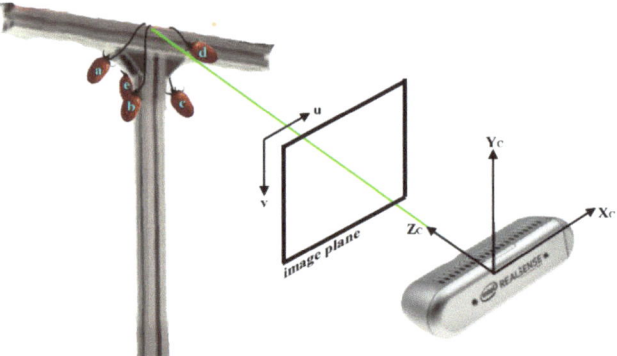

Figure 10. Artificial cherry tomatoes with different poses.

4.1. Point-to-Point Localization for Target Tomato Manipulation

The proposed PBVS, IBVS, and HVSC were tested for point-to-point localization of a target tomato. The artificial cherry tomato was laid out with the pose angles $\theta = \varphi = 0$. The position of the centroid in the image plane is located at (222, 141) pixels, and the initial depth from the image is 322 mm. The operation location is denoted at the location (320, 240) pixels in the image plane and at a depth of 370 mm. Due to the presumed pose angles, the robotic manipulator will be controlled to reach the operation position without considering the orientations of the end effector.

The errors e_1, e_2, and e_3 by the three visual feedback controllers are presented in Figure 11. It is shown that the three controllers can effectively align the target and reach the operation position. Their performances were compared as shown in Figure 12. The PBVS has larger errors in e_1, e_2, and e_3 because of the camera parameters' uncertainty and measurement errors that lead to inaccuracy in the coordinates of the target in space. However, the PBVS has a shorter execution time because the PBVS need not frequently capture images to serve as feedback information.

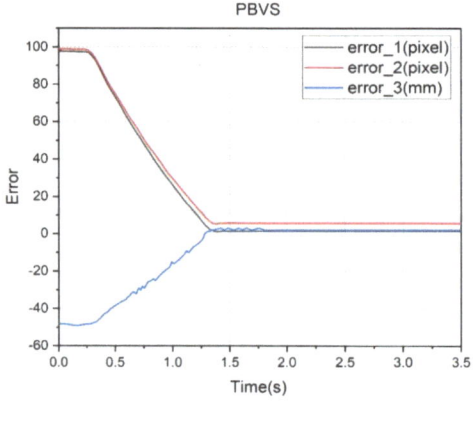

(a)

Figure 11. *Cont.*

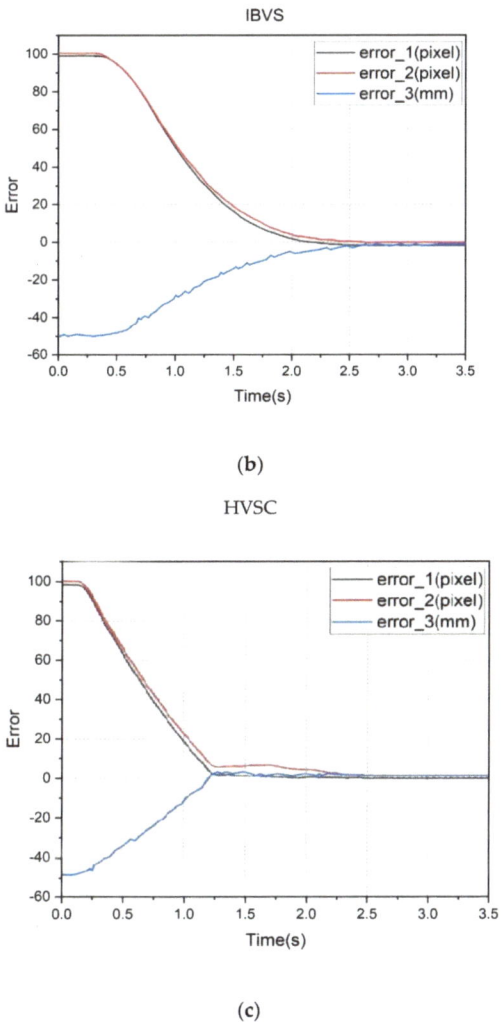

(b)

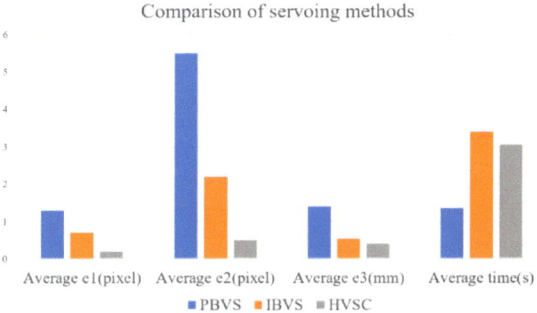

(c)

Figure 11. Error trajectories of (**a**) PBVS, (**b**) IBVS, and (**c**) HVSC for point-to-point localization.

Figure 12. Performance comparisons for PBVS, IBVS, and HVSC.

4.2. HVSC with Constant and Fuzzy Feedback Gains

In this subsection, the HVSC with a separation constant and fuzzy feedback, respectively, were performed and compared for target localization with varied poses. The results for reaching the operation position are presented in Figure 13 with $\theta = 10°$, $\varphi = 30°$ and Figure 14 with $\theta = 15°$, $\varphi = 45°$. Even when the target has a far distance from the end effector, it is seen that the HVSC with fuzzy feedback gains has better stabilization than the constant gains, due to robusticity against disturbances. In addition, the performance for larger pose angles may engender a larger localization deviation because the larger pose angles are difficult to compute and identify accurately.

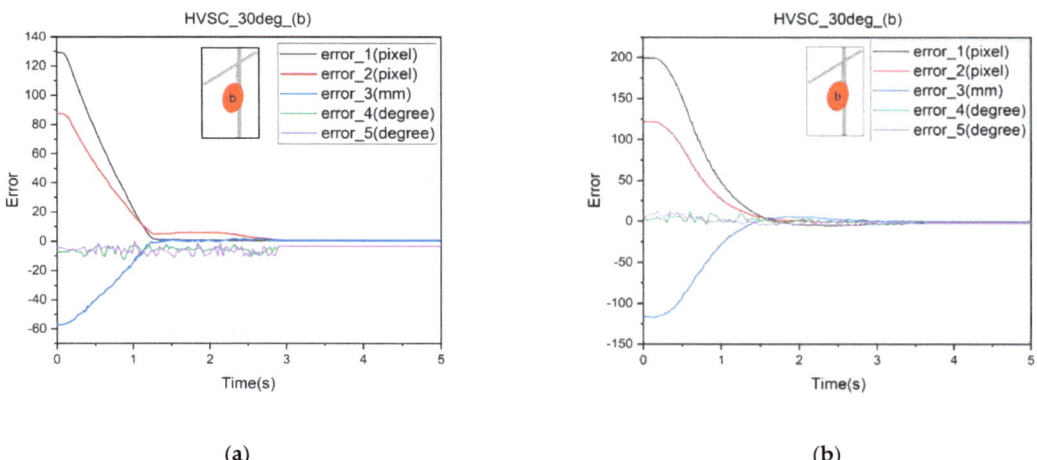

Figure 13. Error trajectories of HVSC for $\theta = 10°$, $\varphi = 30°$ with (**a**) constant, (**b**) fuzzy feedback gains.

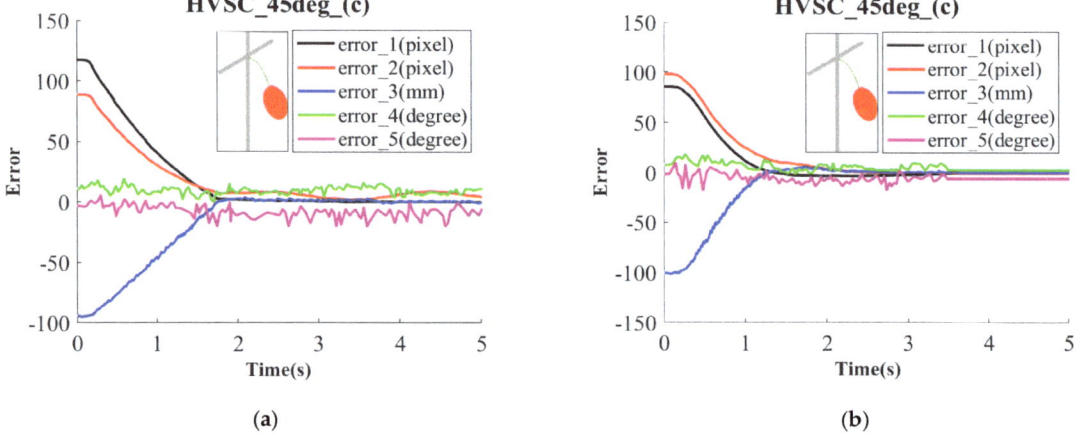

Figure 14. Error trajectories of HVSC for $\theta = 15°$, $\varphi = 45°$ with (**a**) constant, (**b**) fuzzy feedback gains.

4.3. Application to Cherry Tomato Picking

Finally, the artificial target cherry tomatoes were picked by the proposed robotic manipulator with the fuzzy-based HVSC. After identifying the tomato and determining the corresponding position and orientation, the harvesting mechanism moves to the operation position using the HVSC. According to the harvesting mechanism design, if the rectangle sleeve can successfully capture the target cherry tomato, the object must be picked without needing accurate positioning. Also, PBVS has a comparatively fast execution speed, so the visual control was switched to PBVS to pick the target following HVSC.

Figures 15–17 depict the harvesting trajectories in space for target tomatoes with growth orientations $\varphi = 30°$, $45°$, and $60°$. Initially, the surface of the rectangle frame is parallel to the ground. For the growth pose $\varphi = 30°$ and $45°$, the orientation of the end effector does not adjust very much while moving for picking. However, in the case of $60°$ of growth pose, it is apparent that the orientation of the end effector must be varied to pick the cherry tomato successfully. Moreover, based on numerous tests for each case, it is demonstrated that the picking success rate is 100% for $30°$ of growth pose and 94.5% for $45°$ of growth pose, while the picking success rate for $60°$ is the lowest with 89.2%. The reason results from the large computational errors for a target cherry tomato with a large angle for growth orientation.

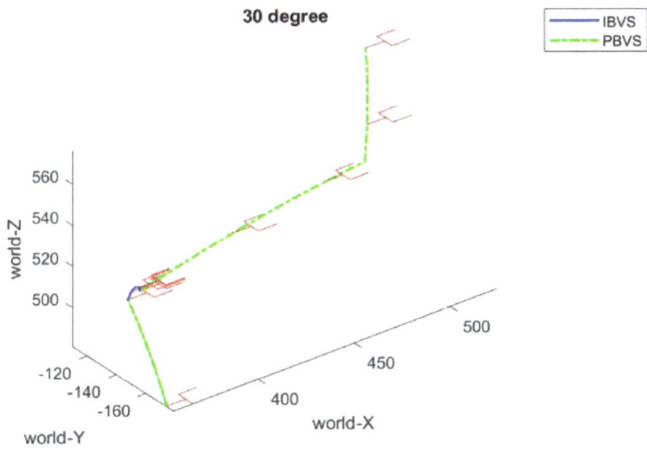

Figure 15. Trajectory of cherry tomato harvesting for $\varphi = 30°$.

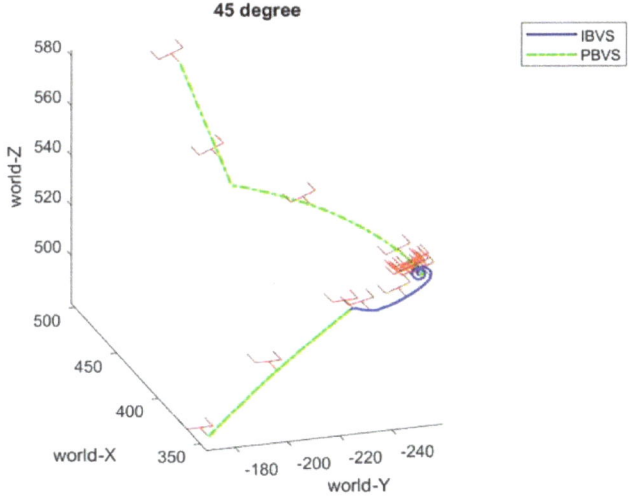

Figure 16. Trajectory of cherry tomato harvesting for $\varphi = 45°$.

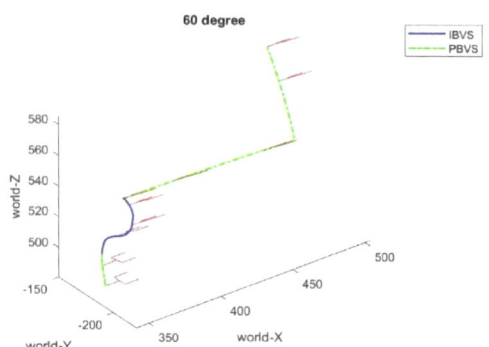

Figure 17. Trajectory of cherry tomato harvesting for $\varphi = 60°$.

5. Conclusions

This paper concludes with the realization of a robotic manipulator for cherry tomato harvesting. To perform smooth and accurate localization tasks, the fuzzy-based HVSC was used to implement the point-to-point localization and picking tasks, in which the PBVS was first performed for the coarse localization of the end effector, and the IBVS was then executed to drive the end effector to the desired operation position. Finally, the robotic manipulator was again switched to the PBVS to perform the cherry tomato picking using our developed cutting and clipping integrated mechanism. The laboratory experiments for different poses of artificial cherry tomatoes demonstrate the feasibility of the proposed robotic manipulator and visual servo control for cherry tomato harvesting. The overall results show that the developed robotic manipulator using fuzzy-based HVSC has an average harvesting time of 9.40 s/per and an average harvesting success rate of 96.25% in picking cherry tomatoes with random pose angles. The picking failures always result from the noise on the measured depth values and the associated computational pose errors such that the sleeve cannot successfully capture the target cherry tomatoes.

In the future, more investigations of factors such as the picking order, occlusion, overlapping, and environmental lighting problems are to be conducted for practical field applications. Further comparative analyses and comprehension of the proposed system in real field tests will be thus evaluated.

Author Contributions: Conceptualization: Y.-R.L. and C.-T.C.; Investigation: Y.-R.L., W.-Y.L. and Z.-H.H.; Methodology: Y.-R.L., W.-Y.L. and C.-T.C.; Software: Y.-R.L. and W.-Y.L.; Supervision: C.-T.C.; Validation: Y.-R.L. and C.-T.C.; Writing—original draft: Y.-R.L., W.-Y.L. and C.-T.C.; Writing—review & editing: Y.-R.L. and C.-T.C. All authors have read and agreed to the published version of the manuscript.

Funding: This work was supported by the Ministry of Science and Technology of Taiwan under the Grant No. MOST 108-2221-E-003 -024 -MY3, MOST 110-2623-E-003-001 and MOST 111-2221-E-003-021.

Data Availability Statement: The data that support the findings of this research are available from the corresponding author, [C.T. Chen], upon reasonable request.

Conflicts of Interest: The authors declare no conflict of interest.

References

1. Wu, H.C. Study on human power structure of current agriculture. *ATTS Q.* **2019**, *118*, 36–39.
2. Bac, C.W.; van Henten, E.J.; Hemming, J.; Edan, Y. Harvesting robots for high-value crops: State-of-the-art review and challenges ahead. *J. Field Robot.* **2014**, *31*, 888–911. [CrossRef]
3. Barnett, J.; Duke, M.; Au, C.K.; Lim, S.H. Work distribution of multiple Cartesian robot arms for kiwifruit harvesting. *Comput. Electron. Agric.* **2020**, *169*, 105202. [CrossRef]
4. Zahid, A.; Mahmud, M.S.; He, L.; Heinemann, P.; Choi, D.; Schupp, J. Technological advancements towards developing a robotic pruner for apple trees: A review. *Comput. Electron. Agric.* **2021**, *189*, 106383. [CrossRef]
5. Ji, W.; Pan, Y.; Xu, B.; Wang, J. A real-time apple targets detection method for picking robot based on ShuffleNetV2-YOLOX. *Agriculture* **2022**, *12*, 856–873. [CrossRef]

6. Xu, B.; Cui, X.; Ji, W.; Yuan, H.; Wang, J. Apple grading method design and implementation for automatic grader based on improved YOLOv5. *Agriculture* **2023**, *13*, 124–141. [CrossRef]
7. Sa, I.; Ge, Z.; Dayoub, F.; Upcroft, B.; Perez, T.; McCool, C. DeepFruits: A fruit detection system using deep neural networks. *Sensors* **2016**, *16*, 1222–1244. [CrossRef]
8. Liang, X.; Wang, H.; Liu, Y.H.; Chen, W.; Jing, Z. Image-based position control of mobile robots with a completely unknown fixed camera. *IEEE Trans. Autom. Control* **2018**, *63*, 3016–3023. [CrossRef]
9. Gans, N.; Hutchinson, S.; Corke, P. Performance tests for visual servo control systems with application to partitioned approaches to visual servo control. *Int. J. Robot. Res.* **2003**, *22*, 955–981. [CrossRef]
10. Dewi, T.; Risma, P.; Oktarina, Y.; Muslimin, S. Visual servoing design and control for agriculture robot; a review. In Proceedings of the 2018 International Conference on Electrical Engineering and Computer Science (ICECOS), Pangkal, Indonesia, 2–4 October 2018; pp. 57–62.
11. Jun, J.; Kim, J.; Seol, J.; Kim, J.; Son, H.I. Towards an efficient tomato harvesting robot: 3D perception, manipulation, and end-effector. *IEEE Access* **2021**, *9*, 17631–17640. [CrossRef]
12. Edan, Y.; Rogozin, D.; Flash, T.; Miles, G.E. Robotic melon harvesting. *IEEE Trans. Robot. Autom.* **2000**, *16*, 831–835. [CrossRef]
13. Zhao, D.; Lv, J.; Ji, W.; Zhang, Y.; Chen, Y. Design and control of an apple harvesting robot. *Biosyst. Eng.* **2011**, *110*, 112–122.
14. Lehnert, C.; English, A.; McCool, C.; Tow, A.W.; Perez, T. Autonomous sweet pepper harvesting for protected cropping systems. *IEEE Robot. Autom. Lett.* **2017**, *2*, 872–879. [CrossRef]
15. Chaumette, F.; Hutchinson, S. Visual servo control. I. Basic approaches. *IEEE Robot. Autom. Mag.* **2006**, *13*, 82–90. [CrossRef]
16. Yoshida, T.; Kawahara, T.; Fukao, T. Fruit recognition method for a harvesting robot with RGB-D cameras. *ROBOMECH J.* **2022**, *9*, 15. [CrossRef]
17. Mehta, S.S.; Burks, T.F. Vision-based control of robotic manipulator for citrus harvesting. *Comput. Electron. Agric.* **2014**, *102*, 146–158. [CrossRef]
18. Li, T.; Yu, J.; Qiu, Q.; Zhao, C. Hybrid uncalibrated visual servoing control of harvesting robots with RGB-D cameras. *IEEE Trans. Ind. Electron.* **2023**, *70*, 2729–2738. [CrossRef]
19. Barth, R.; Hemming, J.; van Henten, E.J. Design of an eye-in-hand sensing and servo control framework for harvesting robotics in dense vegetation. *Biosyst. Eng.* **2016**, *146*, 71–84. [CrossRef]
20. Li, S.; Xie, W.; Gao, Y. Enhanced IBVS controller for a 6DOF manipulator using hybrid PD-SMC method. In Proceedings of the 43rd Annual Conference of the IEEE Industrial Electronics Society (IECON), Beijing, China, 29 October–1 November 2017; pp. 2852–2857.
21. Singh, A.; Kalaichelvi, V.; Karthikeyan, R. A survey on vision guided robotic systems with intelligent control strategies for autonomous tasks. *Cogent Eng.* **2022**, *9*, 2050020. [CrossRef]
22. Malis, E.; Chaumette, F.; Boudet, S. 2 1/2 d visual servoing. *IEEE Trans. Robot. Autom.* **1999**, *15*, 238–250. [CrossRef]
23. Machkour, Z.; Ortiz-Arroyo, D.; Durdevic, P. Classical and deep learning based visual servoing systems: A survey on state of the art. *J. Intell. Robot. Syst.* **2022**, *104*, 11. [CrossRef]
24. Li, Y.R.; Lian, W.Y.; Liu, S.H.; Huang, Z.H.; Chen, C.T. Application of hybrid visual servo control in agricultural harvesting. In Proceedings of the International Conference on System Science and Engineering, Taichung, Taiwan, 26–29 May 2022; pp. 84–89.
25. Hannan, M.W.; Burks, T.F.; Bulanon, D.M. A machine vision algorithm combining adaptive segmentation and shape analysis for orange fruit detection. *Agric. Eng. Int. CIGR J.* **2009**, *6*, 1–17.
26. Hayashi, S.; Shigematsu, K.; Yamamoto, S.; Kobayashi, K.; Kohno, Y.; Kamata, J.; Kurita, M. Evaluation of a strawberry-harvesting robot in a field test. *Biosyst. Eng.* **2010**, *105*, 160–171. [CrossRef]
27. Suzuki, S. Topological structural analysis of digitized binary images by border following. *Comput. Vision Graph. Image Process.* **1985**, *30*, 32–46. [CrossRef]
28. Ghosal, S.; Mehrotra, R. A moment-based unified approach to image feature detection. *IEEE Trans. Image Process.* **1997**, *6*, 781–793. [CrossRef] [PubMed]
29. Shih, C.-L.; Lee, Y. A simple robotic eye-in-hand camera positioning and alignment control method based on parallelogram features. *Robotics* **2018**, *7*, 31. [CrossRef]
30. Dong, J.; Zhang, J. A new image-based visual servoing method with velocity direction control. *J. Frankl. Inst.* **2020**, *357*, 3993–4007. [CrossRef]
31. Chiang, Y.F.; Liu, Y.H.; Chen, C.T. Hybrid visual servo control for point-to-point localization of an autonomous wheeled mobile robot. *Int. J. iRobot.* **2022**, *5*, 20–28.

Disclaimer/Publisher's Note: The statements, opinions and data contained in all publications are solely those of the individual author(s) and contributor(s) and not of MDPI and/or the editor(s). MDPI and/or the editor(s) disclaim responsibility for any injury to people or property resulting from any ideas, methods, instructions or products referred to in the content.

Article

An Accurate Dynamic Model Identification Method of an Industrial Robot Based on Double-Encoder Compensation

Xun Liu [1,*], Yan Xu [2], Xiaogang Song [3], Tuochang Wu [4], Lin Zhang [1] and Yanzheng Zhao [1]

1. School of Mechanical Engineering, Shanghai Jiao Tong University, Shanghai 200240, China; linzhang-sjtu@sjtu.edu.cn (L.Z.); yzh-zhao@sjtu.edu.cn (Y.Z.)
2. College of Information and Computer Engineering, Northeast Forestry University, Harbin 150036, China; xuyan@nefu.edu.cn
3. Mechanical Engineering and Automation, Harbin Institute of Technology, Shenzhen 518055, China; 19b953014@stu.hit.edu.cn
4. College of Intelligent Science and Technology, National University of Defense Technology, Changsha 410073, China; wutuochang21@nudt.edu.cn
* Correspondence: liux_robot@sjtu.edu.cn

Abstract: Aiming at the challenges to accurately simulate complex friction models, link dynamics, and part uncertainty for high-precision robot-based manufacturing considering mechanical deformation and resonance, this study proposes a high-precision dynamic identification method with a double encoder. Considering the influence of the dynamic model of the manipulator on its control accuracy, a three-iterative global parameter identification method based on the least square method and GMM (Gaussian Mixture Model) under the optimized excitation trajectory is proposed. Firstly, a bidirectional friction model is constructed to avoid using residual torque to reduce the identification accuracy. Secondly, the condition number of the block regression matrix is used as the optimization objective. Finally, the joint torque is theoretically identified with the weighted least squares method. A nonlinear model distinguishing between high and low speeds was established to fit the nonlinear friction of the robot. By converting the position and velocity of the motor-side encoder to the linkage side using the deceleration ratio, the deformation quantity could be calculated based on the discrepancy between theoretical and actual values. The GMM algorithm is used to compensate the uncertainty torque that was caused by model inaccuracy. The effectiveness of the proposed method is verified by a simulation and experiment on a 6-DoF industrial robot. Results prove that the proposed method can enhance the online torque estimation performance by up to 20%.

Keywords: dynamic identification; double encoder; block regression matrix; weighted least squares; GMM algorithm; industrial robot

1. Introduction

Traditional industrial robots are widely used in manufacturing due to their high speed and precision, such as in welding, spraying, polishing, and cutting, where stable processing trajectories are required. The performance of these tasks highly depends on accurate dynamic models, which must account for factors such as friction and unknown disturbance torques. Although the robot dynamics model can be obtained from CAD models, the parameters obtained through this method may not accurately reflect the actual dynamic parameters. Consequently, researchers have proposed various methods to decompose and analyze the robot dynamics model to improve its accuracy. Vandanjon [1] used a method that independently considers each part of the robot dynamics, identifies the inertial forces, centrifugal forces, inertial integrals, and gravity separately, and designs various exciting trajectories. While this method can enhance the model's accuracy, excessive subdivisions will augment the model's uncertainty and diminish the accuracy of the identification outcomes. Currently, the widely used method for robot model identification is the global identification method [2], which can comprehensively incorporate various factors in the dynamic modeling process, and collect and process data for all joints, before applying identification algorithms to calculate all dynamic parameters simultaneously. Due to its easy implementation, this has become a commonly practiced approach to robot dynamics modeling. The mainstream process is to determine the minimum identifiable parameter set and regression matrix [3]; on this basis, design the optimal exciting trajectory to fully stimulate the dynamic characteristics of the robot [4]. Solving for kinetic parameters, design the parameter identification algorithm through the data collected along the exciting trajectory. Designing the optimal exciting trajectory [5] and parameter identification algorithm is a current research hotspot. In the case of industrial robots lacking joint torque sensors, conventional identification methods [6–8] typically leverage proprioceptive signals, such as kinematic states and motor current. Direct access to the torques applied to the robot links is unavailable, as these values are affected by errors in friction modeling and the limited precision of torque constants. Consequently, the identification results are susceptible to disturbances. Identification methods based on current measurements depend on precise prior knowledge of joint drive gains [9]. Unfortunately, the calibration scenarios for drive gains provided by manufacturers often differ from the identification scenarios [10].

Because of the high-load joint rigidity of industrial robots, the theoretical modeling of dynamics is sufficient to approximate the real joint driving torque due to the separation design of driving and joints. However, small industrial manipulators have lightweight structures and components, such as harmonic reducers, double encoders, and torque sensors, resulting in a highly integrated servo drive and motor in a single joint. This structure reduces the rigidity of the joint. Thus, theoretical dynamics modeling can only establish the link dynamics on the load side, and the motor-side dynamics need to compensate for the flexible error. In order to address the challenges posed by flexible systems, robots often rely on torque sensors, as exemplified in study [11]. The introduction of joint torque sensors transforms the system into a passive control configuration, ensuring system stability. However, traditional industrial robots typically lack joint torque sensors, necessitating the use of flexible deformation from dual encoders for approximate torque compensation. The primary application scenario for this approach is dynamic identification, aiming to acquire more accurate models. This, in turn, facilitates applications such as drag teaching or collision detection. Spong [12] introduced a modeling approach for the flexible joints of the manipulator, equivalent to a spring model with only stiffness and damping between the motor end and the connecting-rod load end. However, because of the large stiffness of the operating arm, unless correspondingly large external forces are acting on the connecting rod, the error of the double encoder cannot fully capture the physical characteristics of the flexible joint. Therefore, researchers have proposed methods to improve the model, such as static parameter identification and neural network model fitting [13].

Linear identification methods typically model frictional forces as Coulomb and linear viscous forces. Coulomb and linear viscous forces are directly identified through the method discussed in [14]. However, some studies demonstrate a nonlinear relationship between the viscous frictional force and joint velocity [15,16]. Several identification methods have been widely used, including the least squares method [17], the weighted least squares method [18], and the maximum likelihood method [19]. The least squares method is a classic algorithm used in a linear regression analysis that is easy to understand and implement; however, it is vulnerable to noise and has poor robustness. For this reason, many scholars at home and abroad have conducted further research [20–24]. Recently, some researchers have utilized neural networks to establish dynamic models, but the results are unreliable due to the networks' high sensitivity to noise and tendency to overfit. Thus, despite advancements in dynamic modeling, improving dynamic model accuracy is still an active area of research [25].

The motor and the load are not directly coupled but are essentially an elastic system. This paper presents an algorithm for the identification and compensation of dynamic model parameters for industrial robots, based on compensating for information with double encoders. This approach is exemplified using a 6-DOF industrial robot, where the minimum parameter set is derived by constructing the dynamic model and employing QR decomposition. The optimization parameters for the incentive trajectory are determined using a trajectory optimization algorithm, resulting in the acquisition of the optimized incentive trajectory. Subsequently, upon obtaining the trajectory, joint torque is identified using the iterative weighted least squares method. The nonlinear friction force of the manipulator is modeled by constructing a nonlinear model that distinguishes between high and low speeds. The WLS (weighted least squares) method is used for the identification of the dynamic parameters. Finally, the information of the nonlinear residue is fitted using a double encoder to complete the identification and compensation of dynamic parameters of the industrial manipulator. To address uncertainties in torque components that cannot be precisely modeled, the GMM (Gaussian Mixture Model) algorithm is applied for compensation. This improves the accuracy and robustness of the identification results. The entire dynamic identification process in this study is conducted in an offline mode, with only torque estimation being performed in real-time online. Figure 1 illustrates the functional flowchart of the offline identification method.

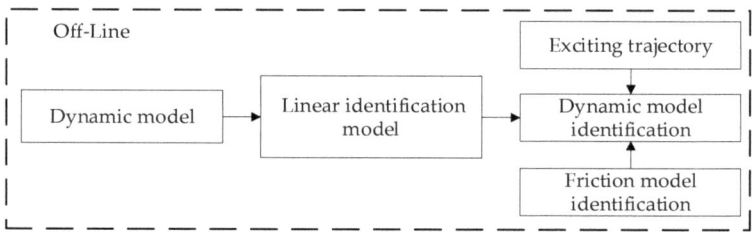

Figure 1. Flowchart of the offline identification method.

The rest of the article is arranged as follows: Section 2 introduces the linearization of the dynamic model and the identification of friction. Section 3 proposes the block regression matrix, which is used as the index to optimize the trajectory parameters for obtaining a relatively ideal exciting trajectory. Section 4 provides the identification method of the dynamic parameters based on WLS. Section 5 proposes the GMM algorithm, which is used to compensate for the uncertain torque component that cannot be accurately modeled. The simulation and experimental results are demonstrated in Section 6. At last, Section 7 concludes the article.

2. Linearization of Dynamic Model and Identification of Friction

The expression of the joint torques of a serial robot can be obtained using the Newton–Euler iterative method base on MDH [26] and can be represented as follows: "∈".

$$\tau_m = M(\theta)\ddot{\theta} + C(\theta, \dot{\theta}) + G(\theta) + \tau_f + J(\theta)^T F_{ext} + \tau_u \qquad (1)$$

In this equation, $M(\theta) \in \mathbb{R}^{n \times n}$ and n represent the positive definite symmetric inertia matrix and the number of joints, respectively. $C(\theta, \dot{\theta}) \in \mathbb{R}^{n \times n}$ and $G(\theta) \in \mathbb{R}^{n \times 1}$ represent the Coriolis and gravitational torques, respectively. $\theta, \dot{\theta}, \ddot{\theta}$ are displacement, velocity, and acceleration vectors of the joint's $n \times 1$ vector space. $\tau_m \in \mathbb{R}^{n \times 1}$ and $\tau_f \in \mathbb{R}^{n \times 1}$ represent the driving torque and frictional torque of the joint in the dynamic model, respectively. F_{ext} and $J(\theta)$ represent the external force acting on the robot endpoint and jacobian matrix, and $\tau_u \in \mathbb{R}^{n \times 1}$ represents the unmodeled and disturbance torques of the joint. Since the external force, unmodeled part, and disturbance torque are independent of the parameters of the robot dynamic model itself, the dynamic model becomes a link dynamic model. Therefore, in this paper, Equation (1) is linearized [3] to obtain the link dynamic model. The link dynamic model is built solely on the characteristics of the link, without accounting for joint influences. On the other hand, the joint dynamic model incorporates the effects of joint parameters. The torque in this part will be studied in the following text. After rearranging the link dynamic parameters of the robot to be identified and removing and integrating the columns that do not affect the identification process, the basic dynamic parameters, namely the minimum parameters, will be obtained.

$$\tau_{link} = \begin{bmatrix} \Gamma_{link} & \Gamma_f \end{bmatrix} \begin{bmatrix} \Phi_{link} \\ \Phi_f \end{bmatrix} = \Gamma \Phi \tag{2}$$

The dynamic model becomes a dynamic model of a link without a motor. Where $\tau_{link} \in \mathbb{R}^{n \times 1}$ is link torque, $\Gamma_{link} \in \mathbb{R}^{n \times rank(\Gamma_{link})}$ is link parameters, $\Phi_{link} \in \mathbb{R}^{rank(\Gamma_{link}) \times 1}$ is the link regression matrix, $\Gamma_f \in \mathbb{R}^{n \times 8}$ is friction parameters, and $\Phi_f \in \mathbb{R}^{8 \times 1}$ is the friction regression matrix. Γ is solely dependent on the mechanical arm's motion state and independent of its structural parameters. The regression matrix can be obtained using the kinematic formula. Φ represents the minimum set of expressions corresponding to the dynamic structural parameters. Bidirectional torque detection is an accurate method to obtain frictional force data. It can be derived that the Coriolis/centrifugal matrix satisfies

$$C(\theta, -\dot{\theta})(-\dot{\theta}) = C(\theta, \dot{\theta})\dot{\theta} \tag{3}$$

Given that a majority of industrial robots lack joint torque sensors, directly acquiring joint friction torque becomes impractical. Nevertheless, it is feasible to deduce the joint friction torque by examining the characteristics and design of the robot's configuration and kinematic state. Industrial robots typically feature encoders on each joint's motors, enabling the direct reading of joint velocity. Subsequently, [27] introduces a bidirectional friction estimation method for extracting joint friction torque from the overall joint torques. For simplicity, it can be assumed that the friction torque is only related to the joint velocity:

$$\tau_f(-\dot{\theta}) = -\tau_f(\dot{\theta}) \tag{4}$$

Assume two industrial robot configurations θ_1, θ_2 meet the following conditions:

$$\begin{aligned} \theta_1 &= \theta_2 = q \\ \dot{\theta}_1 &= -\dot{\theta}_2 = \dot{q} \\ \ddot{\theta}_1 &= -\ddot{\theta}_2 = \ddot{q} \end{aligned} \tag{5}$$

Substituting Equation (5) into Equation (3), it can be obtained that

$$\begin{aligned} \tau_1 &= M(q)\ddot{q} + C(q, \dot{q})\dot{q} + G(q) + \tau_{estf}(\dot{\theta}) \\ \tau_2 &= M(q)(-\ddot{q}) + C(q, -\dot{q})(-\dot{q}) + G(q) + \tau_{estf}(-\dot{\theta}) \end{aligned} \tag{6}$$

Substituting Equation (5) into Equation (6), it can be found that

$$\tau_1 - \tau_2 = 2M(q)\ddot{q} + 2\tau_{estf}(\dot{q}) \tag{7}$$

According to Equation (7), when the robot moves slowly, friction is typically obtained under low-speed and constant-speed conditions, where joint accelerations are exceedingly small and can be approximated as negligible. This method provides a straightforward and easily implementable approach for acquiring frictional force data in robotic arm systems. The inertia force/torque $M(q)$ can be ignored, and it can be obtained that

$$\tau_{estf}(\dot{\theta}) = \frac{\tau_1 - \tau_2}{2} \tag{8}$$

A common approach for dynamic model identification involves assuming the friction model as Coulomb friction plus viscous friction linear to the joint velocity; this is often inadequate in practical scenarios. Recognizing the nonlinearity of friction, several advanced friction models have been proposed in relevant literature [28,29]. However, these models are typically isolated from mass-inertial parameters and identified independently using nonlinear optimization methods. In recent developments, a unified approach has been introduced for dynamic model identification that incorporates nonlinear friction. This paper adopts the following nonlinear friction model for each joint:

$$\tau_{fi} = \begin{cases} \tau_{fci} & |\dot{\theta}_i| < \lambda \\ \tau_{fci} + \tau_{fvi} & |\dot{\theta}_i| \geq \lambda \end{cases} \quad (9)$$

$$\tau_{fci} = \begin{cases} k_{ci1} \dfrac{\tanh\left(\frac{\dot{\theta}_i}{eps}\right)\left(\tanh\left(\frac{\dot{\theta}_i}{eps}\right)+1\right)}{2} & \dot{\theta}_i \geq 0 \\ k_{ci2} \dfrac{\tanh\left(\frac{\dot{\theta}_i}{eps}\right)\left(1-\tanh\left(\frac{\dot{\theta}_i}{eps}\right)\right)}{2} & \dot{\theta}_i < 0 \end{cases} \quad (10)$$

$$\tau_{fvi} = \begin{cases} k_{vi1} \dfrac{\tanh\left(\frac{\dot{\theta}_i}{eps}\right)+1}{2} \dot{\theta}_i + k_{vi1}^2 \dot{\theta}_i^2 + k_{vi1}^3 \dot{\theta}_i^3 & \dot{\theta}_i \geq 0 \\ k_{vi2} \dfrac{1-\tanh\left(\frac{\dot{\theta}_i}{eps}\right)}{2} \dot{\theta}_i + k_{vi2}^2 \dot{\theta}_i^2 + k_{vi2}^3 \dot{\theta}_i^3 & \dot{\theta}_i < 0 \end{cases} \quad (11)$$

Coulomb and viscous friction models are adopted in this paper, where k_{ci1} and k_{vi1} represent the friction coefficients during forward motion, and k_{ci2} and k_{vi2} represent the friction coefficients during backward motion. However, accurately defining static and low-speed friction poses a significant challenge. A suitable threshold λ is set with Equation (9) to make the joint's low-speed and high-speed movements smoother, and friction model accuracy is ensured by considering the joint velocity squared and velocity cubed in the calculations. $\tanh(\cdot)$ is the hyperbolic tangent function, and eps is the transition accuracy typically set to 0.0001. This method overcomes the discontinuity problem of the $\text{sign}(\cdot)$ function near the switching point at 0 and avoids the estimation errors in friction force caused by identification errors or switching the direction of movement. τ_{estfi} is the friction torque of the i-th joint determined through measuring torque and inner-layer identification, while τ_{fi} is the estimated friction torque of the i-th joint obtained through the friction model.

$$\arg(\lambda, k_{ci1}, k_{ci2}, k_{vi1}, k_{vi2}) \min(\sum_{i=1}^{6} \|\tau_{estfi} - \tau_{fi}\|) \quad (12)$$

3. Optimization Index Based on the Condition Number of Block Regression Matrix

Matrix calculation sensitivity to errors can be reflected by a matrix's condition number. A smaller condition number of the regression matrix, viewed physically by a robot, results in an exciting trajectory, allowing higher velocity and acceleration over the entire workspace, thereby collecting more information for parameter identification. The dynamic model used in this paper suggests that a smaller condition number of the regression matrix results in higher joint acceleration, stimulating the robot's inertia tensor matrix. Higher joint velocities better stimulate the centrifugal force and Coriolis force terms. Significant joint position changes create larger torque differences, thus better stimulating the gravitational force term. Thus, the condition number $\text{cond}(\Gamma)$ can be used as an index for the regression matrix's influence on inertial parameter identification. The condition number of the regression matrix Γ generally serves as the optimization index for the exciting trajectory. However, research indicates that optimizing only the condition number of the regression matrix fails to meet accurate dynamic model requirements. Therefore, optimizing the condition number of submatrices is also necessary during the optimization process to constrain the internal structure of the regression matrix Γ. The text introduces the weight matrix based on the least squares method and converts the optimization objective to a weighted regression matrix Γ^* following [24].

$$\text{cond}(\Gamma^*) = \text{cond}(\Omega^{-\frac{1}{2}} \cdot \Gamma) \leq \text{cond}(\Omega^{-\frac{1}{2}}) \cdot \text{cond}(\Gamma) \tag{13}$$

The constant matrix in the formula is denoted by Ω. The matrix is calculated by measuring noise throughout this paper. The matrix is utilized to optimize the condition number of Γ^* throughout the paper. It serves as a prerequisite for obtaining a more accurate dynamic model. The observation matrix of frictional force contains many zeros since the exciting trajectory of each joint is independent, which could increase the condition number. Therefore, this paper uses an observation matrix for exciting trajectory that does not have the frictional force part. Optimizing the condition number of Γ^* alone as the optimization objective does not achieve the desired results, as per experimental observations. We also optimize the condition number of the submatrix of Γ^* to constrain its internal structure. The Γ^* matrix is decomposed into sub-regression matrices including the acceleration term Γ_α^* and velocity term Γ_β^* and joint position term Γ_γ^*. The impact of each sub-regression matrix on the total regression matrix varies. As such, assigning varying weights to different sub-regression matrices is necessary. Ω_α, ω_β, and ω_γ, respectively, represent the respective weights of sub-regression matrices. In order to optimize the internal structure more effectively, parts with larger condition numbers are given heavier weights and those with smaller condition numbers are given lighter weights. The weight values are computed by finding the variance of the corresponding columns of the regression matrix. The values show that indicators with greater differences in variation are assigned larger weights, while those with smaller differences are given smaller weights. The larger the weight, the more significant the respective target. To summarize, the optimization objective of the paper is

$$\text{co}\Gamma = (\omega_\alpha + \omega_\beta + \omega_\gamma) \cdot \text{cond}(\Gamma^*) + \omega_\alpha \cdot \text{cond}(\Gamma_\alpha^*) + \omega_\beta \cdot \text{cond}(\Gamma_\beta^*) + \omega_\gamma \cdot \text{cond}(\Gamma_\gamma^*) \tag{14}$$

This paper uses a limited Fourier series trajectory as the identification exciting trajectory.

$$\begin{cases} \theta_i(t) = \sum_{l=1}^{N} \frac{a_{l,i}}{w_f l} \sin(w_f l t) - \frac{b_{l,i}}{w_f l} \cos(w_f l t) + \theta_{i0} \\ \dot{\theta}_i(t) = \sum_{l=1}^{N} a_{l,i} \cos(w_f l t) + b_{l,i} \sin(w_f l t) \\ \ddot{\theta}_i(t) = w_f \sum_{l=1}^{N} -a_{l,i} l \sin(w_f l t) + b_{l,i} l \cos(w_f l t) \end{cases} \tag{15}$$

where N is the number of terms in the Fourier series trajectory, the sampling frequency of the trajectory is ff, and the fundamental frequency is $w_f = 2\pi f_f$; $a_{l,i}$ and $b_{l,i}$ are the amplitudes of the trigonometric functions. Considering joint limits, velocity, and acceleration limits, the following objectives and constraints are given, where t_s and t_e are the start and end times of the sampling time:

$$\min \text{co}\Gamma$$
$$\text{subject to}: \begin{cases} |\theta_i(t)| \leq \theta_{i,\max} \\ |\dot{\theta}_i(t)| \leq \dot{\theta}_{i,\max} \\ |\ddot{\theta}_i(t)| \leq \ddot{\theta}_{i,\max} \\ \theta_i(t_s) = \theta_i(t_e) = 0 \\ \dot{\theta}_i(t_s) = \dot{\theta}_i(t_e) = 0 \\ \ddot{\theta}_i(t_s) = \ddot{\theta}_i(t_e) = 0 \end{cases} \tag{16}$$

To perform optimization and solve the problem, it is necessary to process Equation (16) above and convert it to

$$\begin{cases} |\theta_i(t)| = \left| \sum_{l=1}^{N} \frac{a_{l,i}}{\omega_f l} \sin(\omega_f l t) - \frac{b_{l,i}}{\omega_f l} \cos(\omega_f l t) + \theta_{i0} \right| \\ \leq \sum_{l=1}^{N} \frac{1}{\omega_f l} \sqrt{a_{l,i}^2 + b_{l,i}^2} + |\theta_{i0}| \leq \theta_{i,\max} \\ \left| \dot{\theta}_i(t) \right| = \left| \sum_{l=1}^{N} a_{l,i} \cos(\omega_f l t) + b_{l,i} \sin(\omega_f l t) \right| \\ \leq \sum_{l=1}^{N} \sqrt{a_{l,i}^2 + b_{l,i}^2} \leq \dot{\theta}_{i,\max} \\ \left| \ddot{\theta}_i(t) \right| = \left| \omega_f \sum_{l=1}^{N} b_{l,i} \cos(\omega_f l t) - a_{l,i} l \sin(\omega_f l t) \right| \\ \leq \omega_f \sum_{l=1}^{N} l \sqrt{a_{l,i}^2 + b_{l,i}^2} \leq \ddot{\theta}_{i,\max} \end{cases} \quad (17)$$

$$\begin{cases} \theta_i(t_s) = \theta_i(t_e) = \sum_{l=1}^{N} \frac{b_l}{\omega_f l} - \theta_{i0} = 0 \\ \dot{\theta}_i(t_s) = \dot{\theta}_i(t_e) = \sum_{l=1}^{N} a_l = 0 \\ \ddot{\theta}_i(t_s) = \ddot{\theta}_i(t_e) = \sum_{l=1}^{N} \omega_f l b_l = 0 \end{cases} \quad (18)$$

At the beginning of each iteration, a starting point is randomly selected. During optimization, if the objective function decreases in the current process, the present outcome will be upheld, and the regression matrix will be amplified. Otherwise, a new starting point will be randomly selected, and the regression matrix will remain unchanged. If the objective function value does not decrease after k attempts, the global optimal solution is considered to have been reached, and the search process stops.

4. Dynamic Model Identification of Link Based on WLS

Currently, we have identified the required exciting trajectory coefficients and performed dynamic identification to obtain parameter sets. We must note that the collected torque belongs to the joint driving torque, and we need to identify the link kinematics after excluding the friction torque. Consequently, Equation (2) can be modified to

$$\tau_m = \Gamma_{link} \Phi_{link} + \delta \quad (19)$$

where δ is the torque error and noise error. The cause of these errors is that the collected joint torque does not possess a complete equal relationship with the identified link kinematics. Torque is collected from each joint at different sampling time units, where it is concatenated and combined into the final collected torque set. Furthermore, the torque error collected from each joint has different standard deviation. To mitigate the impact of collected data errors on the accuracy of the identification, we followed the approaches presented in reference [24]. Consequently, errors are defined with the following attributes:

$$E(\delta^T \delta) = o_\delta^2 e \quad (20)$$

$E(\cdot)$ represents mathematical expectation, and o_δ^2 represents the variance of δ. Assuming that each joint's noise error is independent of each other, e is a unit diagonal matrix. $o_\delta^2 e$ represents the variance of the noise error of the driving torque of the six joints. Directly using the traditional standard LS (least squares) method for identification can only minimize the 2-norm of the error between the collected torque and the estimated torque of the linear part, without minimizing the 2-norm error δ. This leads to suboptimal optimization of the minimum parameter set variance during identification. To overcome this limitation, we recommend using the WLS (weighted least square) method. First, calculate the torque error, define the collected torque as τ_{sample}, the data number is 6 m, and estimate the torque through LS as τ_{LS}.

$$E = \tau_{sample} - \Gamma_{sample}\Phi_{LS} \quad (21)$$

$$\begin{cases} \Sigma = \begin{bmatrix} \Sigma_{11} & & & \\ & \Sigma_{22} & & \\ & & \ddots & \\ & & & \Sigma_{nn} \end{bmatrix} \\ \Sigma_{ii} = \mathrm{var}(\mathbb{E}_i) \end{cases} \quad (22)$$

$$\Phi_{WLS} = \left(\Gamma^T \Sigma_{6m}^{-1} \Gamma\right)^{-1} \Gamma^T \Sigma_{6m}^{-1} \tau \quad (23)$$

$\mathbb{E} \in \mathbb{R}^{6m \times 1}$, $\varepsilon_i \in \mathbb{E}$, $\mathrm{var}(\cdot)$ represents the calculation of variance, and $\Sigma_{6m}^{-1} \in \mathbb{R}^{6m \times 6m}$ is a block diagonal matrix consisting of m identical blocks of Σ. There is no unique method to determine the weighting coefficients. Due to the assumption that the joint noise is mutually independent, $\sigma_\delta^2 e$ can be a diagonal matrix. However, in reality, the joint noise is correlated. Hence, the weight can be calculated by computing the non-diagonal covariance matrix.

$$\Omega = \frac{E \cdot E^T}{m - rank(\Gamma_{linkmin})} \quad (24)$$

$$\Phi_{WLS} = \left(\Gamma^T \Omega_{6m}^{-1} \Gamma\right)^{-1} \Gamma^T \Omega_{6m}^{-1} \tau \quad (25)$$

5. Nonlinear Joint Dynamics Compensation

This study proposes a three-stage iterative identification method for dynamic model identification. In the first stage, theoretical identification is conducted for the link dynamics. The second stage focuses on identifying friction. The third stage involves compensating for uncertain components based on flexible error. The sections and functional modules in the paper are as shown in Figure 2. The three-ring identification algorithm proposed in this paper is included in it.

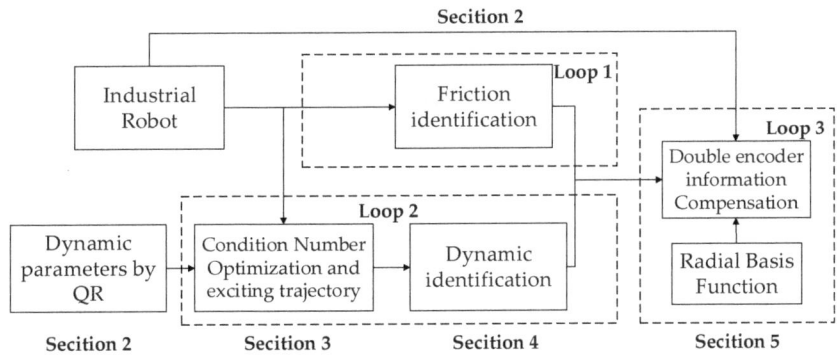

Figure 2. The flowcharts of three-loop dynamic identification.

In the absence of considering factors such as joint vibration and flexibility, the servo motor and the load end are regarded as rigid bodies to improve the basic identification accuracy as much as possible. However, in actual situations, the motor and the load end are not directly coupled, but rather form an elastic system, and the joint bearings and outer frame are not completely rigid. Under the action of motor drive torque, mechanical deformation occurs. Mechanical resonance has a certain effect on the dynamic performance of the system, mainly due to harmonic reducers, and the joint physical model is shown in Figure 3. Therefore, for joint systems that have requirements for accuracy and speed, elastic deformation cannot be ignored. This principle can be used to estimate the torque component during mechanical resonance in reverse. Another significant torque component can be estimated directly through current, while compensating for frictional forces, which will result in a better effect.

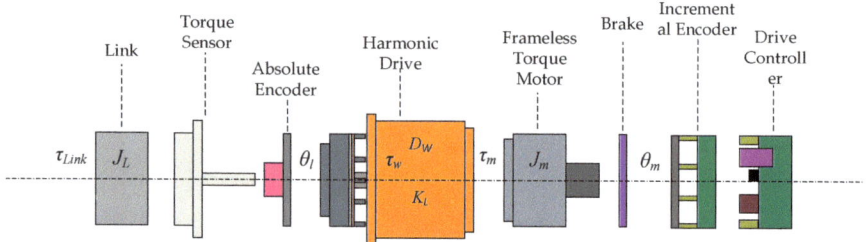

Figure 3. Physical model of joints.

The servo drive section is located on the right; the link end is located on the left, with the harmonic reducer transmission device being in the center. The servo motor drives the entire executing mechanism, and its position is measured with an incremental encoder. The position of the link end is measured with an absolute value encoder. The position and velocity of the motor-side encoder are converted to the link side through the reduction ratio. Thus, the theoretical and actual errors can be calculated to determine the deformation. The joint module is driven by a coupled drive, which results in the servo system becoming a highly coupled and multi-inertia system. For ease of study, the system can be simplified into a flexible connected servo system with two inertias. τ_w represents the harmonic resonance torque component caused by the harmonic reducer, and τ_m is generally directly obtained from the motor without taking this into account. J_m and J_l refer to the rotational inertia of the motor and the link, respectively. K_w is the transmission coupling stiffness coefficient while D_w is the transmission shaft damping coefficient. θ_m and θ_l are, respectively, the theoretical angle calculated to the link side and the actual angle of the link side. Due to the challenging friction modeling and accurate modeling of the harmonic link and torque transmission error, errors occur. Also, the acceleration is measured inaccurately prone to fluctuations, and filtering causes errors. Therefore, the torque generated by $J_m \ddot{\theta}_m$ is not taken into account, and the following text will incorporate it into the error. Based on the traditional dynamic model, this paper calculates the difference between the model torque and the actual torque, and analyzes it using an error model and data model analysis. To utilize dynamic compensation, the torque τ_w is no longer used, and the torque τ_u is defined.

$$\tau_m = \tau_f + \tau_u + \tau_{link} \tag{26}$$

An effective and relatively fast method, which is designed as a unified three-loop iterative scheme [21], is proposed to acquire an accurate dynamic model, but does not compensate for unmodeled torque or harmonics. This paper proposes an improved version of the three-loop method. The first loop identifies the friction force, the second loop identifies the τ_{link} of the link dynamic, and the third loop compensates for uncertain components using flexible error to improve accuracy. If the torque component τ_u cannot be accurately modeled, direct application of the learning method will lead to over-reliance on data. Furthermore, the linear modeling method suggested in [6] shows poor accuracy in some states, and estimation is impossible without flexible deformation. This paper proposes a method based on double-encoder information for identifying the residual torque. This method is designed to enable nonlinear approximation and smoothing of the residual torque, effectively solving the problems mentioned above. For the nonlinear $f(\cdot)$ part, it is inspired by the methods similar to DMP (Dynamic Movement Primitives) and RBF (radial basis function) in [30] to fit the nonlinear part. The radial basis function is

$$\psi_j(u) = \exp\left(-\frac{1}{2\sigma_j^2}(u-c_j)^2\right), \quad j = 1, 2, \ldots, N \tag{27}$$

σ represents the width of the radial basis function, c represents its center position, and the number of basis functions is N. Since offline identification is used to obtain weight coefficients, it is hoped that the fitting parameters have high compatibility, and the error limit of $\theta_m - \theta_l$ is defined as $\Delta\theta_{max}, \Delta\theta_{min}$; then, the calculation method of σ_i and c_i is

$$\begin{cases} c_j = \Delta\theta_{\min} + (j-1)\left(\frac{\Delta\theta_{\max} - \Delta\theta_{\min}}{N-1}\right) \\ \sigma_j = \frac{1}{(0.5(c_{j+1}-c_j))^2} \end{cases} \tag{28}$$

In the same way, operating on $\dot{\theta}_m - \dot{\theta}_l$, σ_j only needs to expand one data value at the end to satisfy data synchronization, so that the values of c_j and σ_j can be determined. Ensure that the Gaussian function covers the entire flexible error space of the industrial robot, and has an effective mapping for the input u of the RBF network. If the traditional RBF network is used to fit the data, it is necessary to update the center position, width, and weight of the radial basis function. However, increasing the amount of program also modifies the coverage interval of the radial basis function, which is unfavorable for dynamic identification, so $[\theta_m - \theta_l, \dot{\theta}_m - \dot{\theta}_l]$ needs to be identified uniformly, so as to reflect the uncertain torque components under different joint states. Assume that the number of radial basis functions is N, and the data have M groups:

$$f(u) = \frac{\sum\limits_{i=1}^{N} w_{rj}\psi_j(u)}{\sum\limits_{j=1}^{N} \psi_j(u)} \tag{29}$$

It is necessary to identify all the data at one time, and the expected fitting data are $f_{target} = [f_{target}(1), f_{target}(2), \ldots f_{target}(M)]$; then, the weight identification is

$$\begin{bmatrix} \psi_1(1) & \psi_1(2) & \cdots & \psi_N(M) \\ \psi_2(1) & \psi_2(2) & \cdots & \psi_N(M) \\ \vdots & \vdots & \cdots & \vdots \\ \psi_N(1) & \psi_N(2) & \cdots & \psi_N(M) \end{bmatrix} \cdot \begin{bmatrix} f_{target}(1) \\ f_{target}(2) \\ \vdots \\ f_{target}(m) \end{bmatrix} = \begin{bmatrix} w_{r1}\sum\limits_{j=1}^{M}\psi_1(i) \\ w_{r2}\sum\limits_{j=1}^{M}\psi_2(i) \\ \vdots \\ w_{rN}\sum\limits_{j=1}^{M}\psi_N(i) \end{bmatrix} \tag{30}$$

Directly calculate the network weight $W_r = [w_{r1}, w_{r2}, \ldots, w_{rN}]$; the actual process needs to $[\theta_m - \theta_l, \dot{\theta}_m - \dot{\theta}_l]$ as the input data, and torque τ_u as the expected training.

Due to the multi-degree-of-freedom serial structure of robotic manipulators, employing a single neural network alone cannot adequately capture the coupling between joints. Therefore, a GMM (Gaussian Mixture Model) is employed to model each joint of the multi-degree-of-freedom robotic manipulator. Subsequently, GMR (Gaussian Mixture Regression) is applied to fit the data for each joint individually. This approach is essential for accurately characterizing the intricate interdependencies among the joints in multi-degree-of-freedom robotic arms.

GMM (Gaussian Mixture Model) assumes that data are composed of multiple Gaussian distributions, each referred to as a component, and a data point may originate from any one of these components. Model parameters include the mean, covariance matrix, and weight for each component. These parameters are estimated using the EM (Expectation–Maximization) algorithm, where the E-step calculates the probability of each data point belonging to each component, and the M-step updates the model parameters. The EM algorithm is an iterative optimization algorithm used to estimate parameters in models with latent variables. It comprises two steps: the Expectation step and Maximization step. The E-step computes the expectation of latent variables for observed data given the current parameters. The M-step maximizes the expectation calculated in the E-step, updating model parameters. GMR (Gaussian Mixture Regression) is a regression model that uses GMM to model the conditional probability distribution, capturing the relationship between input and output. GMM parameters are estimated using the EM algorithm. Given input data, conditional probability distribution is computed using GMM, followed by the calculation of the expected value and variance of the output. This is commonly used for modeling complex nonlinear relationships. In summary, the fundamental idea of the GMM EM GMR algorithm is to model data using GMM, iteratively optimize model parameters with the EM algorithm, and then apply these parameters in GMR to establish the relationship between input and output for predictive purposes.

The modeling and regression process for the i-th joint is as follows: In the first step, data from joint i are $\xi_i = \{u_i, f(u_i)\}$, with $\xi_i \in \mathbb{R}^{3 \times m}$, $u_i \in \mathbb{R}^{2 \times m}$ comprising an input vector consisting of joint position and velocity, where $f(u_i) \in \mathbb{R}^m$ represents the output vector composed of joint torque residuals, and m represents the number of sampled points in the dataset. The second step involves modeling the dataset ξ using a GMM consisting of K Gaussian components. The joint probability density of the GMM is defined as follows:

$$p(\xi_i) = \sum_{k=1}^{K} \pi_k N(\xi_i; \mu_k, \Sigma_k) \tag{31}$$

In the equation, $\pi_1, \ldots, \pi_K$ represents the mixture coefficient for the k-th Gaussian component, subject to the constraints $\pi_k > 0$ and $\sum_{k=1}^{K} \pi_k = 1$. $\mu_1, \ldots, \mu_K$ denotes the mean of the k-th Gaussian component, and $\Sigma_1, \ldots, \Sigma_K$ is the covariance matrix associated with it. $N(\mu_1, \Sigma_k)$ represents the Gaussian component defined by mean μ_k and covariance Σ_k. Specifically, the k-th Gaussian component is defined as follows:

$$p(\xi_i | \mu_k, \Sigma_k) = \frac{1}{2\pi\sqrt{|\Sigma_k|}} e^{-\frac{1}{2}((\xi_i - \mu_k)^T \Sigma_k^{-1}(\xi_i - \mu_k))} \tag{32}$$

The third step involves utilizing the EM algorithm to compute the parameters π_k, μ_k, and Σ_k for each Gaussian component.

In the fourth step, after obtaining the GMM parameters, GMR is employed to fit the expected function $f(\cdot)$. For each Gaussian component, given the input data u_i, the conditional probability $f(u_i)$ satisfies a Gaussian distribution.

$$p(f(u_i) | u_i, k) = N(f(u_i); \hat{f}(u_i), \hat{\Sigma}_{f(u_i),k}) \tag{33}$$

$$\hat{f}_k(u_i) = \mu_{f(u_i),k} + \Sigma_{f(u_i)u_i,k} \left(\Sigma_{u_i,k}\right)^{-1} (u_i - \mu_{u_i,k}) \tag{34}$$

$$\hat{\Sigma}_{f(u_i),k} = \Sigma_{f(u_i),k} - \Sigma_{f(u_i)u_i,k} \left(\Sigma_{u_i,k}\right)^{-1} \Sigma_{u_i f(u_i),k} \tag{35}$$

$$p(f(u_i) | u_i) = \sum_{k=1}^{K} h_k(u_i) N(f(u_i); \hat{f}_k(u_i), \hat{\Sigma}_{f(u_i),k}) \tag{36}$$

$$h_k(u_i) = \frac{p(k)p(u_i|k)}{\sum_{i=1}^{K} p(i)p(u_i|i)} = \frac{\pi_k N(u_i; \mu_{u_i,k}, \Sigma_{u_i,k})}{\sum_{i=1}^{K} \pi_i N(u_i; \mu_{u_i,k}, \Sigma_{u_i,k})} \tag{37}$$

The estimation of the conditional expectation $f(u_i)$ given u_i under the Gaussian distribution is defined using the linearity property. The parameters of the Gaussian distribution are defined as

$$\hat{f}(u_i) = \sum_{k=1}^{K} h_k(u_i)\hat{f}_k(u_i) \tag{38}$$

$$\hat{\Sigma}_{f(u)} = \sum_{k=1}^{K} h_k^2(u)\hat{\Sigma}_{f(u),k} \tag{39}$$

The above equation represents the torque residual $\hat{f}(u_i)$ fitted under joint position and joint velocity u_i for the i-th joint.

6. Simulation and Experiment

To illustrate the proposed method, several experiments were undertaken on the 6-DoF industrial robot. The experiment system is shown in Figure 4, and the controller's hardware platform is equipped with a SpeedGoat RCP (The SpeedGoat, Bern, Switzerland) real-time simulation platform [31] that has a computation cycle of 1 ms. The control system is developed with the MDH model, utilizing the real-time function of MATLAB Simulink (The MathWorks, Natick, USA). The computer in use is equipped with a CPU: I7-11800H-2.30 GHz, 64 G-3200 MHZ memory, and the MATLAB version is 2022b.

(a)

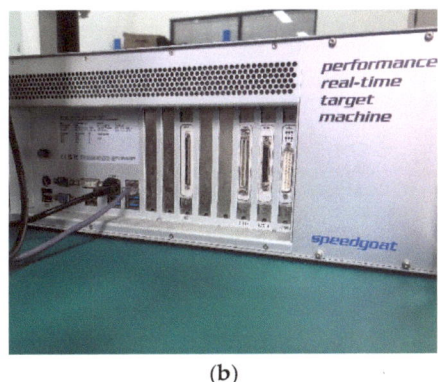
(b)

Figure 4. The experimental robotic system: (**a**) 6-DoF industrial robot; (**b**) SpeedGoat simulation platform.

In this paper, the fifth-order Fourier series is used to design the exciting trajectory. The exciting frequency of the trajectory is $f_s = 0.02$ Hz, and the cycle is 20 s. The displacement, velocity, and acceleration of each joint are calculated using the Fourier series, to obtain 200 discrete points every 0.1 s. The constraint limits of the joint displacements, velocities, and accelerations of the industrial robots used in the experiment are shown in Table 1.

Table 1. The parameters of exciting trajectory limits.

Limits		Joint 1	Joint 2	Joint 3	Joint 4	Joint 5	Joint 6
Joint limit $\theta_{I,max}$ (deg)	Max	120	90	60	120	120	120
	Min	−120	−90	−60	−120	−120	−120
Joint velocity limit $\dot{\theta}_{I,max}$ (deg/s)	Max	100	60	60	80	80	80
	Min	−100	−60	−60	−80	−80	−80
Joint acceleration limit $\ddot{\theta}_{I,max}$ (deg/s^2)	Max	120	120	120	120	120	120

Comparing the condition number of the regression matrix obtained with the optimization index method based on the block regression matrix condition number and the least squares method used in this paper, the results are shown in Table 2 below:

Table 2. Comparison table of condition number optimization results of different methods.

Optimization Method	Condition Number
WLS	189.4012
Ours	162.2440

It can be seen from Table 2 that the condition number of the regression matrix obtained with the method used in this paper is the smallest. At the same time, the method used in this paper optimizes the sub-matrix of the regression matrix and adjusts the internal structure of the regression matrix, so it can better stimulate the characteristics of the dynamic parameters. Because there are six joints, the number of variable coefficients for the total trajectory optimization solution is 66. According to the constraint parameters provided in Table 1, the coefficients of the exciting trajectory are calculated through the pattern search optimization function provided using Matlab, as shown in Table 3. The total system runtime is determined by the maximum integer time. Loop1 corresponds to the friction identification module, taking 1 s to complete. Loop2 represents the dynamics identification module, where the trajectory optimization and execution take 30 s, and the dynamic parameter identification process requires 5 s. Loop3 corresponds to the dynamics compensation module based on dual-encoder deformation. The GMM algorithm within this module has a relatively longer runtime, contributing to the total module time of 2 s.

Table 3. Coefficient results of exciting trajectories.

Optimization Parameters	Joint 1	Joint 2	Joint 3	Joint 4	Joint 5	Joint 6
	0.0046	−0.0406	−0.1772	−0.1704	−0.0210	0.0673
	−0.0074	0.0597	0.0661	0.1161	−0.1317	0.2685
$a_{l,i}$	0.4845	0.2492	0.1019	−0.1703	0.3614	0.1682
	−0.5103	0.0678	0.0189	0.1426	−0.1994	−0.4241
	0.0286	−0.3361	−0.0097	0.0820	−0.0094	−0.0763
	−0.0053	−0.0234	−0.0001	0.0088	0.1460	0.1335
	−0.0182	−0.2122	−0.0583	−0.3357	−0.3491	0.0677
$b_{l,i}$	0.6963	0.1619	0.0450	0.1996	0.3639	0.0942
	−0.3354	−0.1082	0.1273	0.1731	−0.1008	−0.2907
	−0.1411	0.0789	−0.1056	−0.1257	−0.0273	0.1227
θ_{i0} (rad)	0.3361	−0.2762	−0.0112	−0.2369	0.1976	0.4777

The actual running period of the industrial robot is set to 20 s, and the displacement, velocity, and acceleration of each joint exciting trajectory are shown in Figure 5.

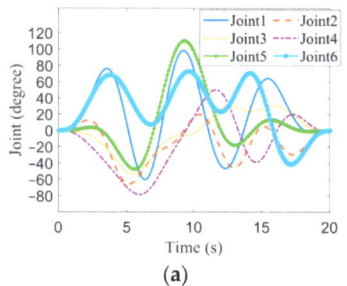
(a)

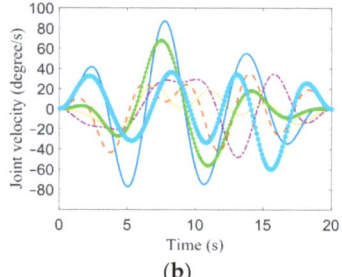
(b)

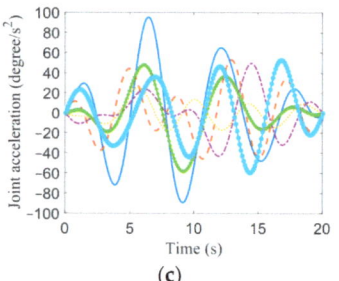
(c)

Figure 5. Exciting trajectories: (a) Joint position; (b) Joint velocity; (c) Joint acceleration.

The rotation axis of Joint 1 is orthogonal to that of Joint 2, with minimal gravitational impact on Joint 1 and maximal gravitational impact on Joint 2. Consequently, selecting the friction between Joint 1 and Joint 2 as the test parameter for the algorithm is a compelling choice. Figure 6 illustrates the ultimate results of Joint 1 and Joint 2 friction estimation along with the fitting parameters. The nonlinear parameters prove to be highly effective in capturing the friction characteristics under conditions close to zero. Tailoring to specific requirements, more sophisticated friction models can be seamlessly integrated into the proposed framework.

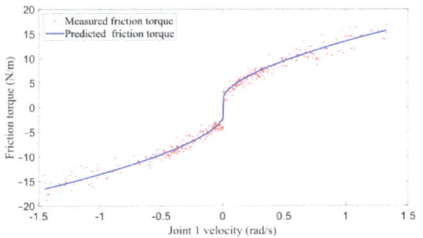 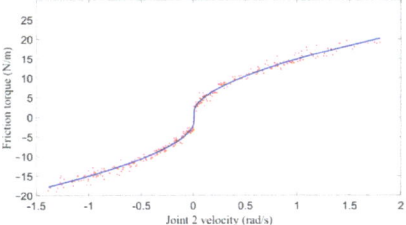

Figure 6. Nonlinear friction torque estimation of robot Joint 1 and Joint 2.

Directly use the traditional WLS method to identify the dynamic model and friction of the link. Observing the results in Figure 7, it can be found that the effect is not ideal, and even when the joint moves smoothly in one direction, a large error occurs. As long as the joints are in motion, there will be joint deformations. At this time, only the encoder on the link side is used for dynamic identification, and there will be errors. When the joints return to the zero position, the industrial robot is in a vertical state at this time, and there is almost no motion deformation in each joint, so the residual torque is almost zero. Finally, we compensate for the residual torque based on the double-encoder information, and Figure 8 shows the results of the identification method proposed in this paper.

Given that the identification framework in this paper is also based on an iterative strategy, our proposed method exhibits a notable improvement in torque estimation, particularly in turning and local positions compared to the WLS method. The collection of torque and joint state data introduces high-frequency noise, originating from joint and mechanical vibrations, as well as friction forces that cannot be precisely estimated. To address this, we introduce low-pass filtering for effective noise reduction, ensuring a more accurate representation of torque information without compromising the signal.

In Figures 7 and 8, it can be seen that the dynamic model identified with the method used in this paper is more accurate, and the measured torque basically agrees with the predicted torque current. The calculated RMSE (root mean square error) between the predicted torque and the actual measured torque of each joint is shown in Table 4. The results show that the three-loop dynamics identification scheme based on double-encoder information compensation proposed in this paper has a significant improvement compared with the WLS, and the RMSE of the residual torque is reduced by more than 20%, which proves the superiority of the method in this paper compared with the traditional method.

Table 4. Result of RMSE.

Identification Method	Joint 1	Joint 2	Joint 3	Joint 4	Joint 5	Joint 6
WLS	2.3376	2.1278	1.5395	0.3361	0.7101	0.2578
Ours	1.9594	1.7345	1.1609	0.3027	0.5852	0.2096

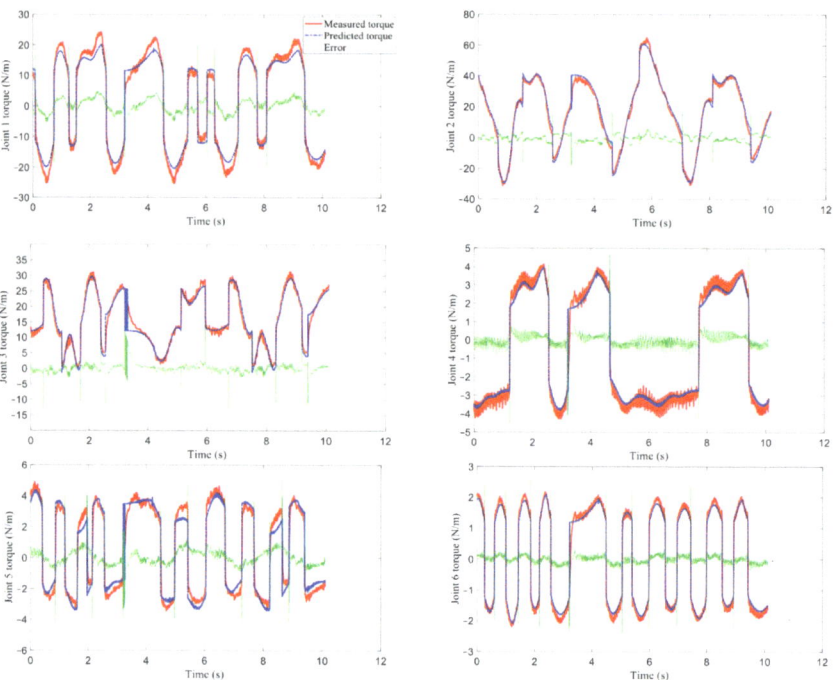

Figure 7. The result of traditional identification method WLS.

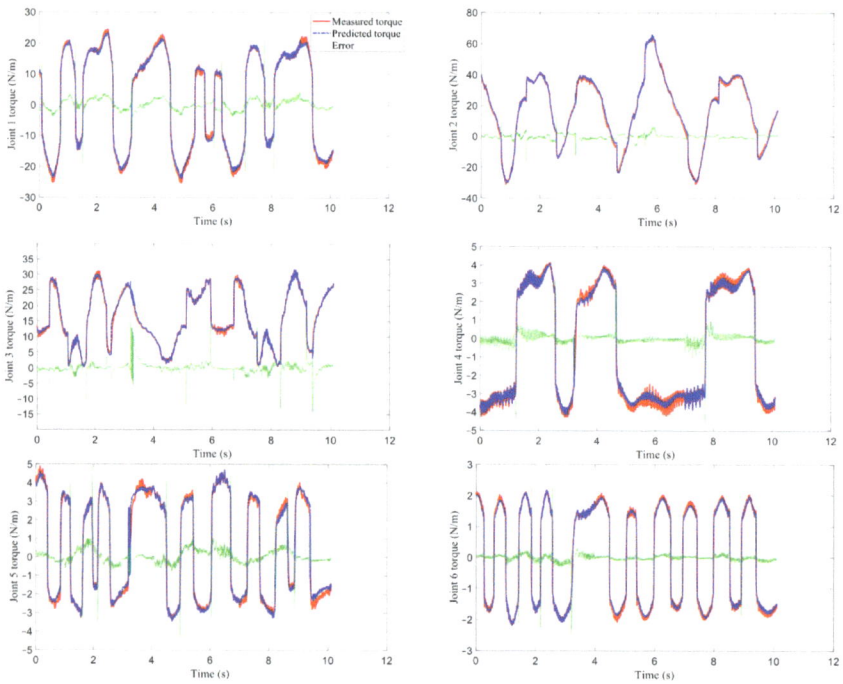

Figure 8. The result of identification method in this paper.

7. Conclusions

The method proposed in this paper firstly identifies the friction force through two directional moments, and then conducts a theoretical identification on the dynamics of the link, in which the block matrix condition number is used as the optimization index for the exciting trajectory. Finally, the deformation moments that cannot be accurately modeled are estimated using double-encoder information, which can reflect the influence of unmodeled parts such as harmonic reducers. The dynamic parameter identification of an industrial robot has been enhanced in two aspects. Firstly, a nonlinear friction force model, distinguishing between high and low speeds, is employed to better fit the dynamic friction effects of the robotic arm. Secondly, the GMM algorithm is introduced into dynamic parameter identification to compensate for the uncertain torque residue arising from nonlinear fitting. However, through an intuitive analysis of the residual torque, it can be found that the accuracy of the torque estimation will decrease in the place where the speed switches direction, and the error will be large. This is also due to the difficulty in estimating the friction force when the movement switches directions. This approach reduces the root mean square of identification residuals by 20%, signifying a significant improvement in the precision of model parameter identification.

Author Contributions: X.L. contributed the central idea and wrote the initial draft of the paper. Y.X., X.S., T.W., L.Z. and Y.Z. contributed to refining the ideas and revised this paper. All authors have read and agreed to the published version of the manuscript.

Funding: This work was supported by the National Key Research and Development Program for Robotics Serialized Harmonic Reducer Fatigue Performance Analysis and Prediction and Life Enhancement Technology Research, Grant No. 2017YFB1300603.

Data Availability Statement: All data generated or analyzed during this study are included in this paper or are available from the corresponding authors on reasonable request.

Conflicts of Interest: The authors declare no conflict of interest.

References

1. Vandanjon, P.; Gautier, M.; Desbats, P. Identification of robot inertial parameters by means of spectrum analysis. In Proceedings of the 1995 IEEE International Conference on Robotics and Automation (ICRA), Nagoya, Japan, 21–27 May 1995; pp. 3033–3038.
2. Wu, J.; Wang, J.; You, Z. An overview of dynamic parameter identification of robots. *Robot. Comput. Integr. Manuf.* **2010**, *26*, 414–419. [CrossRef]
3. Gautier, M.; Khalil, W. Direct calculation of minimum set of inertial parameters of serial robots. *IEEE Trans. Robot. Autom.* **1990**, *6*, 368–373. [CrossRef]
4. Swevers, J.; Verdonck, W.; De Schutter, J. Dynamic model identification for industrial robots. *IEEE Control Syst. Mag.* **2007**, *27*, 58–71.
5. Venture, G.; Ayusawa, K.; Nakamura, Y. A numerical method for choosing motions with optimal excitation properties for identification of biped dynamics-An application to human. In Proceedings of the 2009 IEEE International Conference on Robotics and Automation, Kobe, Japan, 12–17 May 2009; pp. 1226–1231.
6. Huang, Y.; Ke, J.; Zhang, X.; Ota, J. Dynamic parameter identification of serial robots using a hybrid approach. *IEEE Trans. Robot.* **2023**, *39*, 1607–1621. [CrossRef]
7. Zhuang, C.; Yao, Y.; Shen, Y.; Xiong, Z. A convolution neural network based semi-parametric dynamic model for industrial robot. *J. Mech. Eng. Sci.* **2022**, *236*, 3683–3700. [CrossRef]
8. Huang, S.; Chen, J.; Zhang, J.; Zhu, Z.; Zhou, H.; Li, F.; Zhou, X. Robust estimation for an extended dynamic parameter set of serial manipulators and unmodeled dynamics compensation. *IEEE/ASME Trans. Mechatron.* **2022**, *27*, 962–973. [CrossRef]
9. Gautier, M.; Briot, S. New method for global identification of the joint drive gains of robots using a known payload mass. In Proceedings of the 2011 IEEE/RSJ International Conference on Intelligent Robots and Systems, San Francisco, CA, USA, 25–30 September 2011; pp. 3728–3733.
10. Briot, S.; Gautier, M. Global identification of joint drive gains and dynamic parameters of parallel robots. *Multibody Syst. Dyn.* **2015**, *33*, 3–26. [CrossRef]
11. Albu Schäffer, A.; Ott, C.; Hirzinger, G. A unified passivity-based control framework for position, torque and impedance control of flexible joint robots. *Int. J. Robot. Res.* **2007**, *26*, 23–39. [CrossRef]
12. Spong, M.W. Modeling and control of elastic joint robots. *J. Dyn. Sys. Meas. Control* **1987**, *109*, 310–319. [CrossRef]
13. Han, Y.; Wu, J.; Liu, C.; Xiong, Z. Static model analysis and identification for serial articulated manipulators. *Robot. Comput. Integr. Manuf.* **2019**, *57*, 155–165. [CrossRef]

14. Gautier, M. Dynamic identification of robots with power model. In Proceedings of the International Conference on Robotics and Automation, Albuquerque, NM, USA, 25 April 1997; pp. 1922–1927.
15. Wolf, S.; Iskandar, M. Extending a dynamic friction model with nonlinear viscous and thermal dependency for a motor and harmonic drive gear. In Proceedings of the 2018 IEEE International Conference on Robotics and Automation (ICRA), Brisbane, QLD, Australia, 21–25 May 2018; pp. 783–790.
16. Iskandar, M.; Wolf, S. Dynamic friction model with thermal and load dependency: Modeling, compensation, and external force estimation. In Proceedings of the 2019 International Conference on Robotics and Automation (ICRA), Montreal, QC, Canada, 20–24 May 2019; pp. 7367–7373.
17. Ji, Y.; Jiang, X.; Wan, L. Hierarchical least squares parameter estimation algorithm for two-input Hammerstein finite impulse response systems. *J. Frankl. Inst.* **2020**, *357*, 5019–5032. [CrossRef]
18. Kammerer, N.; Garrec, P. Dry friction modeling in dynamic identification for robot manipulators: Theory and experiments. In Proceedings of the 2013 IEEE International Conference on Mechatronics (ICM), Vicenza, Italy, 27 February–1 March 2013; pp. 422–429.
19. Swevers, J.; Verdonck, W.; Naumer, B.; Pieters, S.; Biber, E. An experimental robot load identification method for industrial application. *Int. J. Robot. Res.* **2002**, *21*, 701–712. [CrossRef]
20. Zhang, L.; Wang, J.; Chen, J.; Chen, K.; Lin, B.; Xu, F. Dynamic modeling for a 6-DOF robot manipulator based on a centrosymmetric static friction model and whale genetic optimization algorithm. *Adv. Eng. Softw.* **2019**, *135*, 102684. [CrossRef]
21. Deng, J.; Shang, W.; Zhang, B.; Zhen, S.; Cong, S. Dynamic Model Identification of Collaborative Robots Using a Three-Loop Iterative Method. In Proceedings of the 2021 6th IEEE International Conference on Advanced Robotics and Mechatronics (ICARM), Chongqing, China, 3–5 July 2021; pp. 937–942.
22. Shi, X.; Han, Y.; Wu, J.; Xiong, Z. Servo system identification based on curve fitting to phase-plane trajectory. *J. Dyn. Sys. Meas. Control* **2020**, *142*, 031001. [CrossRef]
23. Shi, X.; Han, Y.; Wu, J.; Xiong, Z. An FFT-based method for analysis, modeling and identification of kinematic error in Harmonic Drives. In Proceedings of the International Conference on Intelligent Robotics and Applications (ICIRA), Shenyang, China, 8–11 August 2019; pp. 191–202.
24. Han, Y.; Wu, J.; Liu, C.; Xiong, Z. An Iterative Approach for Accurate Dynamic Model Identification of Industrial Robots. *IEEE Trans. Robot.* **2020**, *36*, 1577–1594. [CrossRef]
25. Herzog, A.; Righetti, L.; Grimminger, F.; Pastor, P.; Schaal, S. Momentum-based balance control for torque-controlled humanoids. *Comput. Res. Repos.* **2013**, *1*, 1–7.
26. Niku, S.B. *Introduction to Robotics: Analysis, Systems, Applications*; Prentice Hall: Upper Saddle River, NJ, USA, 2001.
27. Lu, Y.; Shen, Y.; Zhuang, C. External force estimation for industrial robots using configuration optimization. *Automatika* **2023**, *64*, 365–388. [CrossRef]
28. Hamon, P.; Gautier, M.; Garrec, P.; Janot, A. Dynamic modeling and identification of joint drive with load-dependent friction model. In Proceedings of the 2010 IEEE/ASME International Conference on Advanced Intelligent Mechatronics, Montreal, QC, Canada, 6–9 July 2010; pp. 902–907.
29. Hamon, P.; Gautier, M.; Garrec, P. New dry friction model with load and velocity-dependence and dynamic identification of multi-DoF robots. In Proceedings of the 2011 IEEE International Conference on Robotics and Automation (ICRA), Shanghai, China, 9–13 May 2011; pp. 1077–1084.
30. Ijspeert, A.J.; Nakanishi, J.; Hoffmann, H.; Pastor, P.; Schaal, S. Dynamical Movement Primitives: Learning Attractor Models for Motor Behaviors. *Neural Comput.* **2013**, *25*, 328–373. [CrossRef] [PubMed]
31. Performance Real-Time Test System. Available online: https://www.speedgoat.com/products-services/real-time-target-machines/performance-real-time-target-machine (accessed on 2 November 2022).

Disclaimer/Publisher's Note: The statements, opinions and data contained in all publications are solely those of the individual author(s) and contributor(s) and not of MDPI and/or the editor(s). MDPI and/or the editor(s) disclaim responsibility for any injury to people or property resulting from any ideas, methods, instructions or products referred to in the content.

Article

Cartesian Stiffness Shaping of Compliant Robots—Incremental Learning and Optimization Based on Sequential Quadratic Programming

Nikola Knežević, Miloš Petrović and Kosta Jovanović *

School of Electrical Engineering, University of Belgrade, 11000 Belgrade, Serbia; knezevic@etf.rs (N.K.); petrovic.milos@etf.rs (M.P.)
* Correspondence: kostaj@etf.rs

Abstract: Emerging robotic systems with compliant characteristics, incorporating nonrigid links and/or elastic actuators, are opening new applications with advanced safety features, as well as improved performance and energy efficiency in contact tasks. However, the complexity of such systems poses challenges in modeling and control due to their nonlinear nature and model variations over time. To address these challenges, the paper introduces Locally Weighted Projection Regression (LWPR) and its online learning capabilities to keep the model of compliant actuators accurate and enable the model-based controls to be more robust. The approach is experimentally validated in Cartesian position and stiffness control for a 4 DoF planar robot driven by Variable Stiffness Actuators (VSA), whose real-time implementation is supported by the Sequential Least Squares Programming (SLSQP) optimization approach.

Keywords: physical human–robot interaction; variable stiffness actuators; Cartesian stiffness shaping; incremental learning; locally weighted projection regression

1. Introduction

Compliant robots constitute a paradigm shift in the field of robotics, characterized by the deliberate integration of pliable materials designed to emulate the inherent flexibility and adaptability observed in natural organisms. Unlike their rigid counterparts, even ones with active compliant control strategies, compliant robots have the unique capability to undergo deformation and reconfiguration, allowing them to adapt and conform to their environment. The compliant nature of these robots imparts a level of dexterity and versatility, making them well suited for tasks that require interaction with delicate objects or for navigating complex, dynamic environments. As the field continues to advance, compliant robots have the potential to revolutionize various industries by providing innovative solutions to problems that were once deemed impossible for traditional robotic systems.

Compliant robots have an elastic element between the actuator and the link, which enables diverse variants of compliant actuators to be systematically designed and engineered by varying actuator configurations and associated elastic elements. They are capable of absorbing sudden impacts and adapting to them [1–3]. Furthermore, robots with flexible joints can outperform rigid robots in repetitive tasks [4], or where a high energy impact is needed to perform tasks such as throwing or nailing [5,6].

The two main types of compliant actuators that have been developed are actuators with constant or variable compliance. Constant compliance actuators or Series Elastic Actuators (SEA) have one elastic element in series to the motor shaft. To accurately control this type of actuator, the characteristics of elastic elements need to be known. The precise joint stiffness of SEA can be acquired either via Finite Element Method analysis or experimentally [7]. SEA exhibits some low-pass filter properties [8] and improves force accuracy by turning the force control problem into a position control problem [9]. However, in some tasks, constant compliance does not lead to the desired behavior—therefore, higher-precision path following might not be possible. Conversely, VSAs have a mechanical structure capable of changing the stiffness properties of the actuator. These types of actuators are mainly composed of two motors coupled with elastic springs—a bidirectional antagonistic setup [10], or they use one motor for position change and another for stiffness variation (independent motor setup) [9]. Basic VSA control methods have generally been founded upon the actuator model: feedback linearization, decoupled control, cascade control, adaptive control, etc. [11–15]. All these methods deal with a nonlinear model or its approximation. Generating the correct representation of the dynamical model is not a trivial task. The characteristics of the springs, as a source of compliance in compliant actuators, are often nonsymmetrical, and the geometry of the actuator itself cannot always be represented correctly. Furthermore, compliant elements are often susceptible to degradation with wear and time, which reduces model accuracy further.

Modeling motor or actuator transfer functions based on the characteristics provided by the manufacturer can be a very challenging task. Moreover, two motors from the same batch with the same declared characteristics do not have exactly matching transfer functions. Furthermore, VSAs have two motors that both work to shape the actuator characteristics, making them even more demanding to model. Consequently, the actuator model needs to be exploited from raw data. Initial approaches to learning models were based on applying step excitation and measuring actuator response. With that information, ARX, ARMAX, or other algorithms can be implemented for transfer function learning. Developing more sophisticated algorithms like neural networks, machine learning techniques, and iterative learning provides easier ways for model learning of actuators [16–19].

Many researchers in the past have implemented different learning techniques to map the relation between system inputs and outputs, tune the dynamics, or control the parameters of a system. In [20], feed-forward control was designed in the form of a PI controller, which gains updates via iterative learning. Iterative learning control was used in [21] for feed-forward control in a decentralized manner, where the feedback control part has a low-gain structure. Generalized iterative learning control for VSA trajectory tracking is presented in [22]. Furthermore, iterative learning was deployed to balance feed-forward and feedback elements described in [23], showing better results than conventional feedback control. Some papers depict a neural network-based adaptive control strategy designed for controlling VSAs [24,25]. Additionally, neural networks can be applied to predict human motion, in order to create the desired robot motion and control the robot in physical human–robot interaction [26]. Reinforcement learning is used for goal-oriented tasks and model-free control. In [27], the authors report accomplishing variable impedance control with reinforcement learning algorithms that are model-free. Furthermore, a model-based policy learning algorithm for closed-loop predictive control of soft robots was implemented in [28], where feed-forward dynamics are represented by a neural network.

The present research considered bidirectional antagonistic actuators. This type of actuator has two DC motors linked to an output shaft with springs. The output position and the stiffness of the actuator can be controlled by changing the position of the two motors. To construct the required system model, the nonlinear relation between inputs (DC motor positions) and outputs (joint position and stiffness) needs to be presented. The complexity of the model depends on the spring's characteristics. The mathematical model almost always assumes that the system is symmetric. Since there are no two identical DC motors or two identical springs, learning algorithms can be applied to learn the model of a system. Constructing accurate models and executing control over compliant robotic systems encounter complexities due to unmodeled friction, asymmetry in springs and motors, and spring nonlinearity.

A key feature of novel compliant robots is advanced and safer physical interaction with the environment. The performance and capabilities of a robot in physical interaction are defined by the mechanical impedance of its End Effector (EE) in Cartesian space or simply by its static component—mechanical stiffness. This property is described by the

Cartesian stiffness matrix—K_c. Variations in stiffness components can be a tradeoff between the accuracy of rigid robots and the safety of compliant robots in different directions. The Cartesian stiffness matrix depends on the configuration of the robot (q) and the stiffness of each joint, which formulated the diagonal joint stiffness matrix (K_j), see Equation (10). Therefore, accurate and fast information on the position and stiffness of compliant actuators is essential for planning and controlling the physical interaction of compliant robots that rely on models.

When it comes to the control of robots driven by rigid joints, Cartesian stiffness is mainly determined by the robot's posture. The stiffness of classical industrial robots with rigid joints is affected by their geometry, material characteristics, actuator and transmission properties, and the robots' posture. In order to control the Cartesian stiffness of an industrial robot and thus make it compliant, researchers have developed different control strategies like impedance and admittance control [29?,30]. In addition, it is possible to control Cartesian stiffness if the robot has more degrees of freedom than the task space via reconfiguration in the null space [32–36].

However, robots with VSAs can control Cartesian stiffness via robot reconfiguration or null space variation, as well as by changing the stiffness on the joint level [37]. Using standard Cartesian stiffness-shaping techniques (active compliance, optimization algorithms), combined with robots that have flexible joints, can provide better control and a wider range of achievable Cartesian stiffness [38].

The above-mentioned collaborative approaches enhance efficiency and flexibility in production processes, reducing the risk of injury by absorbing external forces, which showcase real-world applications where compliant robots excel in industrial settings. Robots with compliant features are employed on assembly lines where they can work safely alongside human workers in the automotive industry [39,40]. Some authors propose the use of impedance control for collaborative human–robot chamfering and polishing applications [41], as well as a null-space search for torque-effective drilling [42]. Collaborative assembly via robot behavior shaping with active and passive compliance was introduced in [43]. Furthermore, a notable real-world application where compliant robots outperform rigid robots is in surgery, specifically in minimally invasive procedures. Compliant robotic systems, such as the da Vinci Surgical System [44], demonstrate superiority over rigid counterparts due to their ability to navigate and manipulate soft and delicate tissues with greater precision and dexterity [45]. The compliance of the robotic arms allows for more natural and adaptive movements, reducing the risk of tissue damage and improving surgeon's control [46].

The contribution of this paper is twofold. The LWPR learning algorithm [47] substitutes the traditional way of modeling actuators and maps VSA characteristics, including the possibility to examine nonlinear phenomena which are often considered as unmodeled dynamics (frictions or drive asymmetry). Additionally, incremental learning features of the LWPR algorithm were used to track model parameter changes due to wear and tear. New measurements are used to expand the learning dataset and incrementally update the actuator model. The proposed methodology improves simultaneous control of both the position and passive stiffness of VSAs. Secondly, SLSQP [48–51] optimization was implemented to shape the Cartesian stiffness of compliant robots with VSAs. This algorithm exploits all the features of quadratic programming, which is used when fast optimization with constraints is needed. Furthermore, SLSQP can optimize functions that have nonlinear criteria with nonlinear constraints. The proposed methodology enables Cartesian stiffness shaping on the EE level to meet the desired robot behavior without concerning stability issues by leveraging compliant behavior via passive stiffness and robot reconfiguration in the null space. Correspondingly, combining joint-level stiffness control and reconfiguration extends the achievable Cartesian stiffness range.

To exploit the full potential of the proposed methodology, the following pipeline was defined through several steps: (1) learning VSA model parameters using LWPR; (2) continuous parameters relearning via incremental learning; (3) utilization of the learned robotic system model (in our case, 4 DoF planar robot with VSAs); and (4) SLSQP algorithm to effectively control behavior by determining the optimal configuration and stiffness on the joint level. The flow chart of the proposed pipeline is presented in Figure 1.

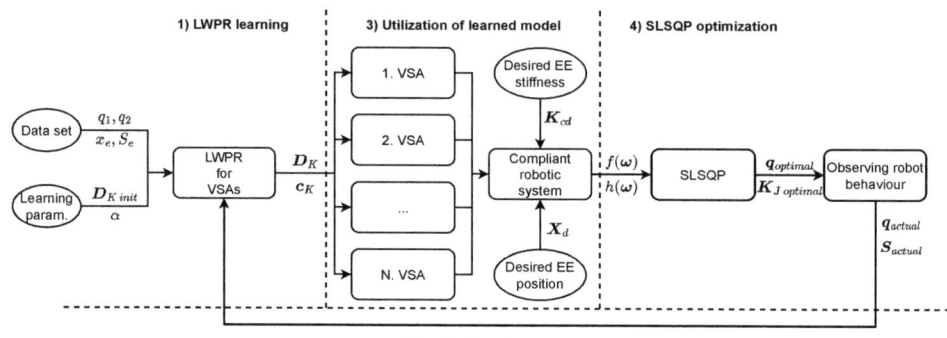

Figure 1. Flowchart presenting pipeline of the proposed methodology. (1) Learning VSA model using LWPR phase. (2) Iterative learning phase. (3) Robotic system model building phase. (4) Optimization phase.

The rest of the paper is organized as follows. Section 2 introduces a general method of LWPR utilization for learning the model of compliant joints and then presents its use case on the QB actuator—the bidirectional antagonistic drive. Section 3 describes a general SLSQP optimization method for Cartesian stiffness shaping, as well as a use case on computing positions and stiffnesses for each of the 4 DoF compliant robot joints for the desired Cartesian stiffness. Finally, Section 4 validates the theory and use cases from Sections 2 and 3 by introducing external disturbances via a compliant robot equipped with an F/T sensor that exposes the motion to the 4 DoF complaint robot and measures the deviation from the desired position. The paper ends with concluding remarks and future work prospects.

2. Learning a Variable Stiffness Actuator Model

LWPR is used as a learning technique in order to deal with the uncertainties of the actuator model parameters, as well as the nonlinearity of the actuator and its susceptibility to change due to wear. The LWPR method is designed to overcome the issue of sparse data because it is effective in learning when a small amount of data is available or when the data are noisy. In [47], the authors describe in detail the features of LWPR compared to other state-of-the-art algorithms, like the Gaussian Process and Support Vector Machine. Furthermore, the complexity of the LWPR algorithm increases linearly with problem dimensionality. Regarding computational efficiency, a 70Hz learning rate has been achieved for a high-dimensional learning problem (90 inputs and 30 outputs). In [52], this technique is used to map the input/output characteristics of SEA. Paper [53] presents a learning algorithm to acquire the inverse dynamics of a 7 DoF manipulator. In the present paper, the LWPR algorithm is used to map the input/output characteristics of VSA, enabling nonlinear function mapping in high-dimensional space, which is very suitable for learning the behavior of robotic systems. Its main idea is to fit a nonlinear function using local linear models. It is shown that locally linear models can be an appropriate substitute for nonlinear and complex models. The essence of LWPR is to determine the validity region of each local model. The validity region can be represented in the form of a Gaussian kernel

$$\omega_k = \exp(-\frac{1}{2}(x - c_k)^T D_k (x - c_k)) \quad (1)$$

where c_k is the center of kth linear model, and D_k corresponds to a positive semi-definite distance metric that determines the size and shape of the validity region of the linear model. Algorithms update the distance matrix D_k ($D_k = M_k^T M_k$) by incorporating gradient descent

$$M_k^{n+1} = M_k^n - \alpha \frac{\partial J}{\partial M_k}, \quad (2)$$

where J is the criteria function for minimizing the prediction error of all linear models.

For each input query data, the local linear model calculates the prediction $\hat{y}_k$. The total output of the learning system is the normalized weighted mean of all K linear models

$$\hat{y} = \sum_1^K \omega_k \hat{y}_k / \sum_1^K \omega_k. \quad (3)$$

To successfully incorporate the LWPR approach into a learning problem, the learning rate parameter α and the initial values for the distance matrix D_k need to be set properly. The typical approach, which can be applied to various VSAs, involves configuring the parameter $D_k = rI$ with a small number for the variable r (e.g., $r = 0.05$). Then, the model is retrained by gradually increasing r until the model achieves satisfactory performance. Also, α can be tuned to improve algorithm performance. This methodology might be demanding and time-consuming until satisfactory performance is achieved.

QB Move Maker Pro [54] was used in this research as a bidirectional antagonistic actuator. Figure 2 shows the QB actuator and its functional scheme. This actuator is a low-cost and open-source variable stiffness actuator. It can be represented as a system with two inputs (q_1 and q_2) and two outputs (x and S), where q_1 and q_2 are the positions of the primal mover motors (DC motors), and x and S are the output shaft position and joint stiffness, respectively. The static relation between the position of the QB actuator primal movers and the equilibrium position and stiffness is given by Equations (4) and (5)

$$x_e = (q_1 + q_2)/2, \quad (4)$$

$$S_e = a \cdot k \cdot cosh(a \cdot (x - q_1)) + a \cdot k \cdot cosh(a \cdot (x - q_2)). \quad (5)$$

Here, $cosh$ is the cosine hyperbolic function, while $a = 6.8465$ and $k = 0.0223$ are spring parameters obtained via identification.

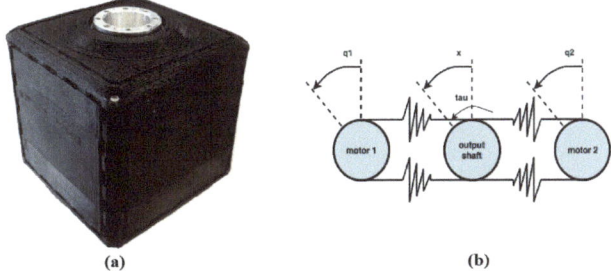

Figure 2. (a) QB actuator. (b) Functional scheme of QB actuator–bidirectional antagonistic actuator.

A proper training set needs to be collected to map the static relation between the inputs and outputs of the bidirectional antagonistic actuator. In the proposed application case, a 4 DoF planar manipulator with VSAs, the input dimension is 2×4 (shaft position and joint stiffness of each actuator), and the output dimension is also 2×4 (primal movers position of each actuator). Data collection is performed on a QB actuator. The authors of [55] suggest five different patterns of the input/output signals). In the first pattern, the reference signals for the primal movers are assumed to have a constant difference between them (0, 20, 40,

60, and 70°), achieving constant stiffness in each subpattern. The difference between the primal movers keeps the constant position of each primal mover from changing from −90 to 90 by 5°. In the following four patterns, the position of the primal movers changed increasingly from −90 to 90° and the difference between their positions changed up and down from −60 to 60°. The dataset is presented in Figure 3.

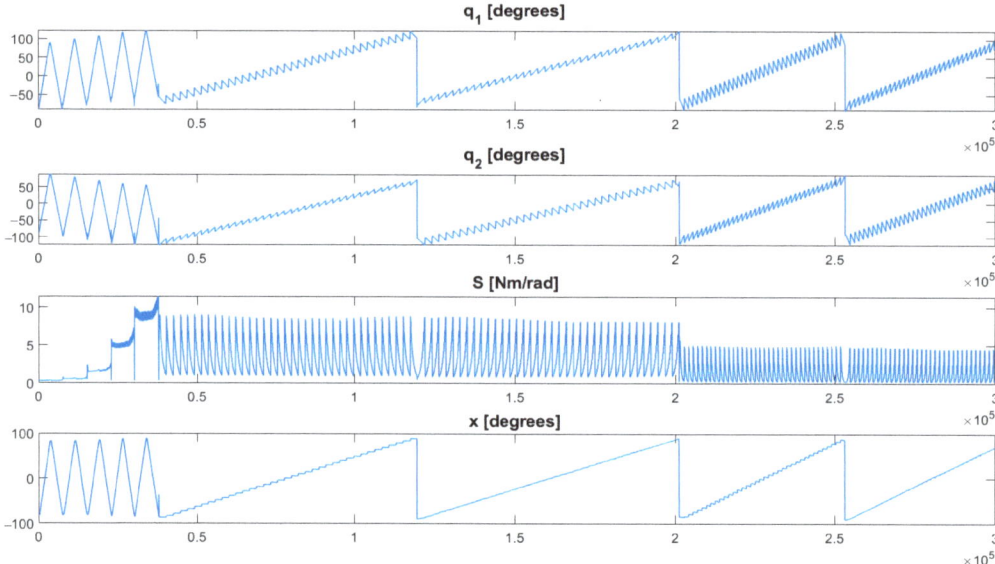

Figure 3. Training patterns designed to cover the entire range of actuator positions and stiffness.

After designing and collecting the training data, the actuator model was learned and the feed-forward control method was implemented. Feed-forward control was used to faithfully represent the accuracy of the learned model. The following diagram represents the control functional scheme (Figure 4).

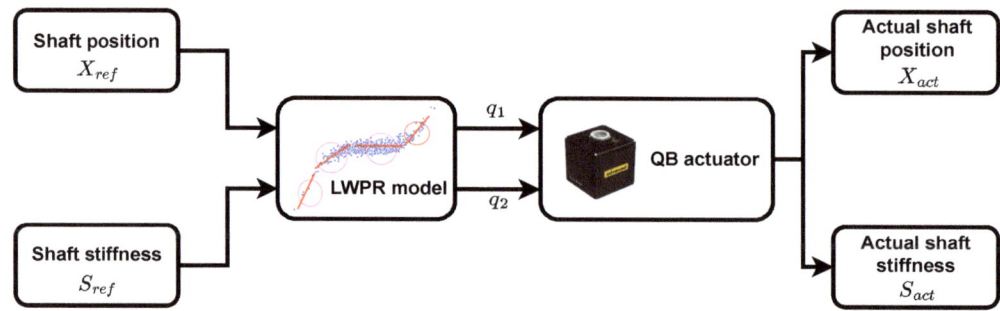

Figure 4. LWPR feed–forward control scheme that maps the reference actuator position and stiffness to the motors' position.

The results from the learned model and the mathematical representation of the static actuator model are shown in Figure 5, to illustrate how well this model can track reference motion. The static model was formed using Equations (4) and (5).

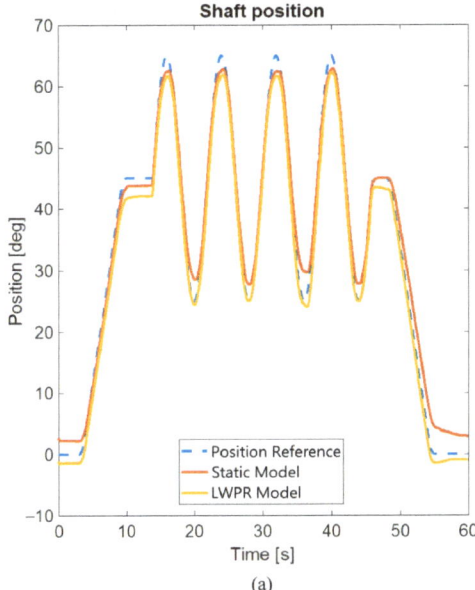
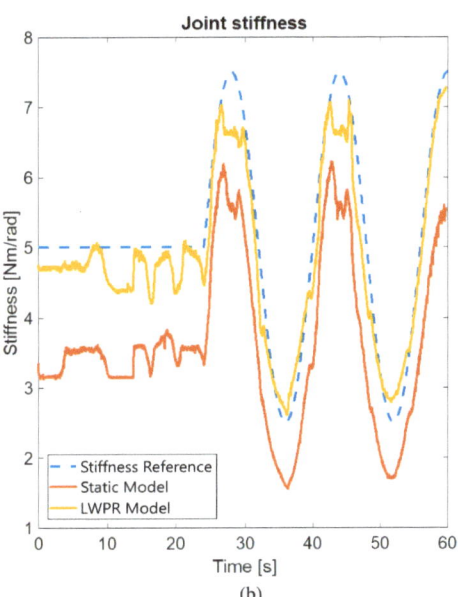

Figure 5. Achieved position (**a**) and stiffness (**b**) tracking results using standard mathematical model (red) and LWPR model (yellow).

It is apparent in Figure 5 that the LWPR and static models of the actuator yield similar position tracking results, but the LWPR model is better in stiffness tracking. However, more effort needs to be put into conducting the detailed mathematical model (including spring and motor asymmetries and friction) compared to the LWPR method, since machine learning techniques simplify model development.

To represent the incremental learning features of the LWPR algorithm, a series of experiments were performed in a simulation environment, using the same learning methodology. After data collection, learning of the actuator model proceeded with LWPR. The initial results from the feed-forward control are shown in Figure 6, light red). Changing the characteristics of one spring on this simulated actuator led to undesired behavior. It is obvious that with new spring parameters, the previously learned model did not consistently track the reference path and stiffness. The reason for this is evident since the actuator model was learned for the initial model parameters (Figure 6, yellow).

When the classical mathematical model is used, a robust controller needs to be developed to suppress the disturbance due to the change in actuator parameters. Designing a robust controller for this type of highly nonlinear system can be challenging. Consequently, it is more convenient to use the incremental learning features of the LWPR. The same model that was learned at the beginning can be used in the process of relearning. Due to model uncertainties introduced by drastic parameter changes (not likely to happen in real-life scenarios, where parameter degradation occurs gradually), new measurements are introduced in the learning set. The new actuator model was learned successfully after only four iterations (Figure 6). In the case of a large deviation from the initial parameters, the model can be relearned an arbitrary number of times.

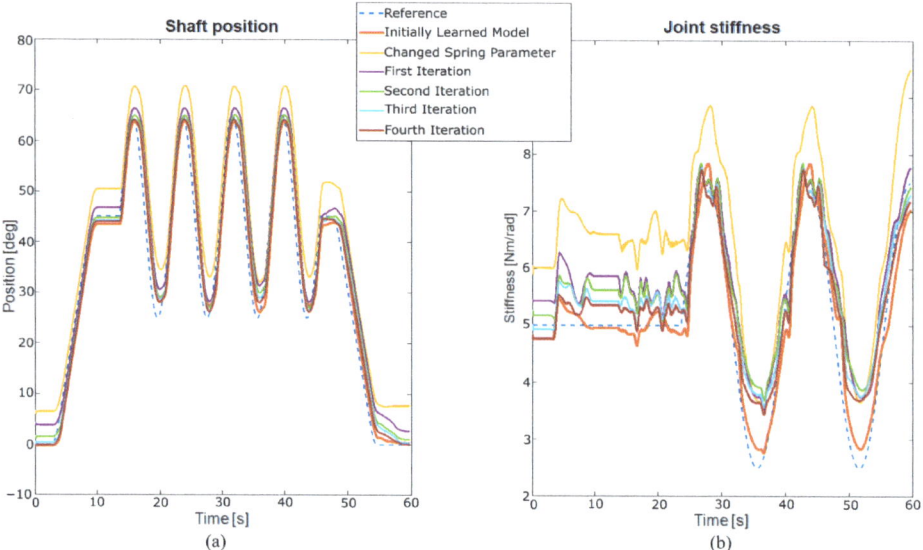

Figure 6. Achieved position (**a**) and stiffness (**b**) tracking results after changing spring parameters (yellow) and after four incremental learning iterations (dark red).

3. Planning End-Effector Cartesian Stiffness

This section presents an optimization algorithm for shaping a robot's EE Cartesian stiffness. To achieve the desired stiffness, the nonlinear function needed to be minimized. SLSQP optimization was used to shape the Cartesian stiffness. This method is an iterative procedure for minimizing nonlinear functions with nonlinear constraints. In each iteration, SLSQP was reduced to a quadratic programming (QP) subproblem by transforming nonlinear functions into quadratic approximation. Furthermore, the result of one QP iteration was used as the starting point for another SLSQP iteration. The problem statement can be formulated as follows

$$\begin{aligned} \min f(\boldsymbol{\omega}), \text{over } \boldsymbol{\omega} &\in \mathbb{R}^n, \\ \text{subject to } h(\boldsymbol{\omega}) &= 0, \\ g(\boldsymbol{\omega}) &\leq 0, \end{aligned} \qquad (6)$$

where the objective function is represented as $f : \mathbb{R}^n \to \mathbb{R}$, while functions $h : \mathbb{R}^n \to \mathbb{R}^m$ and $g : \mathbb{R}^n \to \mathbb{R}^z$ are the equality and inequality constraints for an optimization problem. The value n represents the number of variables in vector $\boldsymbol{\omega}$ (robot's joints position vector and joint stiffnesses) for which optimization is performed, and m and z are the number of equality or inequality constraints, respectively.

For redundant compliant robots, there is an infinite number of robot configurations for one EE position. EE Cartesian stiffness can be alternated by changing the configuration and joint stiffness of the robot. The primary focus of the authors' previous research was the EE Cartesian stiffness control of task-redundant robots with SEAs [56]. The SLSQP algorithm was used to optimize the robot configuration via the null space search, in order to achieve the desired EE Cartesian stiffness. This paper provides an extension of the topic by introducing VSAs in the 4 DoF planar manipulator. To run the optimization algorithm, a kinematic model of the proposed robotic system needed to be developed.

Based on Figure 7, the Cartesian position of the robot EE is defined as

$$x = l_1\cos(q_1) + l_2\cos(q_1 + q_2) + l_3\cos(q_1 + q_2 + q_3) + l_4\cos(q_1 + q_2 + q_3 + q_4), \qquad (7)$$

$$y = l_1\sin(q_1) + l_2\sin(q_1+q_2) + l_3\sin(q_1+q_2+q_3) + l_4\sin(q_1+q_2+q_3+q_4), \quad (8)$$

$$\theta = q_1 + q_2 + q_3 + q_4, \quad (9)$$

where q_1, q_2, q_3, q_4 are the joint positions and l_1, l_2, l_3, l_4 are the lengths of the robot links. The robot stiffness in the Cartesian space is in direct relation to the robot configuration and, therefore, directly related to the Jacobian matrix J.

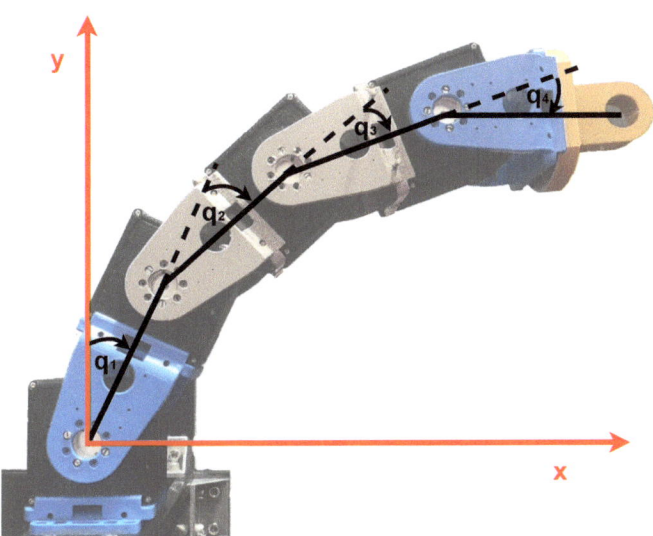

Figure 7. Planar manipulator with 4 DoF consists of variable stiffness actuators.

In the case of VSA-driven robots, Cartesian stiffness is influenced by the robot joint stiffness matrix that has a diagonal matrix form $K_j = \text{diag}(k_i)$, where k_i is the i-th joint stiffness. The Cartesian stiffness matrix can be defined as

$$K_c = (J(q) K_j^{-1} J(q)^T)^{-1}, \quad (10)$$

where K_c is the symmetric 2×2 matrix and q is a 4-dimensional joint position vector.

Only optimization of the $k_{c_{11}}$ and $k_{c_{22}}$ elements will be considered, as they represent stiffness along the X and Y axes in Cartesian coordinates

$$K_{co} = \begin{bmatrix} k_{c_{11}} & 0 \\ 0 & k_{c_{22}} \end{bmatrix}, \quad (11)$$

where $k_{c_{11}}$ is stiffness along the X axis and $k_{c_{22}}$ is stiffness along the Y axis. The desired Cartesian stiffness can be represented as a 2×2 diagonal matrix

$$K_{cd} = \begin{bmatrix} k_{cd_x} & 0 \\ 0 & k_{cd_y} \end{bmatrix}. \quad (12)$$

For the purpose of optimization, the weighted Frobenius norm was used to describe the process performance index (criteria function). The task was to minimize the norm $f(\omega)$ and therefore achieve stiffness tracking.

$$f(\omega) = \sqrt{A(k_{cd_x} - k_{c_{11}})^2 + B(k_{cd_y} - k_{c_{22}})^2}. \quad (13)$$

Coefficients *A* and *B* are weighted factors used to favor one axis over the other. For the optimization process, following the desired EE position can be considered as an optimization constraint. In fact, the optimization process needs to find the robot joint coordinates $q_{o1}, q_{o2}, q_{o3}, q_{o4}$ and joint stiffness $k_{j_{o1}}, k_{j_{o2}}, k_{j_{o3}}, k_{j_{o4}}$ that provides the minimal norm (Equation (13)) and satisfies the constraint that can be described by

$$h(\omega)_1 = -x_d + l_1\cos(q_{o1}) + l_2\cos(q_{o1} + q_{o2}) + l_3\cos(q_{o1} + q_{o2} + q_{o3}) + l_4\cos(q_{o1} + q_{o2} + q_{o3} + q_{o4}), \tag{14}$$

$$h(\omega)_2 = -y_d + l_1\sin(q_{o1}) + l_2\sin(q_{o1} + q_{o2}) + l_3\sin(q_{o1} + q_{o2} + q_{o3}) + l_4\sin(q_{o1} + q_{o2} + q_{o3} + q_{o4}). \tag{15}$$

By repeating this process, the optimization algorithm can lead to a local minimum because the algorithm is based on gradient descent.

A simulated 4 DoF planar manipulator was used to validate the optimization technique. In the simulation, the robot link lengths were set at $l_1 = l_2 = l_3 = l_4 = 0.1$ m. First, the optimization process was simulated over one axis, then over the X axis and Y axis simultaneously. To prove that the optimization technique was working, several cases of the desired robot configuration and stiffness were introduced in the simulation. QB Move Maker Pro parameters were used to achieve more realistic simulation results. The active rotation angle was $\pm 180°$, and the minimal and maximal stiffness were 0.5 Nm/rad and 13 Nm/rad, respectively. At the beginning of each simulation, the initial robot joint stiffness was set to 5 Nm/rad. The time needed to calculate the optimal robot configuration and joint stiffnesses was 0.004 ± 0.001 s.

3.1. Optimization over One Axis

For one-axis optimization, the value of one coefficient, *A* or *B*, in Equation (13) needed to be set to 0. If coefficient *A* is 0, then optimization is performed over the Y axis and vice versa. The robot manipulator is set at some point in the workspace and the algorithm is started. A couple of trials of one-axis optimization are presented in Table 1.

Robot joint stiffness was changed during the simulation, as was joint position, in order to achieve the desired stiffness at a particular position in the workspace. As shown in Table 1, in the case of the one-axis optimization, the algorithm is capable of finding a robot configuration that satisfies the constraints and achieves the desired stiffness along the selected axis.

3.2. Multiple Axis Optimization

Optimization over multiple axes was expected to be more complicated than over one axis, leading to deviation from the desired Cartesian stiffness tracking. In general, the algorithm needs to satisfy the constraints first and then find the robot configuration and joint stiffness that will achieve the desired Cartesian stiffness along multiple axes. To obtain results, coefficients *A* and *B* were set at value 1. Even though the optimization algorithm found the optimal robot configuration and joint stiffness that successfully tracked the desired Cartesian stiffness, as shown in Table 2, the optimization algorithm can fail to find a solution that could track the desired stiffness. Two main reasons can lead to this behavior: (1) in a particular position, the robot cannot physically achieve the desired stiffness, or (2) the optimization algorithm is stuck at the local minimum. This can be overcome by multiple trials of the same desired position and stiffness with different initial positions to find the global minimum. In some scenarios, ideal Cartesian stiffness tracking is not mandatory since in most cases, it is satisfactory to achieve stiff or compliant behavior in a moving direction.

Table 1. One-axis optimization across Y axis.

Sim.	Desired Pos. x, y [m]	Init. Conf. $q_{1,2,3,4}$ [Rad]	Res. Stiff. $k_{j_{1,2,3,4}}$ [Nm/Rad]	Res. Conf. $q_{1,2,3,4}$ [Rad]	Res. Pos. x, y [m]	Norm Val.	Stiffness: Ach. (Des.) [N/m]
1	0.0241 0.3564	1.1472 1.1272 −0.2472 0.7154	7.1004 0.5348 3.5427 0.6301	1.5707 −0.8605 0.9859 0.2777	0.0241 0.3564	2.8055×10^{-5}	235 (235)
2	0.0241 0.3564	0.9472 0.8972 −0.2472 0.9054	6.7482 1.6680 4.5738 2.5607	1.2941 −0.2958 0.5010 0.7727	0.0241 0.3564	6.6554×10^{-6}	440 (440)
3	0.1125 0.3198	1.1472 1.1272 0.0146 −0.8554	2.0157 0.5133 12.0378 3.7201	0.8516 0.0298 1.3616 −1.1421	0.1125 0.3198	6.0982×10^{-5}	728.08 (745)
4	0.1125 0.3198	0.9472 0.8972 0.0146 −0.0783	5.9942 2.7714 9.3229 3.1778	0.9342 −0.2770 1.5074 −0.8967	0.1125 0.3198	1.1922×10^{-5}	1350 (1350)

Table 2. Multiple axes optimization.

Sim.	Desired Pos. x, y [m]	Init. Conf. $q_{1,2,3,4}$ [Rad]	Res. Stiff. x, y [Nm/Rad]	Res. Conf. $q_{1,2,3,4}$ [Rad]	Res. Pos. x, y [m]	Norm Val.	Stiffness: Ach. (Des.) [N/m]
1	0.0241 0.3564	1.1472 1.1272 −0.2472 0.7154	12.9489 13 3.4717 0.5000	0.9865 0.1284 0.7027 0.2841	0.0241 0.3564	1.6819×10^{-4}	60; 350 (60; 350)
2	0.0241 0.3564	0.9472 0.8972 −0.2472 0.9054	12.6546 12.7370 8.5553 5.0958	1.5690 0.3654 −1.2348 1.0421	0.0241 0.3564	9.0114×10^{-4}	75; 2700 (75; 2700)
3	0.1125 0.3198	1.1472 1.1272 0.0146 −0.8554	7.1619 5.9281 8.3310 5.5392	1.4837 0.4001 −0.7289 −0.8127	0.1125 0.3198	1.4×10^{-3}	149.9; 499.9 (150; 500)
4	0.2125 0.2198	0.3224 0.0336 0.3768 1.3664	12.4952 3.5669 8.0491 12.3320	1.4081 0.1878 −1.5175 0.0554	0.2125 0.2198	1.2870×10^{-4}	500; 200 (500; 200)

3.3. Favoring One of the Axes

In the process of multi-axes optimization, in order to favor one axis over another, coefficients A and B need to be set accordingly. To favor stiffness tracking along the X axis, the relation $A > B$ needs to be satisfied and vice versa. This case is different from simple one-axis optimization (where the user has no control over the second axis at all), because control over the non-favored axis is achieved as well. Table 3 shows how changing coefficients A and B affects Cartesian stiffness tracking.

Table 3. Favoring one of the axes.

Sim.	Desired Pos. x, y [m]	A and B	Res. Stiff. $k_{j_{1,2,3,4}}$ [Nm/Rad]	Res. Conf. $q_{1,2,3,4}$ [Rad]	Res. Pos. x, y [m]	Norm Val.	Stiffness: Ach. (Des.) [N/m]
1	0.1172	1	13 0.5 12.99 6.1066	1.5708 −0.4909 −0.7612 1.5040	0.1172	89.86	243; 874
	0.3164	1			0.3164		(330; 850)
2	0.1172	1	13 7.6633 13 5.4937	1.5708 −0.9172 0.0357 1.2859	0.1172	170	159; 847
	0.3164	16			0.3164		(330; 850)
3	0.1172	16	12.9728 5.8234 0.8217 12.9846	1.5653 −0.2659 −1.0942 1.4461	0.1172	72.35	330; 834
	0.3164	1			0.3164		(330; 850)

4. Experimental Validation

For experimental validation, a 4 DoF planar manipulator with QB actuators was used to demonstrate the methodology introduced for compliant actuator model learning (Section 2) and compliant robot Cartesian stiffness shaping (Section 3). To that end, the pipeline presented in Figure 1 was followed. In this process, the joint position and stiffness were obtained from the desired EE position (as a constraint) and Cartesian stiffness. Afterward, the learned LWPR models of each actuator were used to control each joint and achieve the desired joint behavior (position and stiffness). The block diagram of the whole control process is shown in Figure 8. Joint position and stiffness can be calculated from Equations (4) and (5), and the robot EE position and its Cartesian stiffness from Equations (7)–(9) and (10).

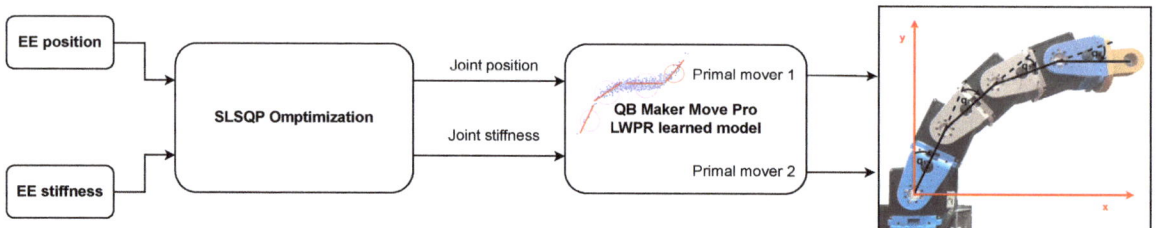

Figure 8. Control block diagram: SLSQP optimization for finding optimal robot configuration and joint stiffness, and LWPR model for controlling QB actuators.

In order to estimate the achieved robot behavior, a contact or disturbance needed to be introduced to the system. This was performed with the Panda robot and the relative deviation from the equilibrium position was measured [57]. The Panda is equipped with an F/T sensor which was used to measure generated contact forces and torques. The experimental setup was composed of the 4 DoF planar robot, the Panda robot, and the F/T sensor (Figure 9), similar to that presented in [58].

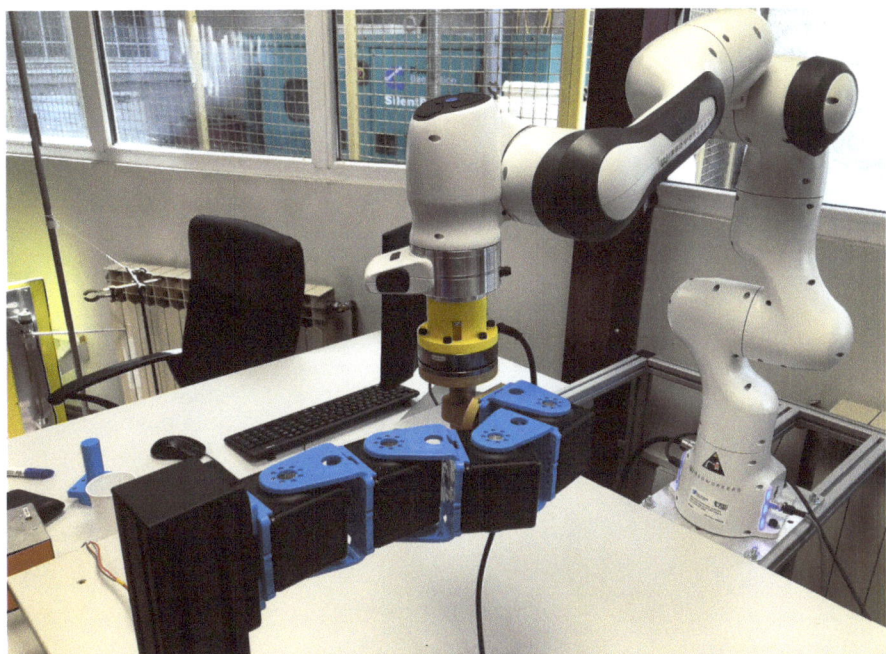

Figure 9. Experimental setup: 4 DoF planar robot with QB actuators, Franka Robotics Panda robot, and Axia80-M8 F/T sensor.

A random perturbation was applied to be able to exploit the achieved behavior of the 4 DoF planar robot. The disturbance was applied in proximity to the equilibrium position. In that way, the robot configuration did not deviate from the optimal configuration, since deviation does not affect Cartesian stiffness due to the infinitesimal change in the Jacobian matrix. In the general case, the Cartesian stiffness matrix of a planar robot is given by

$$\boldsymbol{K_C} = \begin{bmatrix} k_{c_{11}} & k_{c_{12}} \\ k_{c_{21}} & k_{c_{22}} \end{bmatrix}, \tag{16}$$

where $k_{c_{11}}$ and $k_{c_{22}}$ represent stiffness across the X and Y axes, respectively, and $k_{c_{12}} = k_{c_{21}}$ is the coupling stiffness between two axes.

The disturbance or contact in such a system leads to a force generated between the robot and the object in contact (Panda robot). If it is assumed that the behavior of the system is linear in proximity to equilibrium, then the generated force can be expressed as follows

$$\begin{aligned} \boldsymbol{F} &= \boldsymbol{K_C} \times \Delta \boldsymbol{X}, \\ \begin{bmatrix} F_x \\ F_y \end{bmatrix} &= \begin{bmatrix} k_{c_{11}} & k_{c_{12}} \\ k_{c_{21}} & k_{c_{22}} \end{bmatrix} \times \begin{bmatrix} \Delta x \\ \Delta y \end{bmatrix}, \\ F_x &= k_{c_{11}} \Delta x + k_{c_{12}} \Delta y \\ F_y &= k_{c_{21}} \Delta x + k_{c_{22}} \Delta y. \end{aligned} \tag{17}$$

This incomplete system of equations needs to be solved in order to estimate the Cartesian stiffness matrix elements. The measured values are forces and deviation in the XY plane, and the unknown variables are $k_{c_{11}}$, $k_{c_{12}}$, and $k_{c_{22}}$. If there are N randomly applied disturbances, the above equations can be rewritten as

$$\begin{bmatrix} F_{x_1} \\ F_{x_2} \\ \vdots \\ F_{x_N} \\ F_{y_1} \\ F_{y_2} \\ \vdots \\ F_{y_N} \end{bmatrix}_{2N \times 1} = \begin{bmatrix} \Delta x_1 & \Delta y_1 & 0 \\ \Delta x_2 & \Delta y_2 & 0 \\ \vdots & \vdots & \vdots \\ \Delta x_N & \Delta y_N & 0 \\ 0 & \Delta x_1 & \Delta y_1 \\ 0 & \Delta x_2 & \Delta y_2 \\ \vdots & \vdots & \vdots \\ 0 & \Delta x_N & \Delta y_N \end{bmatrix}_{2N \times 3} \times \begin{bmatrix} k_{c_{11}} \\ k_{c_{12}} \\ k_{c_{22}} \end{bmatrix}_{3 \times 1} \quad (18)$$

Using pseudoinverse, the disturbance matrix can be inverted and added to the left side of the equation, providing an estimation of unknown parameters. With this method, the Cartesian stiffness matrix parameters were fitted to minimize the Mean Least Square Error. Figure 10 represents the perturbations and the generated external forces. The applied disturbance was a random movement of the Panda robot in the XY plane, where the maximal movement in each direction was 1.5 cm.

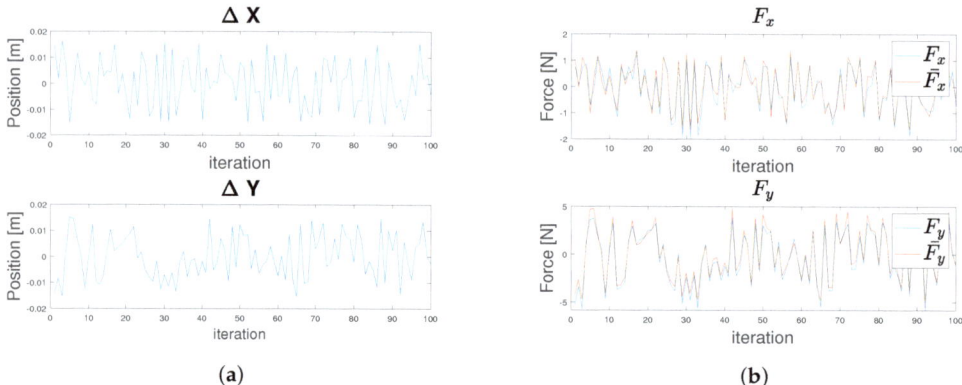

Figure 10. Measurement of applied disturbance. Position of robot EE (**a**). Generated external forces: (**b**) blue. Estimated external force: (**b**) red.

The parameters of the Cartesian stiffness matrix were estimated using the previous equation. The estimated values in this experiment were

$$\hat{K}_c = \begin{bmatrix} 88.6916 & 21.4254 \\ 21.4254 & 326.0492 \end{bmatrix}, \quad (19)$$

while the commanded Cartesian stiffness matrix was

$$K_c = \begin{bmatrix} 77.0128 & 0.0019 \\ 0.0019 & 308.1533 \end{bmatrix}. \quad (20)$$

After estimating the Cartesian stiffness matrix parameters, the estimate of the generated force was calculated by using the newly estimated parameters

$$\begin{aligned} \hat{F}_x &= \hat{k}_{c_{11}} \Delta x + \hat{k}_{c_{12}} \Delta y, \\ \hat{F}_y &= \hat{k}_{c_{21}} \Delta x + \hat{k}_{c_{22}} \Delta y. \end{aligned} \quad (21)$$

Plot (b) in Figure 10 (red) shows the estimated force from the applied disturbance. It is apparent from the estimated force values that using pseudoinverse to minimize the mean least square error can provide good estimation for the Cartesian stiffness parameters.

Interpretation of the Cartesian stiffness matrix can be challenging in some cases. A more convenient way of depicting the Cartesian stiffness matrix is an ellipse representation of the matrix using eigenvalue decomposition. Figure 11 shows the 4 DoF QB robot configuration, the commanded Cartesian stiffness ellipse (black), and the estimated Cartesian stiffness ellipse (red).

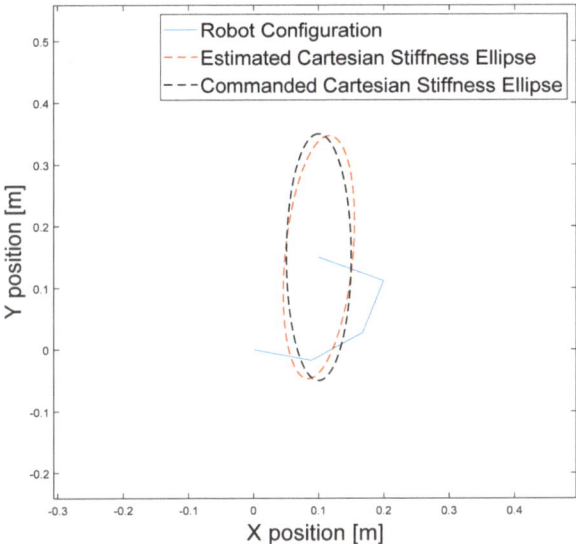

Figure 11. Optimal robot configuration with commanded Cartesian stiffness ellipse (black) and estimated Cartesian stiffness ellipse (red).

The estimated stiffness matrix was approximately equal to the commanded one. The error in the orientation of the estimated Cartesian stiffness ellipse was 2.8% compared to the commanded Cartesian stiffness ellipse.

5. Conclusions

The research aimed to facilitate the physical interaction of a novel compliant robot with the environment by deploying the latest optimization tools and learning methods. The effort reconciled the challenges in modeling actuators of variable stiffness and the need to efficiently determine the position and stiffness of such actuators in order to plan the interaction of the robot EE with the environment. The LWPR iterative learning algorithm demonstrated its efficiency in learning the model parameters of a compliant actuator, which is prone to change due to wear and tear and exploitation time. Based on the model of the robot and its drives, SLSQP efficiently optimized the setting up of the optimal kinematic configuration of the robot and stiffness on the joint level for the desired robot EE Cartesian position and stiffness. Although the proposed methodology was experimentally validated on a 4 DoF planar robot driven by VSAs, the methodology is general and could be exploited by other compliant robots without any additional sensors. Future work will address further improvements of the proposed methodology to allow online Cartesian stiffness shaping beyond discrete points in space (e.g., along a prescribed trajectory), and consequently, its application to real-life in-contact tasks. The proposed approach has several limitations. Finding of the proper learning parameters for the LWPR algorithm can be time consuming on occasion, although parameter finding needs to be performed only once during the initial learning process. EE Cartesian stiffness is limited since it is achieved by exploiting the passive stiffness and kinematics of the manipulator. Cartesian stiffness is shaped using an optimization method that cannot guarantee a global minimum. Although time consuming, this can be overcome by calculating optimal solutions from different initial points. In future work, studies will be conducted on learning techniques that can capture motor dynamics, where a time series dataset will be used. Also, the focus will be on algorithms that combine active and passive stiffness control at the joint level to enhance the algorithm's performance.

Author Contributions: Conceptualization, N.K. and K.J.; methodology, N.K. and K.J.; software, N.K.; validation, N.K. and M.P.; formal analysis, N.K.; investigation, N.K.; resources, K.J.; data curation, N.K. and M.P.; writing—original draft preparation, N.K., M.P., and K.J.; writing—review and editing, N.K. and M.P.; visualization, N.K. and M.P.; supervision, K.J.; project administration, N.K.; funding acquisition, K.J. All authors have read and agreed to the published version of the manuscript.

Funding: This research was supported by the Science Fund of the Republic of Serbia, Modular and Versatile Collaborative Intelligent Waste Management Robotic System for Circular Economy—*CircuBot*, under Grant 6784.

Data Availability Statement: No new data were created or analyzed in this study. Data sharing is not applicable to this article.

Conflicts of Interest: The authors declare no conflicts of interest.

Abbreviations

The following abbreviations are used in this manuscript:

VSA	Variable Stiffness Actuator
LWPR	Locally Weighted Projection Regression
DoF	Degrees of Freedom
EE	End Effector
SLSQP	Sequential Least Squares Programming

References

1. Peshkin, M.A.; Colgate, J.E.; Wannasuphoprasit, W.; Moore, C.A.; Gillespie, R.B.; Akella, P. Cobot architecture. *IEEE Trans. Robot. Autom.* **2001**, *17*, 377–390. [CrossRef]
2. Haddadin, S.; Albu-Schaffer, A.; De Luca, A.; Hirzinger, G. Collision Detection and Reaction: A Contribution to Safe Physical Human-Robot Interaction. In Proceedings of the 2008 IEEE/RSJ International Conference on Intelligent Robots and Systems, Nice, France, 22–26 September 2008; pp. 3356–3363. [CrossRef]
3. Bicchi, A.; Tonietti, G.; Bavaro, M.; Piccigallo, M. Variable Stiffness Actuators for Fast and Safe Motion Control. In *Robotics Research. The Eleventh International Symposium*; Dario, P., Chatila, R., Eds.; Springer: Berlin/Heidelberg, Germany, 2005; pp. 527–536.
4. Visser, L.C.; Stramigioli, S.; Bicchi, A. Embodying Desired Behavior in Variable Stiffness Actuators. *IFAC Proc. Vol.* **2011**, *44*, 9733–9738. [CrossRef]
5. Haddadin, S.; Weis, M.; Wolf, S.; Albu-Schäffer, A. Optimal Control for Maximizing Link Velocity of Robotic Variable Stiffness Joints. *IFAC Proc. Vol.* **2011**, *44*, 6863–6871. [CrossRef]
6. Garabini, M.; Passaglia, A.; Belo, F.; Salaris, P.; Bicchi, A. Optimality principles in variable stiffness control: The VSA hammer. In Proceedings of the 2011 IEEE/RSJ International Conference on Intelligent Robots and Systems, San Francisco, CA, USA, 25–30 September 2011; pp. 3770–3775. [CrossRef]
7. Logozzo, S.; Malvezzi, M.; Achilli, G.; Valigi, M. Characterization of finger joints with underactuated modular structure. *Mater. Res. Proc.* **2022**, *26*, 201.
8. Pratt, G.A.; Williamson, M.M. Series elastic actuators. In Proceedings of the 1995 IEEE/RSJ International Conference on Intelligent Robots and Systems. Human Robot Interaction and Cooperative Robots, Pittsburgh, PA, USA, 5–9 August 1995; Volume 1, pp. 399–406. [CrossRef]
9. Junior, A.G.L.; de Andrade, R.M.; Filho, A.B., Series Elastic Actuator: Design, Analysis and Comparison. In *Recent Advances in Robotic Systems*; IntechOpen: Rijeka, Croatia, 2016; Chapter 10. [CrossRef]
10. Petit, F.; Chalon, M.; Friedl, W.; Grebenstein, M.; Albu-Schäffer, A.; Hirzinger, G. Bidirectional antagonistic variable stiffness actuation: Analysis, design Implementation. In Proceedings of the 2010 IEEE International Conference on Robotics and Automation, Anchorage, AK, USA, 3–7 May 2010; pp. 4189–4196. [CrossRef]
11. Buondonno, G.; De Luca, A. Efficient Computation of Inverse Dynamics and Feedback Linearization for VSA-Based Robots. *IEEE Robot. Autom. Lett.* **2016**, *1*, 908–915. [CrossRef]
12. Trumić, M.; Jovanović, K.; Fagiolini, A. Decoupled nonlinear adaptive control of position and stiffness for pneumatic soft robots. *Int. J. Robot. Res.* **2021**, *40*, 277–295. [CrossRef]

13. Palli, G.; Melchiorri, C.; De Luca, A. On the Feedback Linearization of Robots with Variable Joint Stiffness. In Proceedings of the 2008 IEEE International Conference on Robotics and Automation, Pasadena, CA, USA, 19–23 May 2008; pp. 1753–1759. [CrossRef]
14. Potkonjak, V.; Svetozarevic, B.; Jovanovic, K.; Holland, O. The Puller-Follower Control of Compliant and Noncompliant Antagonistic Tendon Drives in Robotic Systems. *Int. J. Adv. Robot. Syst.* **2011**, *8*, 69. [CrossRef]
15. Lukić, B.; Jovanović, K.; Šekara, T.B. Cascade Control of Antagonistic VSA—An Engineering Control Approach to a Bioinspired Robot Actuator. *Front. Neurorobot.* **2019**, *13*, 69. [CrossRef]
16. Weerasooriya, S.; El-Sharkawi, M. Identification and control of a DC motor using back-propagation neural networks. *IEEE Trans. Energy Convers.* **1991**, *6*, 663–669. [CrossRef]
17. Ismeal, G.A.; Kyslan, K.; Fedák, V. DC motor identification based on Recurrent Neural Networks. In Proceedings of the 16th International Conference on Mechatronics–Mechatronika 2014, Brno, Czech Republic, 3–5 December 2014; pp. 701–705. [CrossRef]
18. Rubaai, A.; Kotaru, R. Online identification and control of a DC motor using learning adaptation of neural networks. *IEEE Trans. Ind. Appl.* **2000**, *36*, 935–942. [CrossRef]
19. Gautier, M.; Jubien, A.; Janot, A. New iterative learning identification and model based control of robots using only actual motor torque data. In Proceedings of the 2013 IEEE/ASME International Conference on Advanced Intelligent Mechatronics, Wollongong, NSW, Australia, 9–12 July 2013; pp. 1436–1441. [CrossRef]
20. Ono, S.; Masuya, K.; Takagi, K.; Tahara, K. Trajectory tracking of a one-DOF manipulator using multiple fishing line actuators by iterative learning control. In Proceedings of the 2018 IEEE International Conference on Soft Robotics (RoboSoft), Livorno, Italy, 24–28 April 2018; pp. 467–472.
21. Angelini, F.; Santina, C.D.; Garabini, M.; Bianchi, M.; Gasparri, G.M.; Grioli, G.; Catalano, M.G.; Bicchi, A. Decentralized Trajectory Tracking Control for Soft Robots Interacting With the Environment. *IEEE Trans. Robot.* **2018**, *34*, 924–935. [CrossRef]
22. Angelini, F.; Mengacci, R.; Santina, C.D.; Catalano, M.G.; Garabini, M.; Bicchi, A.; Grioli, G. Time Generalization of Trajectories Learned on Articulated Soft Robots. *IEEE Robot. Autom. Lett.* **2020**, *5*, 3493–3500. [CrossRef]
23. Della Santina, C.; Bianchi, M.; Grioli, G.; Angelini, F.; Catalano, M.; Garabini, M.; Bicchi, A. Controlling Soft Robots: Balancing Feedback and Feedforward Elements. *IEEE Robot. Autom. Mag.* **2017**, *24*, 75–83. [CrossRef]
24. Huh, S.; Tonietti, G.; Bicchi, A. Neural Network based Robust Adaptive Control for a Variable Stiffness Actuator. In Proceedings of the 2008 16th Mediterranean Conference on Control and Automation, Ajaccio, France, 25–27 June 2008; pp. 1028–1034.
25. Guo, Z.; Pan, Y.; Sun, T.; Zhang, Y.; Xiao, X. Adaptive Neural Network Control of Serial Variable Stiffness Actuators. *Complexity* **2017**, *2017*, 1–9. [CrossRef]
26. Cremer, S.; Das, S.K.; Wijayasinghe, I.B.; Popa, D.O.; Lewis, F.L. Model-Free Online Neuroadaptive Controller with Intent Estimation for Physical Human–Robot Interaction. *IEEE Trans. Robot.* **2020**, *36*, 240–253. [CrossRef]
27. Buchli, J.; Stulp, F.; Theodorou, E.; Schaal, S. Learning variable impedance control. *Int. J. Robot. Res.* **2011**, *30*, 820–833. [CrossRef]
28. Thuruthel, T.G.; Falotico, E.; Renda, F.; Laschi, C. Model-Based Reinforcement Learning for Closed-Loop Dynamic Control of Soft Robotic Manipulators. *IEEE Trans. Robot.* **2019**, *35*, 124–134. [CrossRef]
29. Yang, Q.; Dürr, A.; Topp, E.A.; Stork, J.A.; Stoyanov, T. Variable Impedance Skill Learning for Contact-Rich Manipulation. *IEEE Robot. Autom. Lett.* **2022**, *7*, 8391–8398. [CrossRef]
30. Kronander, K.; Billard, A. Online learning of varying stiffness through physical human-robot interaction. In Proceedings of the 2012 IEEE International Conference on Robotics and Automation, Saint Paul, MN, USA, 14–18 May 2012; pp. 1842–1849. [CrossRef]
31. Keemink, A.Q.; van der Kooij, H.; Stienen, A.H. Admittance control for physical human–robot interaction. *Int. J. Robot. Res.* **2018**, *37*, 1421–1444. [CrossRef]
32. Sadeghian, H.; Villani, L.; Keshmiri, M.; Siciliano, B. Task-Space Control of Robot Manipulators With Null-Space Compliance. *IEEE Trans. Robot.* **2014**, *30*, 493–506. [CrossRef]
33. Guo, Y.; Dong, H.; Ke, Y. Stiffness-oriented posture optimization in robotic machining applications. *Robot. Comput.-Integr. Manuf.* **2015**, *35*, 69–76. [CrossRef]
34. Lukić, B.; Jovanović, K.; Žlajpah, L.; Petrič, T. Online Cartesian Compliance Shaping of Redundant Robots in Assembly Tasks. *Machines* **2023**, *11*, 35. [CrossRef]
35. Ajoudani, A.; Tsagarakis, N.G.; Bicchi, A. On the role of robot configuration in Cartesian stiffness control. In Proceedings of the 2015 IEEE International Conference on Robotics and Automation (ICRA), Seattle, WA, USA, 26–30 May 2015; pp. 1010–1016.
36. Celikag, H.; Sims, N.D.; Ozturk, E. Cartesian Stiffness Optimization for Serial Arm Robots. *Procedia CIRP* **2018**, *77*, 566–569. [CrossRef]
37. Petit, F.; Albu-Schäffer, A. Cartesian impedance control for a variable stiffness robot arm. In Proceedings of the 2011 IEEE/RSJ International Conference on Intelligent Robots and Systems, San Francisco, CA, USA, 25–30 September 2011; pp. 4180–4186.
38. Petit, F.P. Analysis and Control of Variable Stiffness Robots. Ph.D. Thesis, ETH Zurich, Zürich, Switzerland, 2014. [CrossRef]
39. Roveda, L. Adaptive interaction controller for compliant robot base applications. *IEEE Access* **2018**, *7*, 6553–6561. [CrossRef]
40. Masinga, P.; Campbell, H.; Trimble, J.A. A framework for human collaborative robots, operations in South African automotive industry. In Proceedings of the 2015 IEEE International Conference on Industrial Engineering and Engineering Management (IEEM), Singapore, 6–9 December 2015; IEEE: New York, NY, USA, 2015, pp. 1494–1497.

41. Kana, S.; Lakshminarayanan, S.; Mohan, D.M.; Campolo, D. Impedance controlled human–robot collaborative tooling for edge chamfering and polishing applications. *Robot. Comput.-Integr. Manuf.* **2021**, *72*, 102199. [CrossRef]
42. Zanchettin, A.M.; Rocco, P.; Robertsson, A.; Johansson, R. Exploiting task redundancy in industrial manipulators during drilling operations, In Proceedings of the 2011 IEEE International Conference on Robotics and Automation, Shanghai, China, 9–13 May 2011.
43. Cherubini, A.; Passama, R.; Crosnier, A.; Lasnier, A.; Fraisse, P. Collaborative manufacturing with physical human–robot interaction. *Robot. Comput.-Integr. Manuf.* **2016**, *40*, 1–13. [CrossRef]
44. Surgical, I. da Vinci. Surgical System. 2013. Available online: http://www.intusurg.com/html/davinci.html (accessed on 3 January 2024.).
45. Freschi, C.; Ferrari, V.; Melfi, F.; Ferrari, M.; Mosca, F.; Cuschieri, A. Technical review of the da Vinci surgical telemanipulator. *Int. J. Med. Robot. Comput. Assist. Surg.* **2013**, *9*, 396–406. [CrossRef] [PubMed]
46. Burgner-Kahrs, J.; Rucker, D.C.; Choset, H. Continuum robots for medical applications: A survey. *IEEE Trans. Robot.* **2015**, *31*, 1261–1280. [CrossRef]
47. Vijayakumar, S.; D'Souza, A.; Schaal, S. Incremental Online Learning in High Dimensions. *Neural Comput.* **2005**, *17*, 2602–2634. [CrossRef]
48. Kraft, D. *A Software Package for Sequential Quadratic Programming*; Deutsche Forschungs- und Versuchsanstalt fur Luft- und Raumfahrt Koln: Köln, Germany, 1988; Forschungsbericht, Wiss. Berichtswesen d. DFVLR.
49. Boggs, P.T.; Tolle, J.W. Sequential quadratic programming for large-scale nonlinear optimization. *J. Comput. Appl. Math.* **2000**, *124*, 123–137. [CrossRef]
50. Chen, Y.; Ding, Y. Posture Optimization in Robotic Flat-End Milling Based on Sequential Quadratic Programming. *J. Manuf. Sci. Eng.* **2023**, *145*, 061001. [CrossRef]
51. Usevitch, N.S.; Hammond, Z.M.; Schwager, M. Locomotion of Linear Actuator Robots Through Kinematic Planning and Nonlinear Optimization. *IEEE Trans. Robot.* **2020**, *36*, 1404–1421. [CrossRef]
52. Mitrovic, D.; Klanke, S.; Vijayakumar, S. Learning impedance control of antagonistic systems based on stochastic optimization principles. *Int. J. Robot. Res.* **2011**, *30*, 556–573. [CrossRef]
53. Schaal, S.; Atkeson, C.; Vijayakumar, S. Scalable Techniques from Nonparametric Statistics for Real Time Robot Learning. *Appl. Intell.* **2002**, *17*, 49–60. [CrossRef]
54. Catalano, M.G.; Grioli, G.; Garabini, M.; Bonomo, F.; Mancini, M.; Tsagarakis, N.; Bicchi, A. VSA-CubeBot: A modular variable stiffness platform for multiple degrees of freedom robots. In Proceedings of the 2011 IEEE International Conference on Robotics and Automation, Shanghai, China, 9–13 May 2011; pp. 5090–5095.
55. Lukić, B.Z.; Jovanović, K.M.; Kvaščcev, G.S. Feedforward neural network for controlling qbmove maker pro variable stiffness actuator. In Proceedings of the 2016 13th Symposium on Neural Networks and Applications (NEUREL), Belgrade, Serbia, 22–24 November 2016; pp. 1–4.
56. Knezevic, N.; Lukic, B.; Jovanovic, K.; Zlajpah, L.; Petric, T. End-effector cartesian stiffness shaping—Sequential least squares programming approach. *Serbian J. Electr. Eng.* **2021**, *18*, 1–14. [CrossRef]
57. Franka Robotics. Available online: https://franka.de (accessed on 15 November 2023).
58. Deutschmann, B.; Liu, T.; Dietrich, A.; Ott, C.; Lee, D. A Method to Identify the Nonlinear Stiffness Characteristics of an Elastic Continuum Mechanism. *IEEE Robot. Autom. Lett.* **2018**, *3*, 1450–1457. [CrossRef]

Disclaimer/Publisher's Note: The statements, opinions and data contained in all publications are solely those of the individual author(s) and contributor(s) and not of MDPI and/or the editor(s). MDPI and/or the editor(s) disclaim responsibility for any injury to people or property resulting from any ideas, methods, instructions or products referred to in the content.

Article

Using a Robot for Indoor Navigation and Door Opening Control Based on Image Processing

Chun-Hsiang Hsu and Jih-Gau Juang *

Department of Communications, Navigation and Control Engineering, National Taiwan Ocean University, Keelung 202, Taiwan; 10667016@mail.ntou.edu.tw
* Correspondence: jgjuang@mail.ntou.edu.tw

Abstract: This study used real-time image processing to realize obstacle avoidance and indoor navigation with an omnidirectional wheeled mobile robot (WMR). The distance between an obstacle and the WMR was obtained using a depth camera. Real-time images were used to control the robot's movements. The WMR can extract obstacle distance data from a depth map and apply fuzzy theory to avoid obstacles in indoor environments. A fuzzy control system was integrated into the control scheme. After detecting a doorknob, the robot could track the target and open the door. We used the speeded up robust features matching algorithm to recognize the WMR's movement direction. The proposed control scheme ensures that the WMR can avoid obstacles, move to a designated location, and open a door. Like humans, the robot performs the described task only using visual sensors.

Keywords: indoor navigation; image processing; mobile robot; obstacle avoidance; feature matching

1. Introduction

With scientific and technological advancements, advanced machine systems are expected to replace human labor. Robots have always attracted considerable attention in the industrial technology development domain. In recent years, robots have been used to perform a few simple labor tasks. For instance, they are employed in unmanned factories, as vacuum cleaners, for restaurant service, and as tourist guides [1,2]. To allow robots to perform more tasks, it is important to conduct research on how to make robots intelligent and humanized. Diverse types of robots are used in different working environments. In this study, an omnidirectional wheeled robot is used in an indoor working area. Omnidirectional wheeled mobile robots (WMRs) are more flexible than ordinary moving robots as they can move in complex and narrow environments [3]. Omnidirectional robots can move in any direction without turning their heads and have been applied to many tasks [4–6]. In traditional robots, many sensors are installed to detect objects and directions. Obstacle avoidance is mainly achieved using ultrasonic and laser range finders or other distance-measuring instruments. Navigation is performed using a few positioning instruments, such as StarGazer, Bluetooth, or WiFi. In this study, we used cameras to replace traditional obstacle avoidance and positioning sensors to make the robot more human-like.

With advancements in artificial intelligence, intelligent robots have been widely studied. In recent years, omnidirectional wheeled robots with different control systems have been developed. Jia et al. developed an omnidirectional wheeled robot with multiple control Mecanum wheels [7]. Park et al. studied the fuzzy PID steering control structure of a mobile robot prototype [8]. Chung et al. modeled and analyzed an omnidirectional mobile robot with three caster wheels [9]. When robots could move, researchers began to study how to install obstacle avoidance components in them. Ruan et al. used ultrasonic sensors to confer obstacle avoidance capabilities on a two-wheeled self-balancing robot [10]. Jin et al. used a rotating ultrasonic sensor to endow a car with active obstacle-avoidance capabilities [11]. Peng et al. presented a laser-based obstacle avoidance scheme [12]. In addition to ultrasonic sensor–based avoidance, a few researchers have studied image-based

Citation: Hsu, C.-H.; Juang, J.-G. Using a Robot for Indoor Navigation and Door Opening Control Based on Image Processing. *Actuators* **2024**, *13*, 78. https://doi.org/10.3390/act13020078

Academic Editor: Zhuming Bi

Received: 27 December 2023
Revised: 12 February 2024
Accepted: 15 February 2024
Published: 16 February 2024

Copyright: © 2024 by the authors. Licensee MDPI, Basel, Switzerland. This article is an open access article distributed under the terms and conditions of the Creative Commons Attribution (CC BY) license (https://creativecommons.org/licenses/by/4.0/).

obstacle avoidance. Wang et al. used a Kinect depth camera to detect obstacles [13]. They used the Kinect to obtain a depth map and subsequently applied a Gaussian filter and the mean-shift segmentation technique to detect obstacles. Hamzah et al. used a stereo camera to obtain a disparity map and then used the map to compute object distance and the direction of movement [14]. Sharifi et al. used the mean-shift color classification scheme to distinguish obstacles from the ground [15]. AI-Jubouri et al. proposed the use of a set of local features extracted from a sequence of image frames collected using a computer vision system. The extracted features for each free-swimming fish were then compared with pre-extracted sets of features stored in a database using the SURF matching method [16]. Their method yielded a high object recognition rate. Sheu et al. used two cameras to compute a target object's deviation angle and distance and then used an adaptive PID control scheme for real-time target object tracking [17]. In our study, we integrated omnidirectional wheels, an Arduino system, a DC motor, a motor controller, a depth camera, and a robotic arm for indoor navigation. The control scheme involves applying image processing methods and feature matching to detect obstacles and compute the movement direction of a robot. The robot's movement is based on fuzzy control.

This study mainly revised the traditional ranging paradigm. Range detection was realized by using an Intel Realsense depth camera. After computing the distance of an object, fuzzy theory was applied to avoid obstacles. During robot motion on a path, we used the speeded up robust features (SURF) algorithm to compute the robot's self-position and future trajectory. Moreover, the cameras detected the static and dynamic obstacles encountered on the path. After arriving at the designated position, the control scheme used HoughCircles to identify a circular object so that the arm could find the doorknob and claw the door handle. Our experimental results indicate that the proposed visual control scheme can facilitate omnidirectional obstacle avoidance and help the robot move to a designated location and open the desired door. Compared to other relative research, most of them utilized laser and ultrasonic sensors in robot navigation [18,19]. To make the robot human-like, the proposed WMR system only uses visual information to avoid obstacles, navigate indoors, identify the door, and guide the robot arm to reach the doorknob. In addition, a simple fuzzy system is implemented in the control process that can reduce the computing time and is suitable for real-time control.

2. System Description

The proposed control system was realized using an omnidirectional WMR, as shown in Figure 1a. The length, width, and height of the omnidirectional WMR are 600, 400, and 850, respectively. The omnidirectional WMR has two mechanical arms. One of the arms is 570 mm long; it can grasp objects, and its wrist can rotate. The other arm is 470 mm long and equipped with a 150 mm front clip that can hold an object, but the wrist on this arm cannot rotate. The robot's arms have six RX-64 motors; there are two on each shoulder and one on the sides of each elbow. Each RX-64 motor has a length of 40.2 mm, a width of 41 mm, and a height of 61.1 mm. The stall torque of this motor is 5.7 mNm. The left wrist has an RX-28 motor with a length of 35.6 mm, a width of 35.5 mm, and a height of 50.6 mm. Its stall torque is 3.77 mNm. The right wrist has two RX-28 motors and one XM430-W350 motor, which has a length of 46.5 mm, a width of 28.5 mm, and a height of 34 mm. Its stall torque is 4.1 mNm. The waistline of the wheeled robot measures 150 cm. The omnidirectional wheel chassis has a radius of 240 mm. The three omnidirectional wheels are spaced 120° apart, and three 12V-DC motors are installed to provide a rated torque of 68 mNm. The omnidirectional electronic module is installed on the second floor and has three motor encoders and two batteries. Moreover, the electronic module includes a voltage step-down circuit board and a control panel. The battery supply is 12 V, and the capacity is 7 Ah. One of the supply voltages is for the motor controller, and the other is for the other control panels, additional power supply, and voltage step-down board. The voltage step-down board reduces the voltage from 12 V to 9 V and 6 V, providing voltage to the robot arm or other devices. The first panel is the Arduino system. Its main functions are (1) transmitting and receiving signals between the laptop and the receiver and (2) sending control commands to the motor. The second panel is the Arduino I/O extension shield. Its primary function is to connect additional receivers (such as temperature receivers and acceleration/tilt receivers), as shown in Figure 1b. We used an Intel Realsense Depth Camera D415 to detect obstacles and distance, as shown in Figure 1c [20]. The camera has three lenses, two of which are used to measure depth, and the other of which is a red-green-blue (RGB) lens.

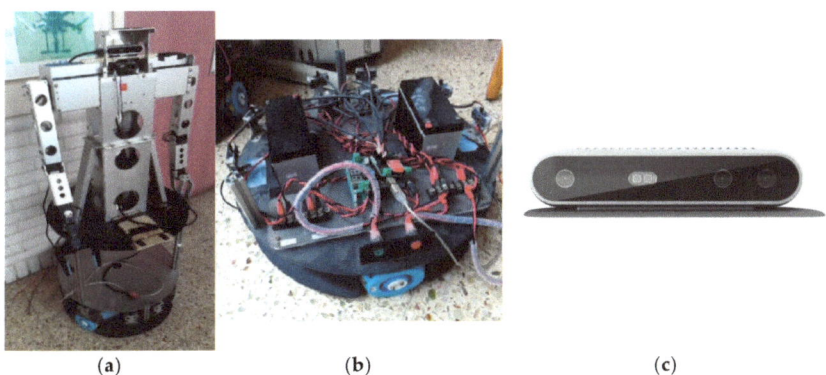

Figure 1. The primary devices of the proposed robot system are (**a**) a mobile robot; (**b**) an omnidirectional wheel, battery, and control components; and (**c**) an Intel Realsense depth camera.

The omnidirectional WMR can move along any angle [21,22]. The structure of the omnidirectional wheels and the coordinate system are shown in Figure 2 [23], respectively. The WMR has three omnidirectional wheels, separated by an angle of 120°. O denotes the WMR center. The length between O and an omnidirectional wheel is L. Counterclockwise and clockwise movements of the wheels are considered positive speed and negative speed, respectively. The road speed of omnidirectional wheel 1 is v_1, that of omnidirectional wheel 2 is v_2, and that of omnidirectional wheel 3 is v_3. The center of the omnidirectional WMR coordinate system is denoted as x_m and y_m, and δ, the angle between v_1 (or v_2) and y_m, is 30°. The angle between v_3 and y_m is 90°.

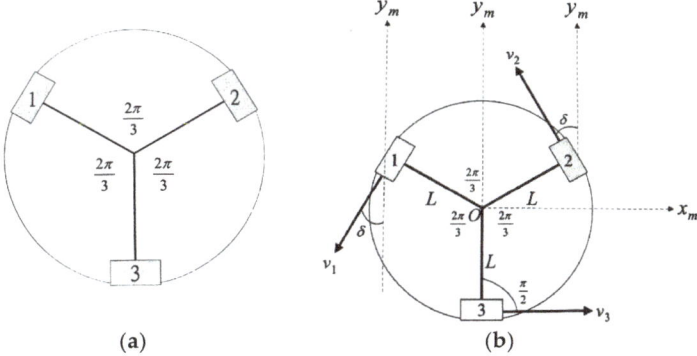

Figure 2. (**a**) Omnidirectional wheel structure. (**b**) Coordinate system.

Based on the wheel radius and angular wheel velocity, we can compute the speed of the omnidirectional wheel. To achieve the desired speed v_m and move along the specified direction, the speed v_i of the omnidirectional wheel i is composed of $\dot{x}_m$ and $\dot{y}_m$, which are the road speeds along the x_m and y_m axes.

The Arduino system is an I/O platform based on an open-source code, and because it uses the Java and C processing and wiring development environment, it has user-friendly features. The Arduino system allows for the rapid development of applications [24]. Figure 3a shows an Arduino Uno R3. It is a microcontroller board based on ATmega328. It has 14 digital input/output pins (6 can be used as PWM outputs), six analog inputs, a 16-MHz ceramic resonator, a USB connection, a power jack, an ICSP header, and a reset button. The device can be connected to a computer through a USB cable or powered with an AC-to-DC adapter or a battery. Figure 3b shows the DFRduino I/O expansion board [25]. Each omnidirectional wheel unit (4202X, KORNYYALK) [26] has three wheels. We added a Microsoft LifeCam studio camera (Figure 3c) attached to the robotic arm to track the doorknob, which is outside the field of view of the depth camera lens, and this camera is only used to track the doorknob.

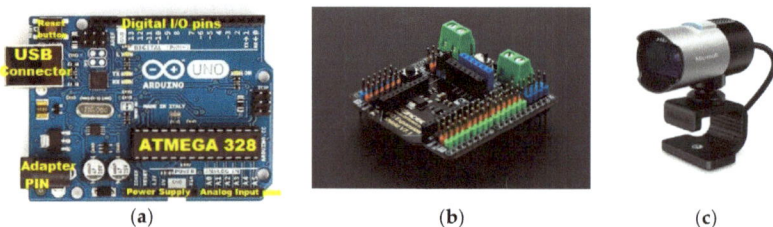

Figure 3. (**a**) Arduino Uno R3 [15], (**b**) DFRduino IO Expansion board, and (**c**) Microsoft LifeCam.

3. Image Processing and Pattern Recognition

The depth camera used in this study has three lenses; two are used to measure depth, and the other is an RGB lens. The color space is mainly three-dimensional, so people can clearly distinguish between colors. Many color spaces exist, such as RGB, YCbCr, and HSV. The representations of the dimensions are different for each of these color spaces. For instance, the RGB color space uses red, green, and blue as the X, Y, and Z axes. The hue, saturation, value (HSV) color space uses chromaticity, saturation, and lightness as the X, Y, and Z axes [27]. The human eye contains several types of cone-shaped photoreceptor cells. Humans see yellow when the stimulation point is slightly larger than green photoreceptor cells. Humans see red when the stimulation point is larger than green photoreceptor cells [28]. Except for white and black, most colors can be obtained by appropriately combining red, green, and blue. The RGB model's red, green, and blue cube coordinates are nonnegative numbers between 0 and 1. The origin (0,0,0) is black, and the intensity of color increases along the coordinate axis direction, with the point (1,1,1) being white. Computer monitors and TV screens mainly use the RGB color space, which combines these three colors in each pixel. Each pixel in a computer monitor is represented by 24 bits, meaning that the color of RGB is represented by 8 bits, and an integer between 0 and 255 represents the intensity of each primary color. A total of 256 such values exist, which can be combined with 16,777,216 colors [18,29]. To measure the distance between an obstacle and the robot, the image depth must be calculated using two cameras [30], and the imaging positions of the left and right cameras can be recognized. Three process steps are required before the camera can be used for depth measurement: camera calibration, stereo rectification, and stereo matching. After completing these steps, the camera can measure the distance from images.

3.1. Camera Calibration

Camera calibration determines a camera's internal parameters, external parameters, and rotation matrix. The external parameters are the transformations describing real-world and camera coordinates. These parameters are used to identify the relationship between the imaging and actual object positions. In this study, we applied the Zhang Zhengyou calibration method [31], which requires numerous samples for position calculation. Some of the samples are shown in Figure 4.

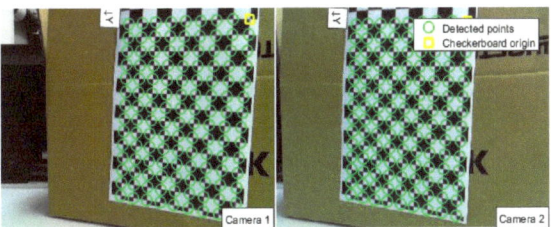

Figure 4. Zhang Zhengyou camera calibration samples.

The stereo rectification step is performed to ensure that the two images correspond after distortion correction. In this step, the epipolar lines of the two images are on the same horizontal line so that one point in one of the images corresponds to the same point in the other image, as shown in Figure 5.

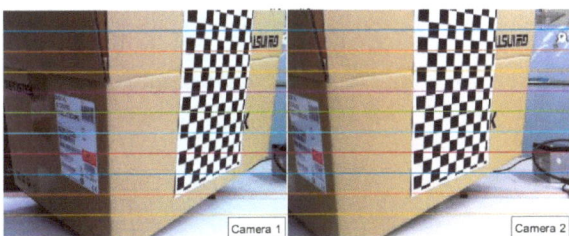

Figure 5. Stereo rectification.

3.2. Depth Map

The parameters obtained after camera calibration and stereo rectification can be used to detect the object's depth. The results of a simulation conducted to detect the distance between a robot and a mug are shown in Figure 6. The actual distance and the detected distance are both 0.83 m.

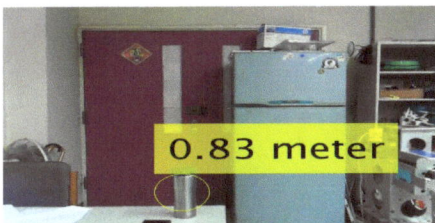

Figure 6. Mug distance.

Because the camera is easily disturbed by light in a place with sunlight, such a disturbance may cause a target to be unrecognizable, or a huge database and extensive calculations may be required for target identification. Therefore, we installed an Intel Realsense D415-type depth camera on the omnidirectional wheel robot. The Intel Realsense depth camera provides a direct depth map estimation. The depth map shows different colors according to the distance, as illustrated in Figure 7. Different colors mean different distances of depth, and the distances are used in the fuzzy control of obstacle avoidance in Section 4.2.

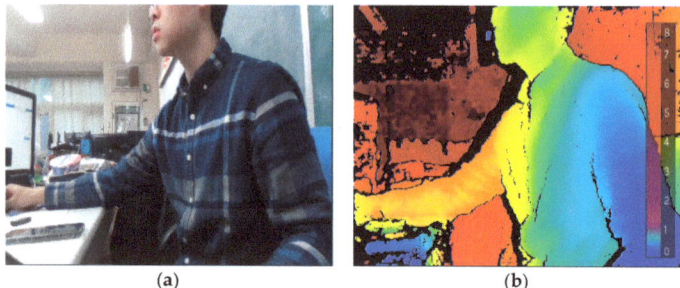

Figure 7. (**a**) is the original image, (**b**) is the depth image where the blue color means the object is near to the camera, and dark red means the object is far from the camera. The distance ranges from 0 m (dark blue) to 8 m (dark red).

3.3. Obstacle Detection

We set up a frame, as indicated by the green square in Figure 8. This frame represents the safe range of the omnidirectional WMR in terms of bumping into an obstacle. Then, we computed the distance of every pixel from the depth webcam for this frame and divided this frame into three parts (left, middle, and right) [14]. After that, we computed the minimum distance between the robot and an obstacle. Given the left distance, middle distance, and right distance, we can determine which obstacle is closer to the robot and identify the position of that obstacle.

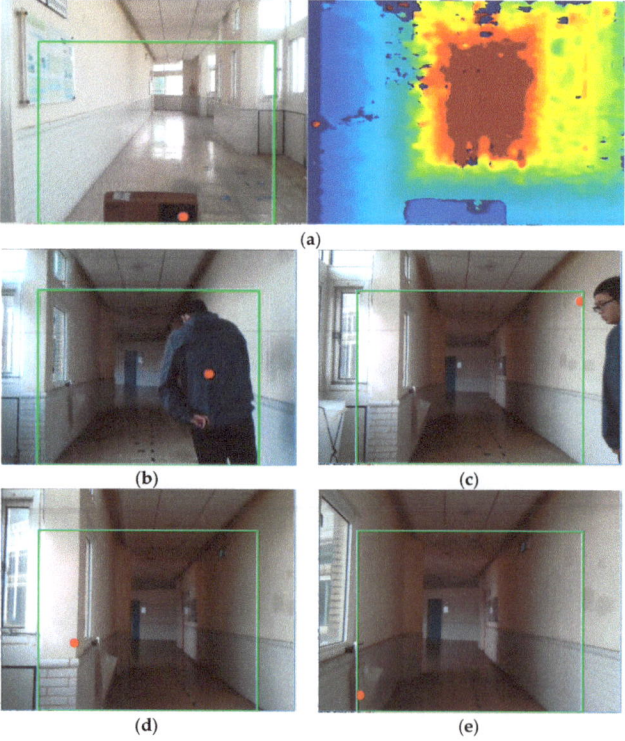

Figure 8. Detection of different obstacles: (**a**) obstacle detection and obstacle depth map, (**b**) nearest obstacle point is on the right side, (**c**) move to the left and the obstacle is outside the safe frame, (**d**) nearest obstacle point is on the left side, (**e**) move to the right and the moving direction is clear.

3.4. Feature Matching

The SURF algorithm was proposed by Herbert Bay [32]. It is a robust algorithm for local feature point detection and description. The SURF algorithm is a modified version of the SIFT algorithm proposed by David Low. The SURF algorithm is faster and more efficient than the SIFT algorithm. The SURF algorithm has three main components: the extraction of local feature points, the description of feature points, and matching of feature points. We used SURF feature matching for route planning. The webcam takes pictures as the robot moves, and the pictures are matched with stored samples, which are images of the known environment. These images provide features of the environment that can help the robot with indoor navigation. We set a threshold for the feature points. When the SURF feature matching feature points exceed the threshold, they represent a proper direction or destination. The feature-matching process is shown in Figure 9.

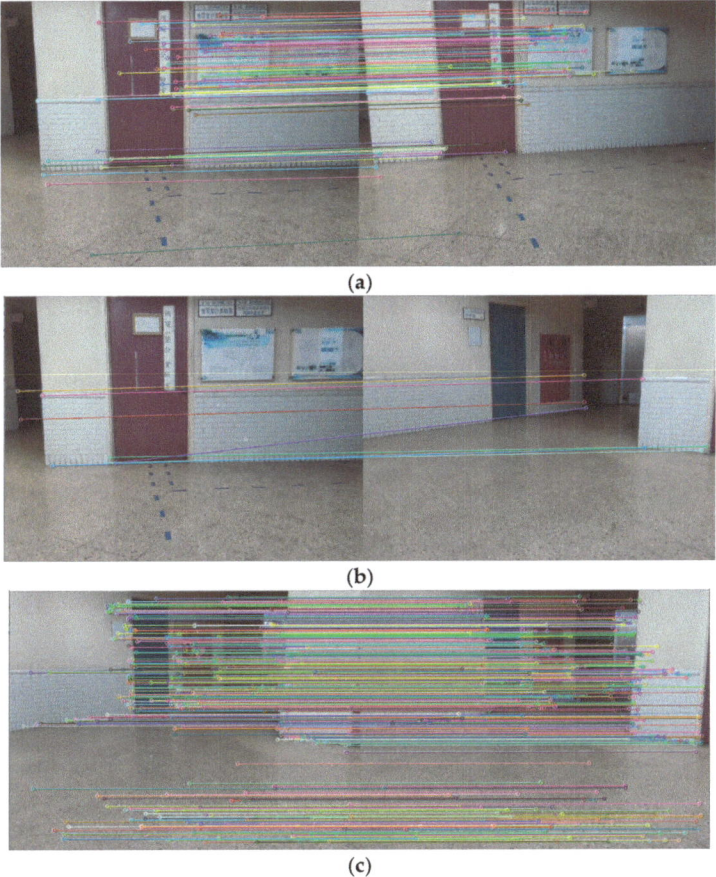

Figure 9. SURF matching: (**a**) robot is heading in the right direction, (**b**) robot is heading in the wrong direction, (**c**) robot is heading in the right direction.

3.5. Circular Doorknob Detection

The target doorknobs are circular. The two-stage HoughCircles transform can identify a circle in an image frame [33]. The first stage involves finding the center of a circle. Given the threshold for an image, edges can be detected, as shown in Figure 10. Then, the gradient line at each nonzero point in the edge image is identified. The greater the number of line intersection points, the greater the likelihood that they are at the center of a circle.

A threshold value is set in the Hough space, which is considered the circle's center if it exceeds the threshold value. The second stage involves detecting the radius of the circle. The Hough transform sets thresholds for the maximum and minimum radii. An object can be extracted using these thresholds, as shown in Figure 10.

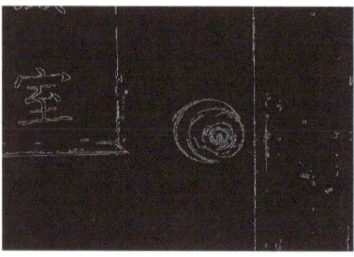 

Figure 10. Edge detection uses a threshold, and the doorknob is detected; the Chinese character on the left part of the figure is the room's name.

4. Control Scheme

In the control system, we used LabVIEW to compile the robot system. The control method and the image processing scheme were written in Python.

4.1. Motion Control

This study integrated obstacle avoidance and route navigation into the omnidirectional wheel robot. The control sequence is presented in Figure 11. The robot will follow a planned path with specified features. From the original starting point, the robot moves straight forward to the next specified feature point. When the robot reaches the desired midway point, it searches the destination point through feature matching and turns to the target's direction. When the destination features are matched, the robot moves forward to the destination point. When the robot reaches the destination point, arm control is activated. A detailed arm control process is shown in Section 4.3. If an obstacle is found on the pathway, the robot performs obstacle avoidance control (Section 4.2), as shown in mark B in Figure 12. After avoiding the obstacle, the robot moves back to the planned path to the desired midway point before mark C in Figure 12.

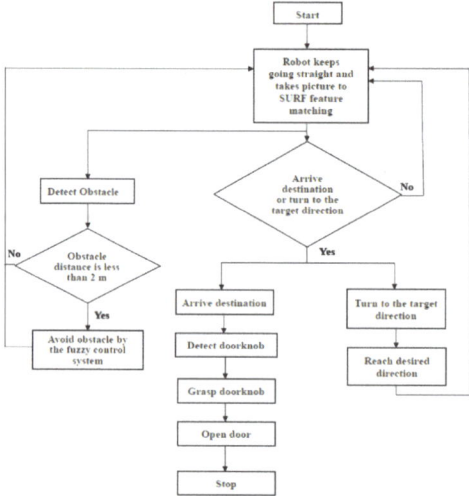

Figure 11. Control flowchart.

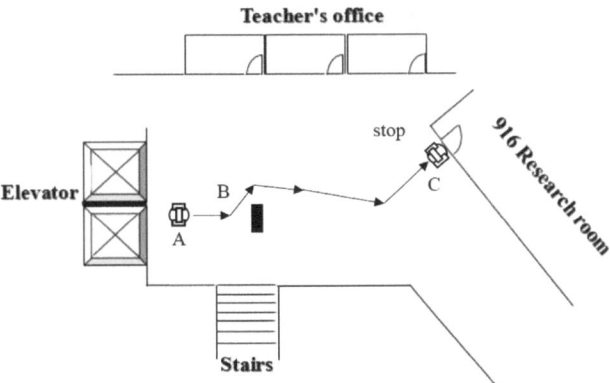

Figure 12. Robot moving path.

After the robot begins operating, it initially uses the SURF feature-matching algorithm to determine whether it has reached a specified position. If the robot is not at the specified position, it estimates the direction to be traveled by matching the features of the SURF algorithm. The obstacle detection feature is always on during the walking process. The obstacle avoidance function is activated if obstacles are found on the path. The robot has fuzzy control installed, so the robot checks for obstacles and avoids them based on its distance from these obstacles. The experimental environment of the omnidirectional wheel robot and the predicted walking path are shown in Figure 12. When the robot arrives at the specified position, doorknob detection is initiated. After detecting the doorknob, the robot tracks the target and opens the door.

4.2. Obstacle Avoidance Control

To simplify the calculation process, other than neural network mobile robot control [34,35], fuzzy control is used in obstacle avoidance to allow the robot to avoid obstacles accurately [36]. The fuzzy control of obstacle avoidance uses three inputs and two outputs. We cut the camera's pixels into three parts: left (L), medium (M), and right (R). Each part detects the nearest obstacle and returns the distance function between the obstacle and the robot. The distances of these three obstacles are the three inputs of the fuzzy control scheme, and the fuzzy sets are near, medium, and far. The outputs are time (T) and pixel (P). The output time (T) is used to control the rotation time of the wheel, and the fuzzy sets are long (LG), medium (MD), and short (ST). The pixel (P) controls the direction of rotation. The fuzzy sets are turn_right (TR), turn_left (TL), and go_straight (TM), as shown in Figure 13, and the fuzzy control models are shown in Figure 14. The fuzzy control scheme is shown in Figure 15.

R1: If L is near and M is near and R is near, then T is LG and P is TR.
R2: If L is near and M is near and R is medium, then T is LG and P is TR.
R3: If L is near and M is near and R is far, then T is LG and P is TR.
R4: If L is near and M is medium and R is near, then T is ST and P is TM.
R5: If L is near and M is medium and R is medium, then T is MD and P is TR.
R6: If L is near and M is medium and R is far, then T is MD and P is TR.
R7: If L is near and M is far and R is near, then T is ST and P is TM.
R8: If L is near and M is far and R is medium, then T is ST and P is TR.
R9: If L is near and M is far and R is far, then T is ST and P is TR.
R10: If L is medium and M is near and R is near, then T is LG and P is TL.
R11: If L is medium and M is near and R is medium, then T is LG and P is TR.
R12: If L is medium and M is near and R is far, then T is LG and P is TR.
R13: If L is medium and M is medium and R is near, then T is MD and P is TL.
R14: If L is medium and M is medium and R is medium, then T is ST and P is TM.

R15: If L is medium and M is medium and R is far, then T is ST and P is TR.
R16: If L is medium and M is far and R is near then, T is ST and P is TL.
R17: If L is medium and M is far and R is medium, then T is ST and P is TM.
R18: If L is medium and M is far and R is far, then T is ST and P is TM.
R19: If L is far and M is near and R is near, then T is LG and P is TL.
R20: If L is far and M is near and R is medium, then T is LG and P is TL.
R21: If L is far and M is near and R is far, then T is LG and P is TL.
R22: If L is far and M is medium and R is near, then T is MD and P is TL.
R23: If L is far and M is medium and R is medium, then T is MD and P is TL.
R24: If L is far and M is medium and R is far, then T is MD and P is TL.
R25: If L is far and M is far and R is near, then T is ST and P is TL.
R26: If L is far and M is far and R is medium, then T is ST and P is TM.
R27: If L is far and M is far and R is far, then T is ST and P is TM.

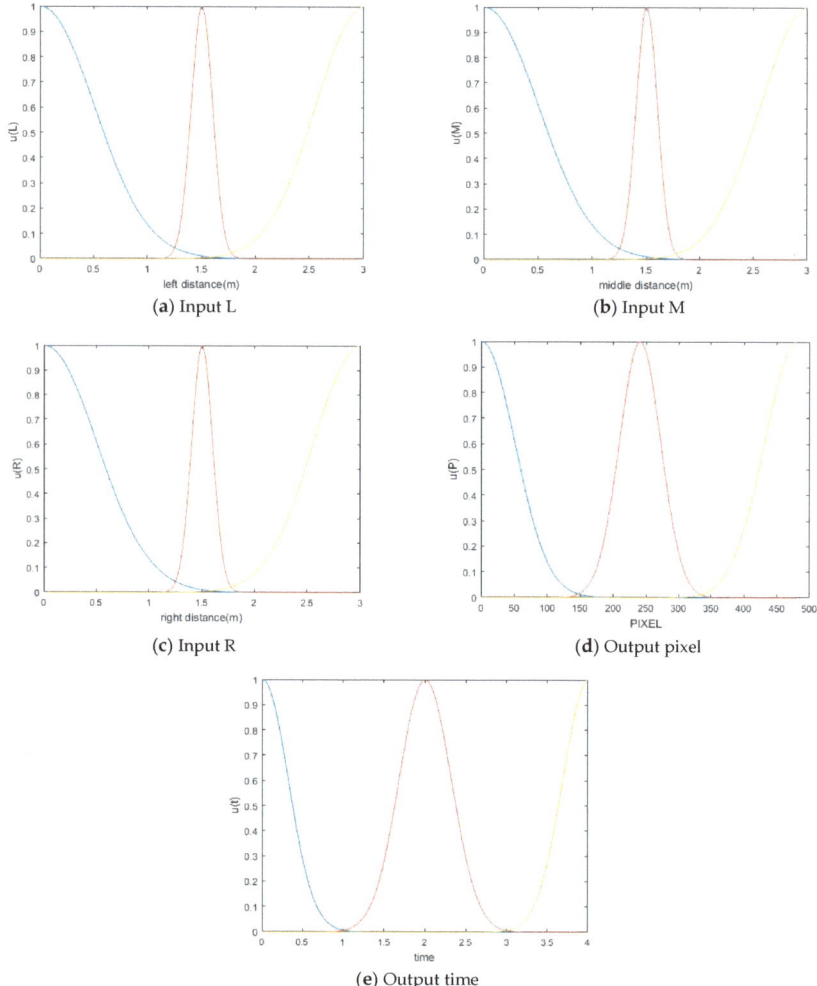

Figure 13. The fuzzy sets of the three inputs are near (blue), medium (red), and far (yellow); the fuzzy sets of the output pixel are turn_left (blue), go_straight (red), and turn_right (yellow); the fuzzy sets of the output time are short (blue), medium (red), and long (yellow).

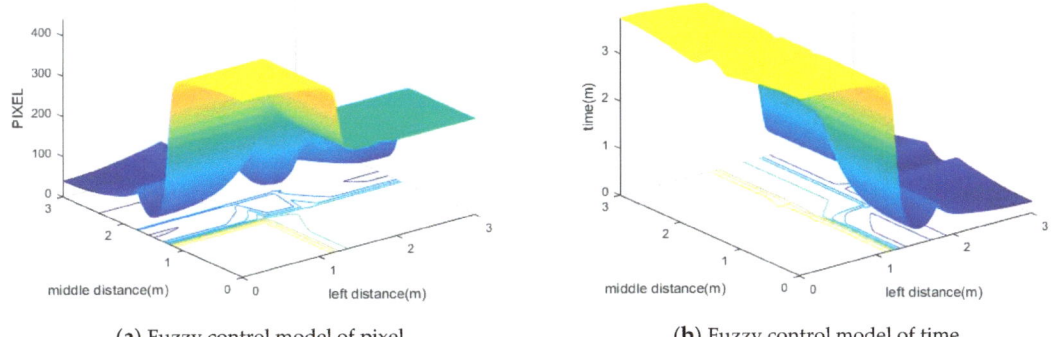

(a) Fuzzy control model of pixel   (b) Fuzzy control model of time

Figure 14. Fuzzy control model where the yellow color means large value and the dark blue means small value.

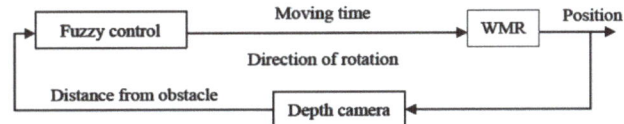

Figure 15. Fuzzy control scheme.

In Figure 12, mark A shows that when the robot detects an obstacle, it inputs the distance between the left, middle, and right to the fuzzy controller and outputs the time and direction of the rotation, as shown in mark B in Figure 12. In Figure 12, mark C shows that the robot stops at the specified position after avoiding an obstacle. When the robot arrives at the specified position, it initiates doorknob detection. The robot tracks the door after detecting the doorknob and then opens the door.

4.3. Arm Control

When the robot reaches the specified position, its arm is automatically lifted to the height of the doorknob, and the camera installed on the arm is activated to start tracking the doorknob, as shown in Figure 16. The robot determines the position of the doorknob and judges the required direction of movement [37]. The directions of movement are left rotation, right rotation, left parallel translation, and right parallel translation. We set two threshold values: the return position is greater than the threshold and lower than the threshold, and the corresponding motions are left and right translations. The times required for each right and left translation movement are 1.2 s and 900 ms, respectively. It is easy to move to the specified range. Then, the left and right rotation movements are performed to finetune the robot's heading angle so that the robot's arm points at the doorknob. The movement of each left and right rotation is 5°. When the robot's position is within these two range values, the robot moves forward. When the distance from the door is shorter than the set distance, the claw of the robot's arm grabs the doorknob and rotates. After the robot claw has turned the doorknob, it pushes the door forward and opens it.

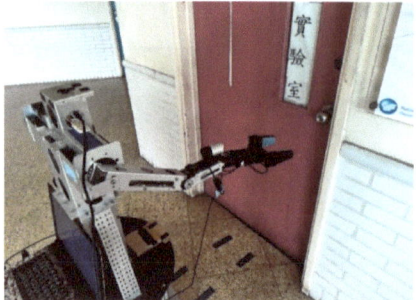

Figure 16. The robot arm lifts to the specified height; the Chinese characters on the figure are the room's name.

5. Experiment Result

We placed two obstacles in front of the robot. The robot was expected to avoid all obstacles without any route planning. After detecting an obstacle, the robot estimated the direction to move, as shown in Figure 17.

(**a**) Robot's starting position

(**b**) Robot avoids box

(**c**) Robot goes straight

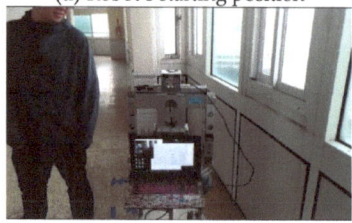
(**d**) Robot avoids human

(**e**) Robot after avoiding human

(**f**) Robot moves out of the obstacle area

Figure 17. Robot obstacle avoidance test.

When the robot is turned on, the onboard computer displays the image captured by the depth camera, as shown in Figure 18. The three circles in the picture represent the nearest left, center, and right distances, respectively. This representation makes it easy for users to identify which objects are detected by the robot. The distance and direction of the action are displayed after the obstacle detection process, as shown in Figure 18. On the depth camera image, the distances of the nearest left, middle, and right objects are greater than 2 m, so the action is GO. At the starting position, the robot does not detect any obstacles, so the control scheme sends the "go" command, which tells the robot to go straight.

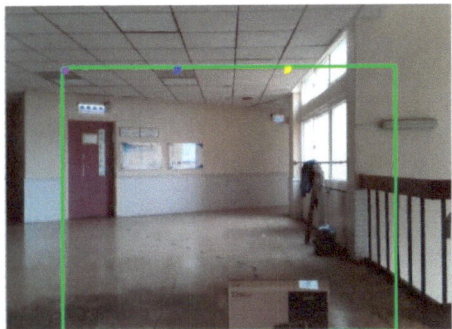

Figure 18. On the depth camera image, the distances of the nearest left, middle, and right objects are greater than 2 m, so the action is GO.

Figure 19 shows the robot detecting an obstacle. Figure 20 shows that the obstacle is detected within 2 m between the middle and the right areas. After detecting the obstacle, the robot executes the fuzzy control scheme to determine the direction to move. The control scheme sends the fuzzy control result "left," which means left rotation. After the robot rotates (Figure 21a), it will go straight and move the same distance parallel to the obstacle and then rotate back in the opposite direction (Figure 21b). The robot can avoid the obstacle successfully, as shown in Figure 21. The robot needs free space to avoid obstacles. If there is not enough space, this means the path has been blocked, and the robot will stop moving until the obstacle is removed.

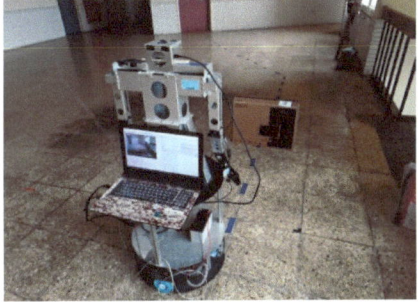

Figure 19. Robot's starting position.

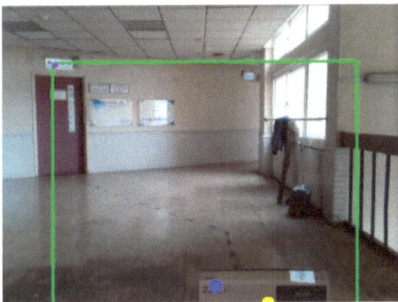

Figure 20. The robot finds an obstacle as the blue mark (in the middle area) and yellow mark (in the right area) on the picture, and the distances are 1.65 m and 1.62 m, respectively; the action is a LEFT turn.

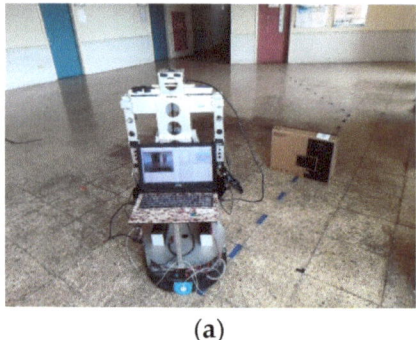 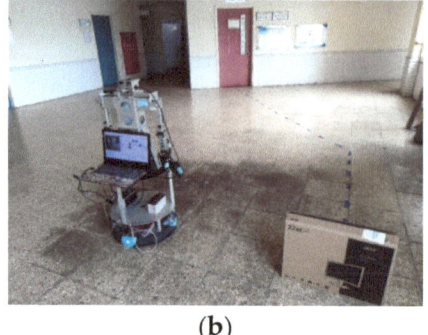

(a) (b)

Figure 21. Robot avoids obstacle (box).

After the rotation is finished (Figure 21b), the robot's movement will be determined again. On the depth camera image (Figure 22), the distances of the nearest left, middle, and right objects are greater than 2 m, so the robot is prompted to "go straight." The robot motion control process always uses SURF to match the target. When it matches the target, the motion control will determine the matching position and the robot's moving direction, as shown in Figures 23 and 24.

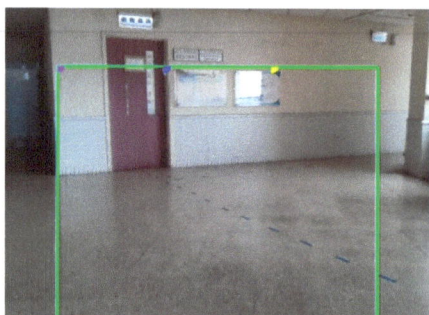

Figure 22. Image on the robot's depth camera.

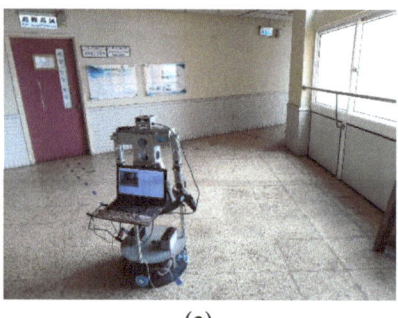 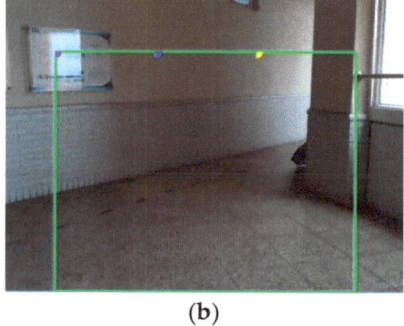

(a) (b)

Figure 23. (a) The robot reaches the midway point; (b) the robot's image.

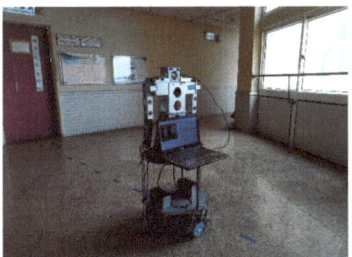

Figure 24. After the SURF matching, the robot turns to the target direction and moves forward.

The robot's walking process always uses SURF for target matching. When the target (predefined door) is matched, the control process shows the matching position and the robot's movement direction. The robot moves forward to the predefined door and stops 1 m ahead of the door, as shown in Figure 25a. The robot stops moving when it arrives at the specified position. According to the image captured by the robot's camera, the control process shows that the robot has arrived at the specified location, as shown in Figure 25b.

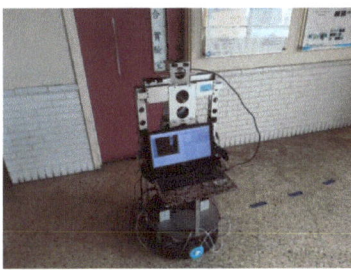 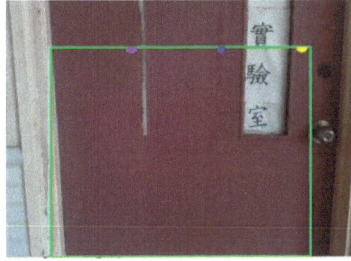

(**a**) Robot arrives at a specified location   (**b**) Robot's camera image

Figure 25. (**a**) The robot arrives at the specified location; (**b**) the Chinese characters on the figure are the room's name.

The arm and arm camera are activated when the robot arrives at the specified location. The arm camera starts to find the doorknob and returns the doorknob position to the robot, as shown in Figure 26. The circle center coordinate of the doorknob is indicated in Figure 26b.

(**a**)   (**b**)

Figure 26. (**a**) Doorknob detection, Activate arm and arm's camera; (**b**) the Chinese characters on the figure are the room's name. Arm's camera image; the object distance is 0.827 m, and the coordinate is (436.5, 51.5).

When the distance between the robot and the door is shorter than the preset threshold, the robot's claw automatically grabs the doorknob and rotates it, and the robot pushes the door forward, opens the door, and stops, as shown in Figure 27.

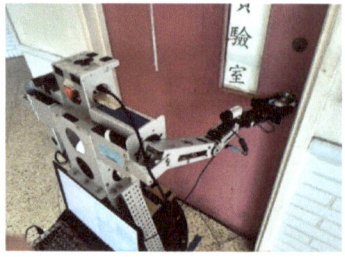 

(**a**) Robot's claw grabs the doorknob position (**b**) Enlarge image of claw grabbing the doorknob

(**c**) Robot claw rotates doorknob   (**d**) Robot pushes the door

Figure 27. The robot opens the door; the Chinese characters on the figure are the room's name.

6. Conclusions

In this study, we proposed the use of real-time images to control the robot's movement. The robot system has many instruments and devices installed. Some devices sense obstacles or measure the targets to be tracked. To ensure our robot was human-like, we only used cameras to make judgments regarding obstacle avoidance and navigation. The Intel Realsense depth camera was used to provide information on the surrounding object's distance for obstacle avoidance usage. The Lifecam was used to identify target objects for doorknob detection. In terms of control, we used LabVIEW to compile the robot system. The control method and the image processing scheme were written in Python. In terms of obstacle avoidance, we used depth images to help the robot avoid obstacles, and the images were more accurate and convenient than those output by conventional ultrasonic obstacle avoidance methods. Compared with other approaches that use laser or ultrasonic sensors, determining which target the robot has detected through images is a superior method. Cameras are relatively inexpensive and easy to maintain compared to laser range finders. Image processing is also more extensive than these other approaches. A fuzzy control system was integrated into the proposed image obstacle avoidance method, and the rotation angles corresponding to different distances were different, allowing the robot to avoid multiple obstacles successfully. Most indoor navigations use SLAM to map and recognize the surroundings. In this study, we assumed the surrounding environments are known; the SURF algorithm was used to inform the robot of the position and direction of the target location. In the experiment, the robot completed actions based only on the results of image processing and recognition; in other words, similar to humans, it used only visual sensors.

Author Contributions: Conceptualization, J.-G.J.; methodology, C.-H.H.; software, C.-H.H.; validation, C.-H.H.; formal analysis, J.-G.J.; Investigation, J.-G.J.; Resources, J.-G.J.; Data curation, C.-H.H.; writing—original draft, C.-H.H.; writing—review & editing, J.-G.J.; visualization, J.-G.J.; supervision, J.-G.J.; project administration, J.-G.J.; funding acquisition, J.-G.J. All authors have read and agreed to the published version of the manuscript.

Funding: This research was funded by the Ministry of Science and Technology (Taiwan), grant number MOST 109-2221-E-019-058.

Data Availability Statement: The original contributions presented in this study are included in this article. Further inquiries can be directed to the corresponding author.

Conflicts of Interest: The authors declare no conflicts of interest.

References

1. Belanche, D.; Casaló, L.V.; Flavián, C.; Schepers, J. Service robot implementation: A theoretical framework and research agenda. *Serv. Ind. J.* **2020**, *40*, 203–225. [CrossRef]
2. Gonzalez-Aguirre, J.A.; Osorio-Oliveros, R.; Rodríguez-Hernández, K.L.; Lizárraga-Iturralde, J.; Menendez, R.M.; Ramírez-Mendoza, R.A.; Ramírez-Moreno, M.A.; Lozoya-Santos, J.d.J. Service Robots: Trends and Technology. *Appl. Sci.* **2021**, *11*, 10702. [CrossRef]
3. Chi, L. Application of Real-Time Image Recognition and Feature Matching to Wheeled Mobile Robot for Room Service. Master's Thesis, National Taiwan Ocean University, Keelung City, Taiwan, 2018.
4. Najim, H.A.; Kareem, I.S.; Abdul-Lateef, W.E. Design and Implementation of an Omnidirectional Mobile Robot for Medi-cine Delivery in Hospitals during the COVID-19 Epidemic. *AIP Conf. Proc.* **2023**, *2830*, 070004.
5. Bernardo, R.; Sousa, J.M.C.; Botto, M.A.; Gonçalves, P.J.S. A Novel Control Architecture Based on Behavior Trees for an Omni-Directional Mobile Robot. *Robotics* **2023**, *12*, 170. [CrossRef]
6. Palacín, J.; Rubies, E.; Clotet, E.; Martínez, D. Evaluation of the Path-Tracking Accuracy of a Three-Wheeled Omnidirectional Mobile Robot Designed as a Personal Assistant. *Sensors* **2021**, *21*, 7216. [CrossRef] [PubMed]
7. Jia, Q.; Wang, M.; Liu, S.; Ge, J.; Gu, C. Research and development of mecanum-wheeled omnidirectional mobile robot implemented by multiple control methods. In Proceedings of the 23rd International Conference on Mechatronics and Machine Vision in Practice, Nanjing, China, 28–30 November 2016.
8. Park, S.; Ryoo, Y.; Im, D. Fuzzy Steering Control of Three-Wheels Based Omnidirectional Mobile Robot. In Proceedings of the International Conference on Fuzzy Theory and Its Applications, Taichung, Taiwan, 9–11 November 2016.
9. Chung, J.H.; Yi, B.-J.; Kim, W.K.; Lee, H. The dynamic modeling and analysis for an omnidirectional mobile robot with three caster wheels. In Proceedings of the IEEE International Conference on Robotics and Automation, Taipei, Taiwan, 14–19 September 2003.
10. Ruan, X.; Li, W. Ultrasonic sensor based two-wheeled self-balancing robot obstacle avoidance control system. In Proceedings of the IEEE International Conference on Mechatronics and Automation, Tianjin, China, 3–6 August 2014.
11. Jin, Y.; Li, S.; Li, J.; Sun, H.; Wu, Y. Design of an Intelligent Active Obstacle Avoidance Car Based on Rotating Ultrasonic Sensors. In Proceedings of the IEEE 8th Annual International Conference on CYBER Technology in Automation, Control, and Intelligent Systems, Tianjin, China, 19–23 July 2018.
12. Peng, Y.; Qu, D.; Zhong, Y.; Xie, S.; Luo, J. The Obstacle Detection and Obstacle Avoidance Algorithm Based on 2-D Lidar. In Proceedings of the IEEE International Conference on Information and Automation, Lijiang, China, 8–10 August 2015.
13. Wang, T.; Bu, L.; Huang, Z. A new method for obstacle detection based on Kinect depth image. In Proceedings of the Chinese Automation Congress, Wuhan, China, 27–29 November 2015.
14. Hamzah, R.A.; Rosly, H.N.; Hamid, S. An Obstacle Detection and Avoidance of a Mobile Robot with Stereo Vision Camera. In Proceedings of the International Conference on Electronic Devices, Systems and Applications, Kuala Lumpur, Malaysia, 25–27 April 2011.
15. Sharifi, M.; Chen, X. Introducing a novel vision based obstacle avoidance technique for navigation of autonomous mobile robots. In Proceedings of the IEEE 10th Conference on Industrial Electronics and Applications, Auckland, New Zealand, 15–17 June 2015.
16. AI-Jubouri, Q.; AI-Nuaimy, W.; AI-Taeeand, M.; Young, I. Recognition of Individual Zebrafish Using Speed-Up Robust Feature Matching. In Proceedings of the 10th International Conference on Developments in eSystems Engineering, Paris, France, 14–16 June 2017.
17. Sheu, J.-S.; Tsai, W.-H. Implementation of a following wheel robot featuring stereoscopic vision. *Multimed. Tools Appl.* **2017**, *76*, 25161–25177. [CrossRef]
18. Tsai, C.-Y.; Nisar, H.; Hu, Y.-C. Mapless LiDAR Navigation Control of Wheeled Mobile Robots Based on Deep Imitation Learning. *IEEE Access* **2021**, *9*, 117527–117541. [CrossRef]
19. Li, C.; Wang, S.; Zhuang, Y.; Yan, F. Deep Sensor Fusion between 2D Laser Scanner and IMU for Mobile Robot Localization. *IEEE Sens. J.* **2019**, *21*, 8501–8509. [CrossRef]
20. Intel Realsense Depth Camera D415. Available online: https://www.intel.com/content/www/us/en/products/sku/128256/intel-realsense-depth-camera-d415/specifications.html (accessed on 21 January 2019).

21. Pin, F.; Killough, S. A new family of omnidirectional and holonomic wheeled platforms for mobile robots. *IEEE Trans. Robot. Autom.* **1994**, *10*, 480–489. [CrossRef]
22. Purwin, O.; D'andrea, R. Trajectory generation and control for four wheeled omnidirectional vehicles. *Robot. Auton. Syst.* **2006**, *54*, 13–22. [CrossRef]
23. Zhong, Q.H. Using Omni-Directional Mobile Robot on Map Building Application. Master's Thesis, National Cheng Kung University, Tainan City, Taiwan, 2009.
24. Arduino Uno R3. Available online: https://electricarena.blogspot.com/ (accessed on 10 January 2019).
25. DFRduino IO Expansion Shield for Arduino. Available online: https://www.dfrobot.com/product-1009.html (accessed on 15 March 2019).
26. Omni Wheel. Available online: http://www.kornylak.com/ (accessed on 15 March 2019).
27. Color Space. Available online: https://en.wikipedia.org/wiki/Color_space (accessed on 20 April 2019).
28. Dragoi, V. Chapter 14: Visual—Eye and Retina. *Neurosci. Online.* 2020. Available online: https://nba.uth.tmc.edu/neuroscience/m/s2/chapter14.html (accessed on 20 April 2019).
29. Color Cube. Available online: https://cs.vt.edu/Undergraduate/courses.html (accessed on 20 April 2019).
30. Zhang, Y.; Xu, X.; Dai, Y. Two-Stage Obstacle Detection Based on Stereo Vision in Unstructured Environment. In Proceedings of the Sixth International Conference on Intelligent Human-Machine Systems and Cybernetics, Hangzhou, China, 26–27 August 2014.
31. Zhang, Z. A Flexible New Technique for Camera Calibration. Available online: https://www.microsoft.com/en-us/research/wp-content/uploads/2016/02/tr98-71.pdf (accessed on 20 January 2019).
32. Bay, H.; Tuytelaars, T.; Gool, L.V. Speed Up Robust Features. In Proceedings of the European Conference on Computer Vision, Graz, Austria, 7–13 May 2006.
33. Liu, H.; Qian, Y.; Lin, S. Detecting Persons Using Hough Circle Transform in Surveillance Video. In Proceedings of the International Conference on Computer Vision Theory and Applications, Angers, France, 17–21 May 2010.
34. Fang, W.; Chao, F.; Yang, L.; Lin, C.-M.; Shang, C.; Zhou, C.; Shen, Q. A recurrent emotional CMAC neural network controller for vision-based mobile robots. *Neurocomputing* **2019**, *334*, 227–238. [CrossRef]
35. Wu, Q.; Lin, C.-M.; Fang, W.; Chao, F.; Yang, L.; Shang, C.; Zhou, C. Self-Organizing Brain Emotional Learning Controller Network for Intelligent Control System of Mobile Robots. *IEEE Access* **2018**, *6*, 59096–59108. [CrossRef]
36. Chao, C.H.; Hsueh, B.Y.; Hsiao, M.Y.; Tsai, S.H.; Li, T.H.S. Real-Time Target Tracking and Obstacle Avoidance for Mobile Robots using Two Cameras. In Proceedings of the ICROS-SICE International Joint Conference, Fukuoka, Japan, 18–21 August 2009.
37. Su, H.-R.; Chen, K.-Y. Design and Implementation of a Mobile Robot with Autonomous Door Opening Ability. *Int. J. Fuzzy Syst.* **2019**, *21*, 333–342. [CrossRef]

Disclaimer/Publisher's Note: The statements, opinions and data contained in all publications are solely those of the individual author(s) and contributor(s) and not of MDPI and/or the editor(s). MDPI and/or the editor(s) disclaim responsibility for any injury to people or property resulting from any ideas, methods, instructions or products referred to in the content.

Article

Fault Detection of Multi-Wheeled Robot Consensus Based on EKF

Afrah Jouili [1], Boumedyen Boussaid [1,2,3,*], Ahmed Zouinkhi [1] and M. N. Abdelkrim [1]

[1] MACS Research Laboratory, Engineering School of Gabes, University of Gabes, Zrig 6029, Tunisia; jouili.afrah.enig@gmail.com (A.J.); ahmed.zouinkhi@enig.rnu.tn (A.Z.); naceur.abdelkrim@enig.rnu.tn (M.N.A.)

[2] Quartz Lab, 95014 Cergy-Pontoise, France

[3] ESEO, Western Higher School of Electronics, Vélizy-Villacoublay, 78140 Paris, France

* Correspondence: boumedyen.boussaid@eseo.fr

Abstract: Synchronizing a network of robots in consensus is an important task for cooperative work. Detecting faults in a network of robots in consensus is a much more important task. In considering a formation of Wheeled Mobile Robots (WMRs) in a master–slave architecture modeled by graph theory, the main objective of this study was to detect and isolate a fault that appears on a robot of this formation in order to remove it from the formation and continue the execution of the assigned task. In this context, we exploit the extended Kalman filter (EKF) to estimate the state of each robot, generate a residual, and deduce whether a fault exists. The implementation of this technique was proven using a Matlab simulator.

Keywords: wheeled mobile robot network; consensus; graph theory; master–slave system; fault detection and isolation (FDI); extended Kalman filter

Citation: Jouili, A.; Boussaid, B.; Zouinkhi, A.; Abdelkrim, M.N. Fault Detection of Multi-Wheeled Robot Consensus Based on EKF. *Actuators* **2024**, *13*, 253. https://doi.org/10.3390/act13070253

Academic Editor: He Chen

Received: 6 May 2024
Revised: 17 June 2024
Accepted: 24 June 2024
Published: 1 July 2024

Copyright: © 2024 by the authors. Licensee MDPI, Basel, Switzerland. This article is an open access article distributed under the terms and conditions of the Creative Commons Attribution (CC BY) license (https://creativecommons.org/licenses/by/4.0/).

1. Introduction

In recent years, there has been a significant shift in focus toward cooperative multi-agent control (MAS) within the realm of control research. This shift is driven by its rapid advancement across various domains, including several robot systems, intelligent transportation, and numerous industrial setups [1–5]. The primary benefit of MAS lies in its ability to enhance system functionality, often overcoming challenges that may prove difficult for humans or even entirely superseding human involvement in executing tasks that are repetitive, hazardous, or beyond human capability [6]. Consequently, the importance of fault detection and isolation has escalated, ensuring safety and maintaining quality standards. Given the inevitability of faults in embedded systems such as wheeled mobile robots (WMRs), it is imperative to ensure that they are promptly identified and addressed. To facilitate and guarantee communication between robots, graph theory is used, which plays a crucial role in modeling the interactions and communication between multiple robots. In representing the robots as nodes and their communication links as edges, graph theory provides a framework for analyzing and designing consensus algorithms. This theoretical foundation is critical in ensuring that all robots in the network can agree on a common set of parameters or states, despite potential faults or communication delays.

Within computer networking, the master–slave model serves as a communication protocol wherein a designated device or process (referred to as the master) governs one or more other devices or processes (the slaves). In a master–slave relationship, the master is often the transmitter of control to the slave.

Master–slave systems often provide data security, improved consistency, and robust fault tolerance. However, their significant drawback arises in the event of master failure, which can lead to the entire system's failure, causing process malfunctions and subsequent degradation in performance [7]. Hence, the necessity of implementing process monitoring and fault detection systems becomes paramount, especially in applications like wheeled robots. This prompts exploration into fault diagnosis techniques to effectively address this challenge. Fault detection and isolation (FDI) has garnered substantial attention in both academic and industrial spheres, as evidenced by recent publications, for example, [8]. According to [9], this methodology is typically categorized into three overarching groups: hardware redundancy methods, model-based methods, and signal analysis methods. There are two other main forks: quantitative approaches and qualitative approaches. In the literature, considerable attention has been directed toward the observer-based technique. For instance, in [10], a method utilizing Kalman filter identification was utilized for the detection and isolation of sensor faults within mobile robotic systems. Also, in [11], the authors presented the integration of a bank of Kalman filters with an expert system for the detection and isolation of sensor faults in mobile robots. Additionally, [12] introduced a Kalman filter designed for joint state prediction and unknown input estimation within linear stochastic discrete-time systems subject to intermittent unknown inputs in measurements. FDI in nonlinear systems has increased significantly in recent years because most of the systems we encounter in practice are nonlinear. For example, in [13], the FDI system is based on a single-model EKF filter that generates residuals as soon as the behavior of the aircraft deviates from the expected trajectory. Also, refs. [14,15] directed their focus toward fault detection in wheeled mobile robots utilizing an EKF filter. The fault detection process typically involves two primary steps: residual generation and subsequent residual evaluation. In our contribution, we will emphasize the observer-based approach, particularly the utilization of the extended Kalman filter.

The main contribution of this article is to detect and isolate faults in a network of robots synchronized to carry out a common task in a master–slave configuration. Robot synchronization is based on graph theory, and fault detection is implemented via the Extented Kalman Filter approach. A new procedure is established to isolate the defective robot from a formation of robots, the majority of which are affected by a defect appearing in one robot.

To present our investigation on fault detection and isolation within a nonlinear system, specifically a wheeled mobile robot, utilizing the extended Kalman filter, this paper is structured as follows. The first section outlines the modeling of a unicycle mobile robot. In the subsequent section, we delve into the extended Kalman filter (EKF) as a corrective predictor technique. This section covers two primary tasks, encompassing prediction and correction phases, alongside the fault diagnosis steps, which include residual generation, evaluation, and decision-making processes. Following this, simulations are presented to illustrate the exploitation of the EKF to detect and generate residues for a faulty master–slave system. Finally, the last section concludes the paper.

2. Multi-Wheeled Robot Consensus

In this section, we delve into the consensus algorithms for multi-wheeled robots, utilizing graph theory to model and analyze the interactions between robots. By employing graph-theoretic concepts, we can design robust consensus protocols that ensure that all robots achieve a unified state.

2.1. Graph Theory

2.1.1. Notation

We denote the set of real numbers as $\mathbb{R}$. The notation $\mathbb{R}^n$ represents the real vector space with n dimensions, while $\mathbb{R}^{n \times n}$ denotes the set of $n \times n$ matrices. A diagonal matrix of size $n \times n$, with diagonal elements $q_1, q_2, ..., q_n$, is denoted as $diag(q_1, q_2, ..., q_n)$. The identity matrix is represented by $I \in \mathbb{R}^{n \times n}$. The Kronecker product of matrices is denoted by $\otimes$. Additionally, we utilize $x = (x_1, ..., x_n)^T$ to signify the vector in $\mathbb{R}^n$. The T in the exponent means transposition.

2.1.2. Graph Theory

Understanding the fundamental concepts of graph theory is indispensable for exploring the dynamics of multi-agent systems [16]. Consider a multi-agent system comprising n mobile robots interconnected through a communication network. The interaction model among agents can be represented by delineating the communication topology in the form of a graph.

Consider $\mathcal{G}_t = (\Lambda, Y)$ an oriented graph, where $\Lambda = [\Lambda_1, \Lambda_2, ..., \Lambda_n]$ represents the set of nodes. Each node i represents the ith agent, and

$$Y = (i,j) \in \Lambda \times \Lambda, j \in N_i$$

represents the collection of edges linking the nodes, where N_i denotes the set of node indexes connected to the ith node. Here, $(i;j)$ signifies a connection between nodes i and j, with j being in N_i, although i may not be in N_j. Therefore, $(i;j)$ and $(j;i)$ do not obligatory denote the same edge. These node connections are consolidated into a matrix $M_d = [a_{ij}]$, known as the adjacency matrix or connection matrix, with dimensions $N * N$. The elements within the matrix indicate whether pairs of vertices are adjacent in the graph or not.

$$a_{kj} = \begin{cases} 1 & \text{if k is connected to j} \\ 0 & \text{if not.} \end{cases}$$

The degree matrix Q is a diagonal matrix of size $N \times N$, providing information regarding the number of edges connected to each node.

$$Q_{ij} = \begin{cases} degree(\Delta_i) & \text{if } i = j \\ 0 & \text{if not.} \end{cases}$$

The Laplacian matrix $H = [l_{ij}]$ is represented by

$$H = Q - M$$

The components within this matrix are as follows::

$$h_{kj} = \begin{cases} -a_{kj} & \text{if } k \neq j \\ \sum_{j=1}^{N} a_{kj} & \text{if } k = j. \end{cases}$$

2.1.3. Closed-Loop System

Consider a collection of $N = 3$ agents of significant scale, wherein the conduct of each agent is delineated by a nonlinear controlled model.

Figure 1 shows the master–slave configuration considered in this study, and the matrices are as follows:

$$A = \begin{bmatrix} 0 & 0 & 0 \\ 1 & 0 & 0 \\ 0 & 1 & 0 \end{bmatrix}, D = \begin{bmatrix} 0 & 0 & 0 \\ 0 & 1 & 0 \\ 0 & 0 & 1 \end{bmatrix} \text{ and } L = \begin{bmatrix} 0 & 0 & 0 \\ 1 & -1 & 0 \\ 0 & 1 & -1 \end{bmatrix}$$

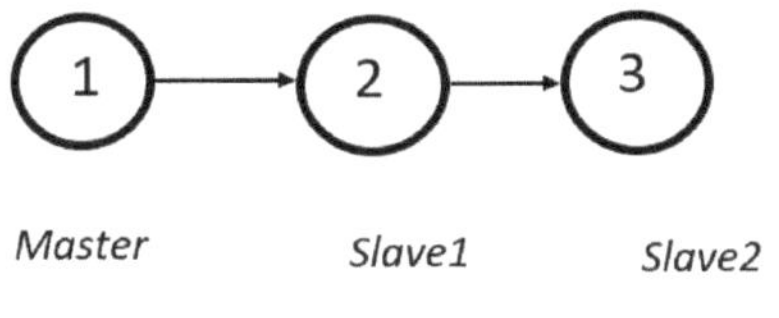

Figure 1. Directed graph.

Remember, the protocol can be written as follows:

$$u_i = -P(X_i)(\sum_{j=1}^{N} a_{ij}(X_i - X_j)) \tag{1}$$

where $P(X_i) \in \Re^{n \times m}$, and a_{ij} are the adjacent elements linked to $\mathcal{G}$.

2.2. Wheeled Mobile Robot Modeling

A unicycle is a robot driven by two independent wheels. It moves in a planar area referenced by an inertial reference $(0, \vec{x}, \vec{y})$. The kinematic model is illustrated in Figure 2.

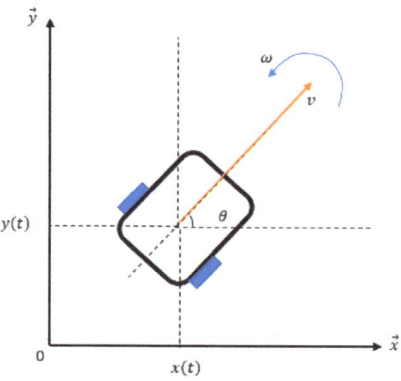

Figure 2. Geometry of a unicycle robot.

The continuous state-space representation is given as follows:

$$\begin{cases} \dot{x}(t) = v \cos(\theta(t)) \\ \dot{y}(t) = v \sin(\theta(t)) \\ \dot{\theta}(t) = \omega(t) \\ \dot{v}(t) = a(t) \end{cases} \tag{2}$$

where x and y denote the position along the x-axis and the y-axis, respectively; $v \in \Re$ represents the linear speed associated with the center of mass; θ is the orientation angle; ω is the angular speed of the center of mass; a is the acceleration; and u represents the input vector $u = \begin{bmatrix} \omega & a \end{bmatrix}^T$. Then, the generalized robot model is given by

$$\begin{cases} \dot{X}(t) = f(X, u, t) \\ Y(t) = CX(t) \end{cases} \tag{3}$$

where $X = \begin{bmatrix} x & y & \theta & v \end{bmatrix}^T$ is the state vector, and $C = \begin{bmatrix} 1 & 0 & 0 & 0 \\ 0 & 1 & 0 & 0 \end{bmatrix}$ is the observation matrix that provides the system its observability condition.

We studied a sinusoidal trajectory, so we developed a regulator that allows us to control the robot.

$$\begin{cases} x_d(t) = t \\ y_d(t) = \sin(t) \end{cases} \tag{4}$$

For this, we used the method of linearizing looping. The command will be, in the end, as follows:

$$\begin{pmatrix} u_1 \\ u_2 \end{pmatrix} = \begin{pmatrix} -v\sin(\theta) & \cos(\theta) \\ v\cos(\theta) & \sin(\theta) \end{pmatrix}^{-1} \begin{pmatrix} (x_d - x) + 2(\dot{x} - v\cos(\theta)) + \ddot{x}_d \\ (y_d - y) + 2(\dot{y} - v\sin(\theta)) + \ddot{y}_d \end{pmatrix} \quad (5)$$

If we define the error vector,

$$e = (e_x, e_y)$$

the dynamics of the error are written as follows:

$$\begin{pmatrix} e_x + 2\dot{e}_x + \ddot{e}_x \\ e_y + 2\dot{e}_y + \ddot{e}_y \end{pmatrix} = \begin{pmatrix} 0 \\ 0 \end{pmatrix}$$

which is stable and converges rapidly to 0. According to Euler's approximation, we discretize state model (1) using a sampling time T_e as follows:

$$\begin{cases} x_{k+1} = x_k + T_e v_k \cos(\theta_k) \\ y_{k+1} = y_k + T_e v_k \sin(\theta_k) \\ \theta_{k+1} = \theta_k + T_e \omega_k \\ v_{k+1} = v_k + T_e a_k \end{cases} \quad (6)$$

2.3. Simulations

NOTATION:

- m : The master robot,
- s_1: The first slave robot,
- s_2: The second slave robot,
- $X_d = \begin{bmatrix} x_d & y_d & \theta_d & v_d \end{bmatrix}^T$: State vector of the desired trajectory,
- $X_m = \begin{bmatrix} x_m & y_m & \theta_m & v_m \end{bmatrix}^T$: State vector of the master robot,
- $X_{s1} = \begin{bmatrix} x_{s1} & y_{s2} & \theta_{s1} & v_{s1} \end{bmatrix}^T$: State vector of the first slave robot,
- $X_{s2} = \begin{bmatrix} x_{s2} & y_{s2} & \theta_{s2} & v_{s2} \end{bmatrix}^T$: State vector of the second slave robot,
- $u_m = \begin{bmatrix} u_{ml} & u_{mr} \end{bmatrix}^T$: The left and right controllers of the master robot,
- $u_{s1} = \begin{bmatrix} u_{s1l} & u_{s1r} \end{bmatrix}^T$: The left and right controllers of the first slave robot,
- $u_{s2} = \begin{bmatrix} u_{s2l} & u_{s2r} \end{bmatrix}^T$: The left and right controllers of the second slave robot.

Consider the desired trajectory in Figure 3.

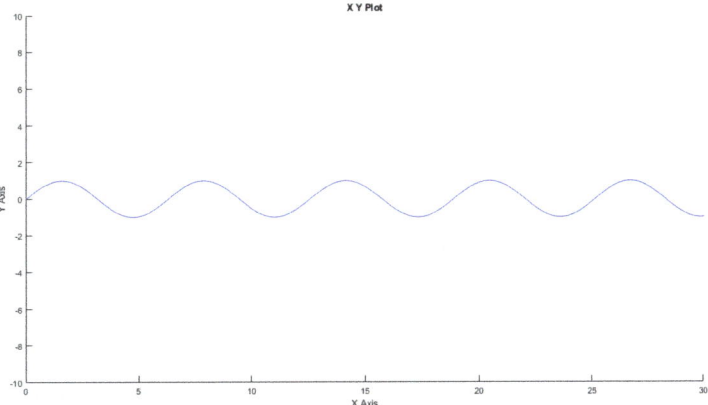

Figure 3. The desired trajectory.

We observe that the robot master accurately follows this trajectory, which is also replicated by the two robot slaves, as depicted in Figures 4–6.

We get the responses of the master and the slaves shown in Figures 7 and 8, also the control is shown in Figures 9–11.

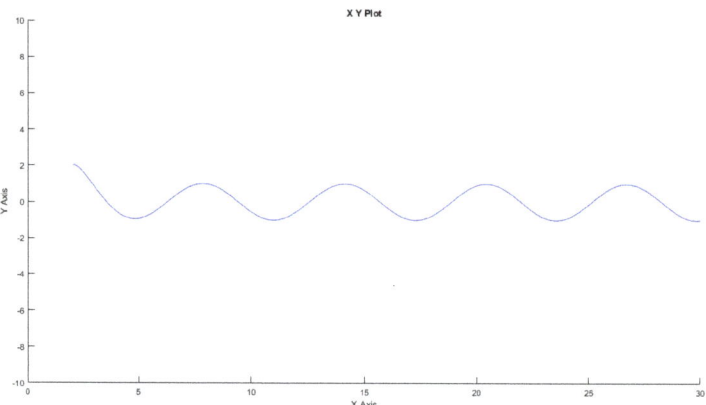

Figure 4. The robot master's trajectory.

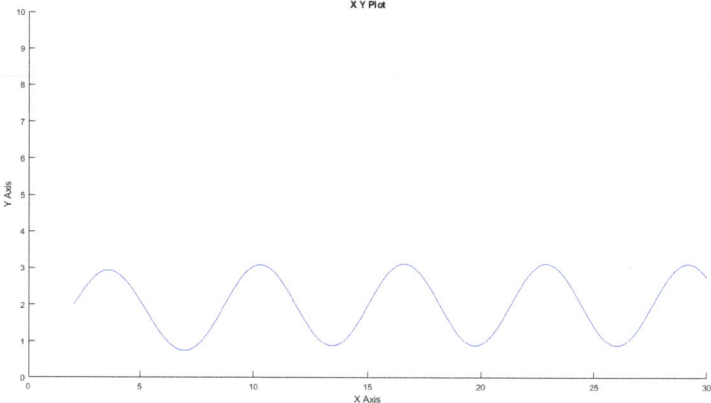

Figure 5. The first robot slave's trajectory.

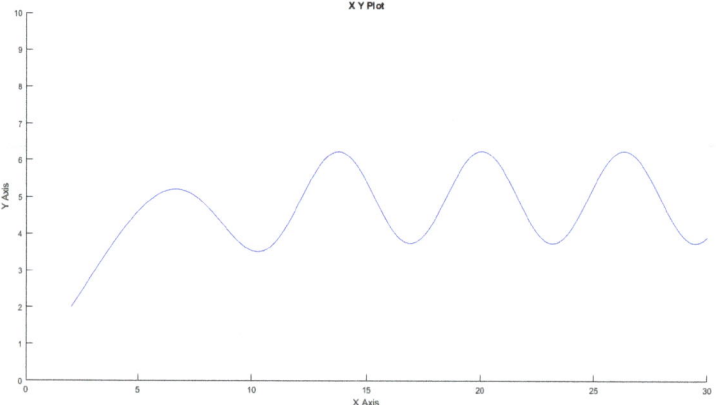

Figure 6. The second robot slave's trajectory.

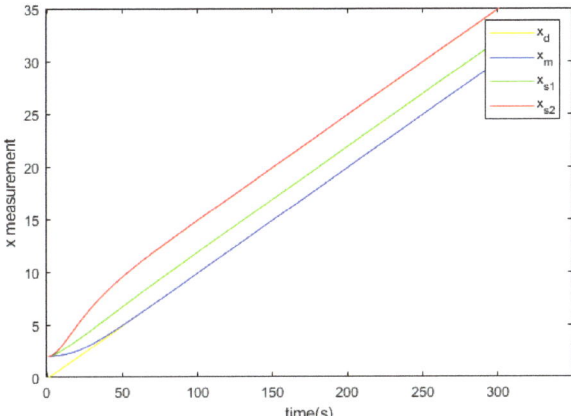

Figure 7. The x coordinate of the desired trajectory of the master and the slaves.

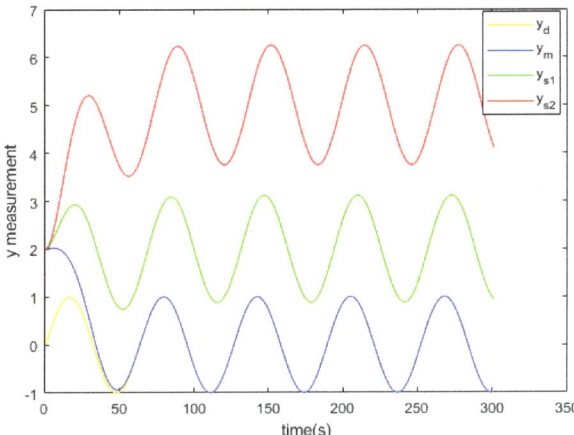

Figure 8. The y coordinate of the desired trajectory of the master and the slaves.

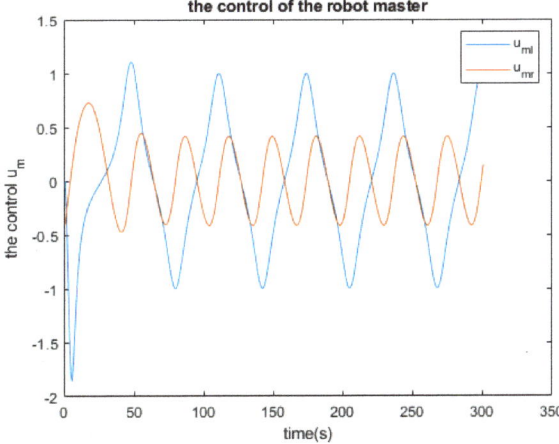

Figure 9. The control of the robot master.

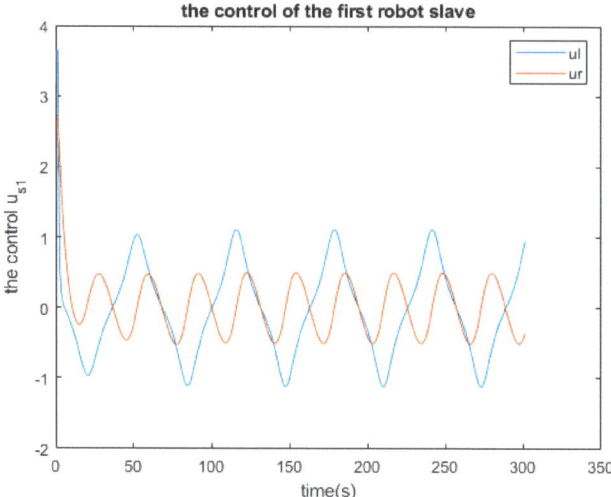

Figure 10. The control of the first robot slave.

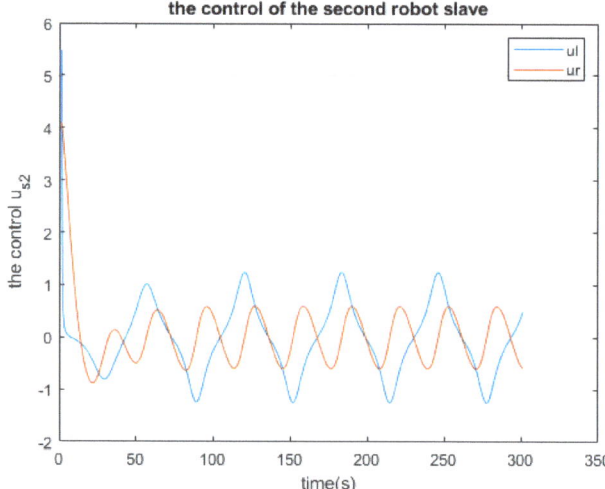

Figure 11. The control of the second robot slave.

3. Fault Detection Based on the Extended Kalman Filter

3.1. Extended Kalman Filter and Fault Diagnosis

3.1.1. The Extended Kalman Filter (EKF)

In fact, fault detection means identifying and diagnosing malfunctions within a system. This involves recognizing deviations from the system's normal or expected behavior, which could be indicative of faults such as sensor errors, actuator failures, process disturbances, or other issues affecting the system's performance. Fault detection typically involves monitoring the system's outputs and comparing them to the predicted values based on a model of the system's normal operation. The extended Kalman filter (EKF) plays a crucial role in this process by providing a dynamic model that can estimate the system's state variables and predict future outputs. Any significant discrepancies between the predicted and actual outputs can indicate the presence of a fault.

Within the extended Kalman filter framework, the nonlinearities inherent in the system dynamics are approximated using a linearized rendition of the nonlinear model based on the most recent state estimate. Let us consider the system characterized by the following state equations:

$$\begin{cases} X(k+1) = f(X, u, k) + \rho(k) \\ Y(k+1) = h(X, u, k) + \gamma(k) \end{cases} \quad (7)$$

$\rho(k)$ and $\gamma(k)$ denote the process and observation noises, which are assumed to have a zero mean with covariances Ψ_{ρ_k} and Ψ_{γ_k}, respectively. Function f, defined as nonlinear according to Equation (7), is employed to compute the predicted state from the previous estimate, as is function h for predicting the measurement from the predicted state. However, due to the nonlinearity of f and h, a Jacobian matrix of partial derivatives is computed, since these functions cannot be directly applied to the covariance.

Hypothesis:

As stated in [17], when the extended Kalman filter (EKF) is applied to a system and placed in a canonical observation form, it gains the following properties associated with global convergence:

- The pair of matrices (A, C) is detectable, which means that there is no unstable mode or no observability in the system.
- The signals $\rho(k)$ and $\gamma(k)$ are central Gaussian white noises. The Density Power Spectral (DSP) covariances Ψ_{ρ_k} and Ψ_{γ_k} mean

$$\begin{cases} \mathbb{E}[\gamma(i)\gamma(j)^T] = \Psi_{\rho_k}, & \text{if } i = j \\ \mathbb{E}[\rho(i)\rho(i)^T] = \Psi_{\gamma_k}, & \text{if } i = j \\ \mathbb{E}[\gamma(i)\rho(j)^T] = 0, & \forall i, j. \end{cases} \quad (8)$$

where $\mathbb{E}[..]$ represents the mathematical expectation.

The latter equation illustrates the stochastic independence of the noises ρ and γ. This assumption is introduced to simplify the subsequent calculations but is not obligatory.

Correction:

- Updated state estimate:
$$\hat{x}_{k/k} = \hat{x}_{k/k-1} + \mathbb{K}_k \hat{y}_k \quad (9)$$

- Updated covariance estimate:
$$\Psi_{k/k} = (I - \mathbb{K}_k C_k) \Psi_{k/k-1} \quad (10)$$

- Measurement residual:
$$\tilde{y}_k = y_k - C_k \hat{x}_{k/k-1} \quad (11)$$

- Innovation covariance:
$$\mathbb{Z}_k = C_k \Psi_{k/k-1} C_k^T + \Psi_{\gamma_k} \quad (12)$$

- Near-optimal Kalman gain:
$$\mathbb{K}_k = \Psi_{k/k-1} C_k^T \mathbb{Z}_k^{-1} \quad (13)$$

Prediction:

- Predicted state estimate:
$$\hat{x}_{k+1/k} = A_k \hat{x}_{k/k} + u_k \quad (14)$$

- Predicted covariance:
$$\Psi_{k+1/k} = A_k \Psi_{k/k} A_k^T + \Psi_{\rho_k} \quad (15)$$

where A_k is the linearized matrix of the function f, which is written as follows:

$$A_k = \frac{\delta f}{\delta X}|_{\hat{x}_{k/k}}$$

Remark 1. *Occasionally, numerical issues can cause the covariance of innovation Z_k to lose its positive definite nature. In such instances, it is advisable to substitute the equation for covariance estimation with*

$$\Psi_{k/k} = \sqrt{(I - K_k C_k)\Psi_{k/k-1}\Psi_{k/k-1}^T(I - K_k C_k)^T}$$

The replacement will always ensure a positive, definite outcome, even if the matrix $\Psi_{k/k-1}$ is not. Consequently, the Kalman equations will exhibit greater stability, as any minor deviations in the positive definite nature of the covariance matrices will be corrected in the subsequent iteration. According to (14),

$$A_k = \begin{bmatrix} 1 & 0 & -T_e v \sin(\theta) & \cos(\theta) \\ 0 & 1 & T_e v \cos(\theta) & \sin(\theta) \\ 0 & 0 & 1 & 0 \\ 0 & 0 & 0 & 1 \end{bmatrix}$$

Suppose that the signals γ_k and ρ_k are Gaussian white noise with a unit covariance matrix; that is to say,

$$\Psi_\gamma = \begin{bmatrix} 1 & 0 \\ 0 & 1 \end{bmatrix} \quad \Psi_\rho = \begin{bmatrix} 1 & 0 & 0 & 0 \\ 0 & 1 & 0 & 0 \\ 0 & 0 & 1 & 0 \\ 0 & 0 & 0 & 1 \end{bmatrix}$$

Also, we take the sample time $Te = 0.01$.

3.1.2. Fault Diagnosis Steps

Figure 12 describes the fault detection procedure in the master–salve robot configuration.

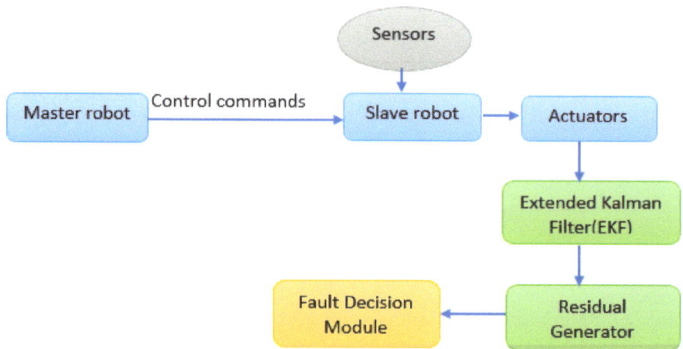

Figure 12. A block diagram illustrating fault detection in a master–slave robot system.

This configuration has the following components:

- **Master Robot**: Sends control commands to the Slave Robot.
- **Slave Robot**: Execute actions based on the received commands.
- **Sensors**: Collect data from the environment and the robots' states, providing inputs for the EKF.
- **Actuators**: Perform the necessary actions in response to control signals.
- **Extended Kalman Filter (EKF)**: Estimates the system's states based on sensor data and predicted models.

- **Residual Generator**: Calculates residuals by comparing the EKF's state estimates with actual measurements.
- **Fault Decision Module**: Analyzes the residuals to detect any discrepancies indicating potential faults, triggering appropriate responses if faults are detected.

Generally, a fault diagnosis system follows three steps: a residual generation, a residue evaluation, and a decision logic [18].

Step 1: Residual generation

The residual generator generates a fault indicator vector, or residual vector, denoted r, based on the measurements of the input and output variables of the system. The nominal value of the residue, excluding transient effects, is theoretically equal to zero under the normal operating conditions of the monitored system. When a fault appears, this residual moves away from zero depending on the fault.

Step 2: Residue evaluation

The residual evaluation module consists of measuring the residue and determining whether the system is functioning properly or not using specific algorithms and methods [18]. Detecting a fault typically involves comparing the residues with a predetermined detection threshold t_{th}. This threshold represents a crucial aspect of residual-based FDI methods, defined as the boundary value for the deviation of a residue from zero. Thus, the process of fault detection proceeds as follows:

$$r = \begin{cases} 1 & if, r > |t_{th}| \\ 0 & if, r < |t_{th}|. \end{cases}$$

Excessively high thresholds may result in failing to detect a fault (missed alarm), while overly low thresholds may lead to false alarms, detecting faults in healthy conditions [19]. There exist two categories of thresholds [8]: the first being a constant threshold, and the second, an adaptive one. The adaptive threshold is employed to accommodate the inevitable parameter uncertainty, disturbances, and noise encountered in practical applications.

Step 3: Decision

This step involves examining the outcome of evaluating a set of residues, and, based on the pattern of activated and non-activated tests, it generates a determination regarding the faulty component within the monitored system.

3.2. Closed-Loop System

The closed-loop system was used in Simulink/Matlab and is shown in the functional diagram in Figure 13.

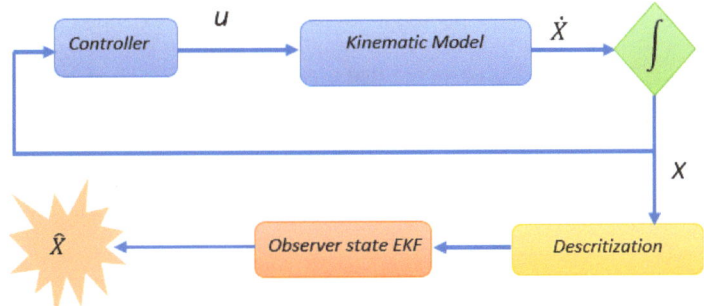

Figure 13. Unicycle mobile robot closed loop.

The robot was controlled in order to follow a desired trajectory given by $P_d = [x_d, y_d]^T$, as shown in Figure 14.

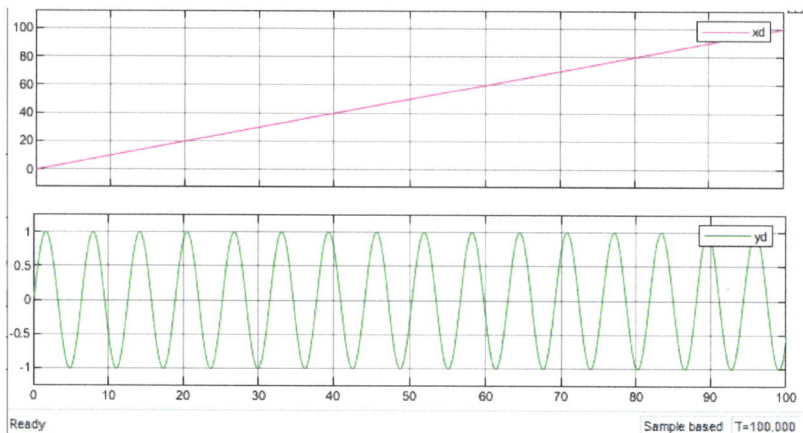

Figure 14. Desired trajectories x_d and y_d.

3.3. Simulation Results

We noticed that the robot master follows the desired trajectory (see Figure 15).

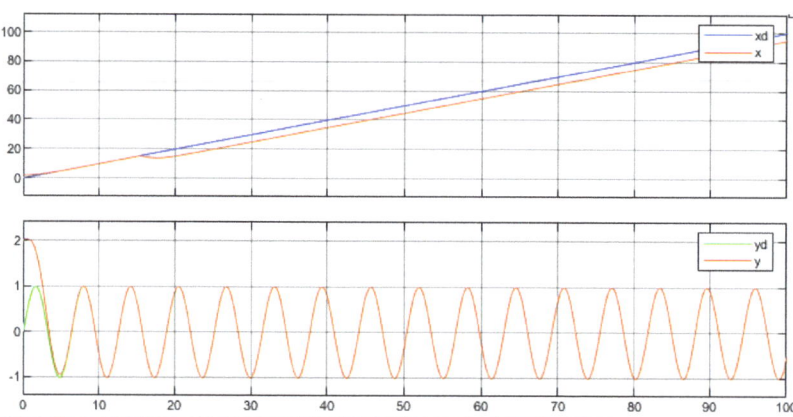

Figure 15. Coordinates x and y of the desired trajectory and the mobile robot.

The fault scenario addressed involves an offset applied to the x measurement at the 15 s mark. Subsequently, we applied our Extended Kalman Filter (EKF), and the resulting robot trajectory is illustrated in Figure 16.

It is observed that there is a slight deviation attributable to the applied fault, yet the robot continues to track the same trajectory consistently.

Furthermore, upon introducing a fault (considered an intermittent fault, such as a switch) to a parameter at time t = 15 s, a minor peak is observed (see Figure 17), after which the robot promptly returns to its intended trajectory. Hence, it can be concluded that the Extended Kalman Filter effectively identifies the fault introduced into the system.

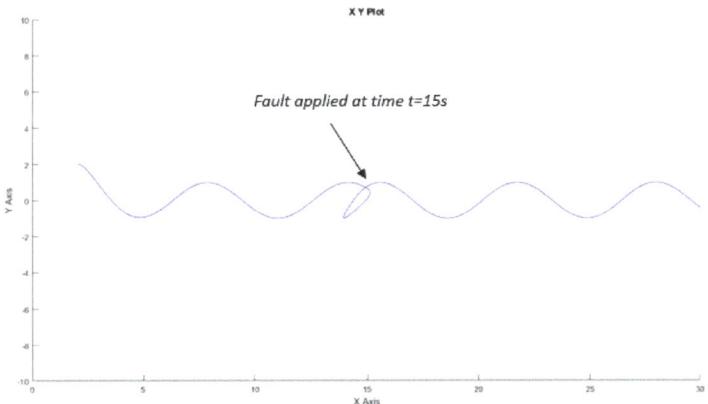

Figure 16. Trajectory of the mobile robot when fault applied at t = 15 s.

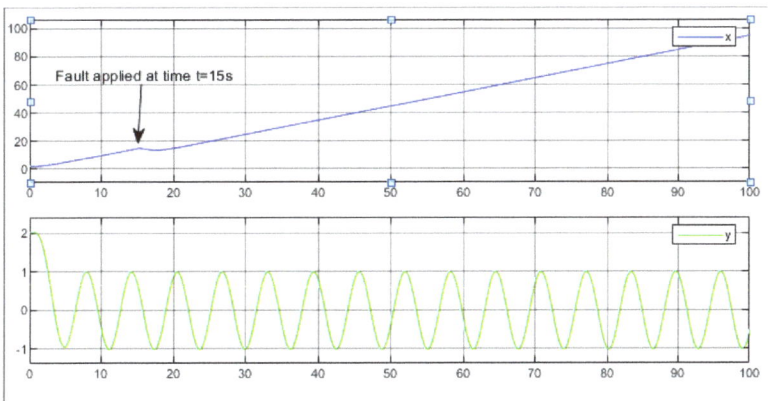

Figure 17. Coordinates *x* and *y* of the mobile robot when fault applied at t = 15 s.

4. Fault Detection in Multi-Robot System

In the preceding section, the system (consisting of a robot master and slaves) executes its trajectory within an ideal environment. However, real-world environments lack such ideal conditions. Thus, we aimed to introduce a fault and observe how the system responds to the challenge. Indeed, the multi-robot system under fault is described in the following figure (see Figure 18). It was observed that the robot master adheres to its trajectory, as depicted in Figure 3.

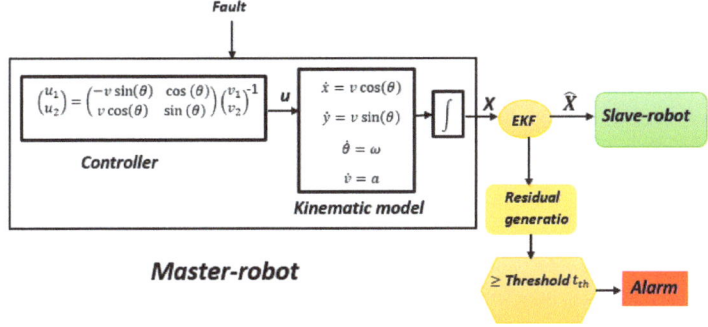

Figure 18. Multi-robot system closed loop.

Subsequently, we induced an offset fault in the master robot's x measurement at 15 s, as illustrated in Figure 19.

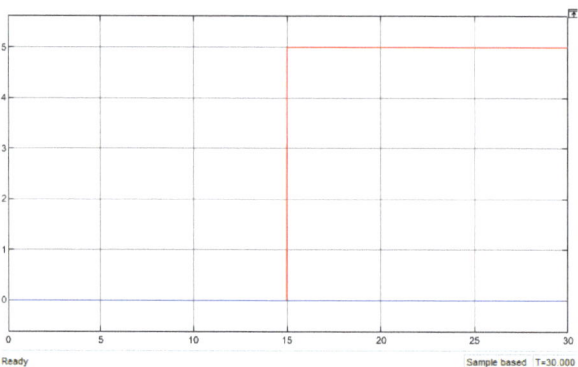

Figure 19. The fault applied at time t = 15 s.

The responses of both the master and the slaves are depicted in the subsequent figures (Figures 20–22). It is observed that the trajectory of the robots experiences a slight disturbance at a specific time, yet it promptly returns to its intended path without deviating from the desired trajectory.

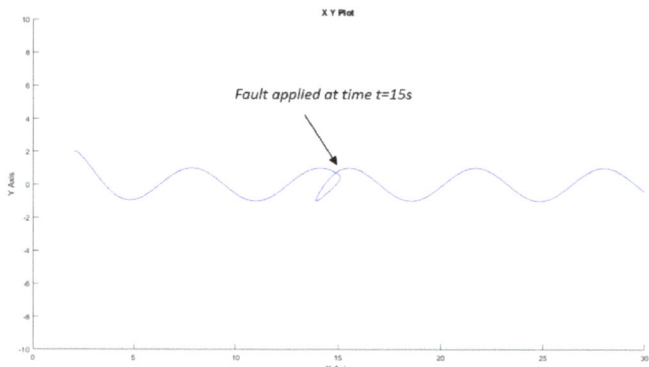

Figure 20. The trajectory of the robot master with fault applied at time t = 15 s.

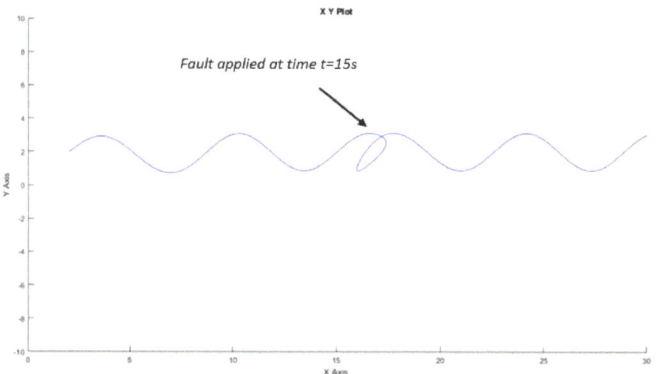

Figure 21. The trajectory of the first robot slave with fault applied at time t = 15 s.

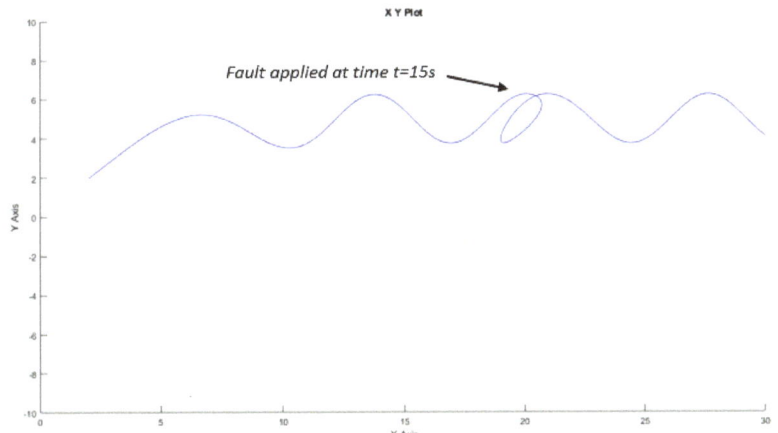

Figure 22. The trajectory of the second robot slave with fault applied at time t = 15 s.

Figures 23 and 24 show the x and y measurements of the desired trajectory, the master and the slaves, respectively.

Furthermore, the simulations of the controllers for both the master and the slaves are illustrated in Figures 25–27, respectively. It is observed that the robot master adheres to the desired trajectory until time t = 15 s, at which point it deviates slightly. However, after a few seconds, it resumes following the trajectory initially set.

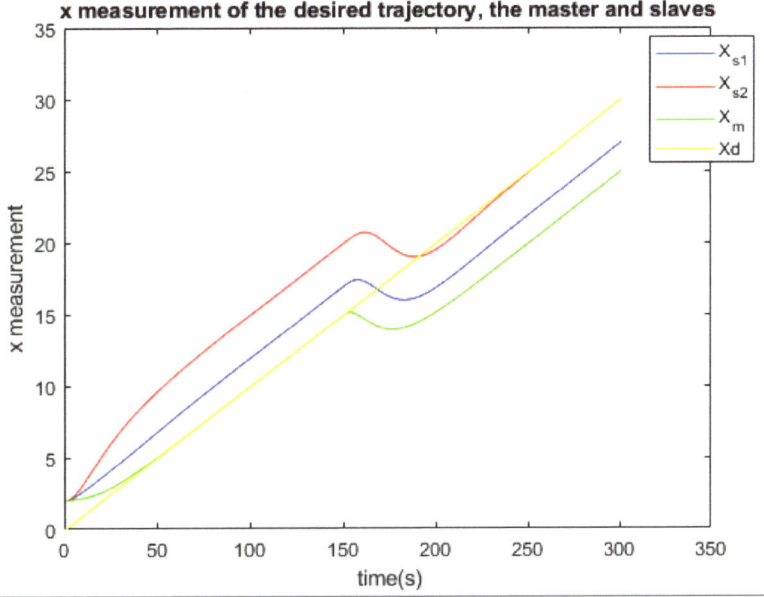

Figure 23. The coordinate x of the desired trajectory of the master and slaves.

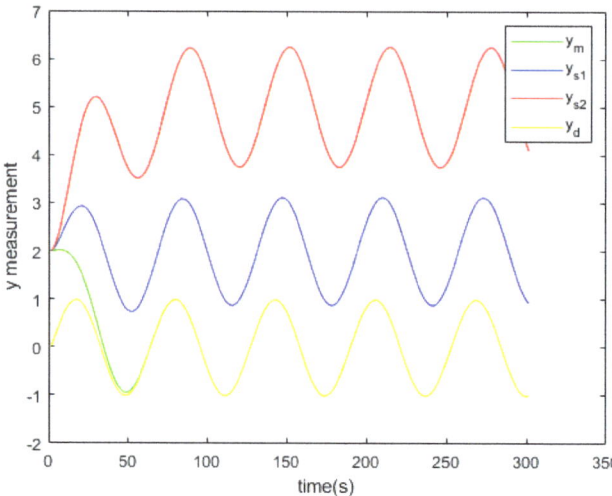

Figure 24. The coordinate y of the desired trajectory of the master and slaves.

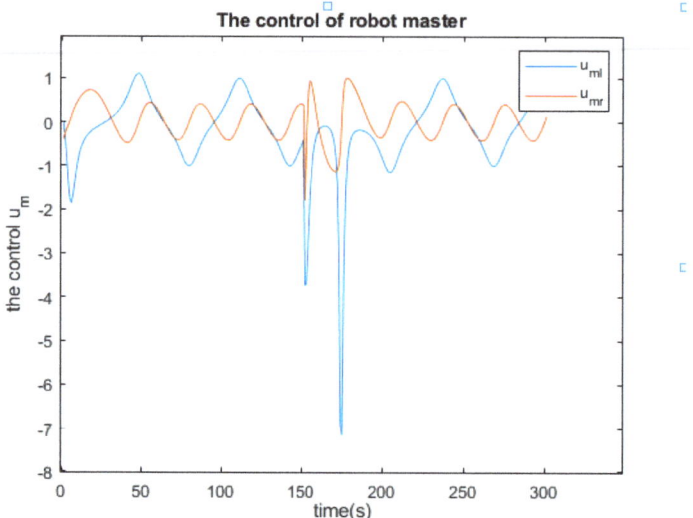

Figure 25. The control of the robot master.

Meanwhile, our attention is directed toward the behaviors of the robot slaves, which entail two cases for investigation: firstly, when a robot slave operates without fault, and secondly, when a robot slave is subjected to a fault injection. The simulations depicted in Figures 21 and 22 revealed the impact on the functionality of the slave robot. Notably, a minor peak is observed in the slave responses (refer to Figures 23 and 24), following which it swiftly resumes its trajectory.

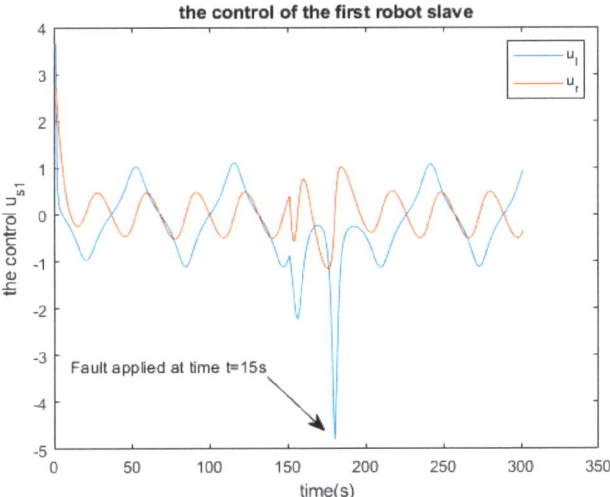

Figure 26. The control of the first robot slave.

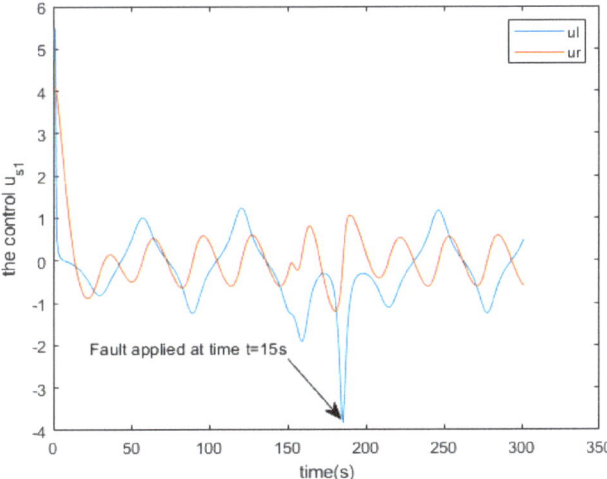

Figure 27. The control of the second robot slave.

Fault Diagnosis

To utilize the extended Kalman filter (EKF) for fault detection in a wheeled mobile robot, the following procedural steps are advised:

1- System Modeling: Develop a comprehensive model that characterizes the robot's behavior under both normal operating conditions and various fault scenarios.

2- State Estimation: Implement the EKF to estimate the current state of the robot utilizing sensor measurements.

3- Fault Models: Define specific fault models detailing how each fault impacts the robot's state and sensor measurements.

4- Residual Calculation: Calculate residuals by computing the disparity between actual measurements and estimated measurements, based on the fault models delineated in the preceding step.

5- Fault Detection: Employ statistical tests or threshold-based techniques on the residuals to identify the occurrence of faults.

6- Fault Diagnosis: Initiate the fault diagnosis phase subsequent to the detection of faults.

For establishing thresholds, the system should first undergo simulations under fault-free conditions, followed by simulations with faults. The fault-free simulations aid in determining thresholds t_{th} through the application of the three-sigma method:

$$\begin{cases} -t_{th} = -3\sigma \\ +t_{th} = +3\sigma \end{cases}$$

where σ denotes the standard deviation of residual r.

Note that σ reflects the EKF accuracy and is calculated based on the EKF error:

$$\sigma = \sqrt{\frac{1}{n} \sum_{i=1}^{n} (e_i^2)} \tag{16}$$

As mentioned in the previous section, the fault is said to be detected when the residual exceeds the threshold more than $|t_{th}|$.

Based on the EKF in our case, we calculated the difference between the values of the robot's state vector and the values of the estimated state vector X_e. So four residuals are generated for each robot:

$$\begin{cases} r_x = x - x_e \\ r_y = y - y_e \\ r_\theta = \theta - \theta_e \\ r_v = v - v_e \end{cases}$$

where the index e means the estimated value. Each residual is compared to the threshold.

Table 1 summarizes the signatures of the residuals under a fault applied to the robot master and the slave robots. The different residual signatures allow for fault isolation. We notice that when the fault appears in the robot master, the other two robots (slaves) are affected. On the other side, if we insert the fault in the robot slaves, the master is not affected, so we conclude that the fault affects only the following robots. Hence, we can say that the fault propagates in one direction only. As shown in the figures (Figures 28–30), a fault appears at time t = 15 s and is detected, and the detection delay does not affect the system performance.

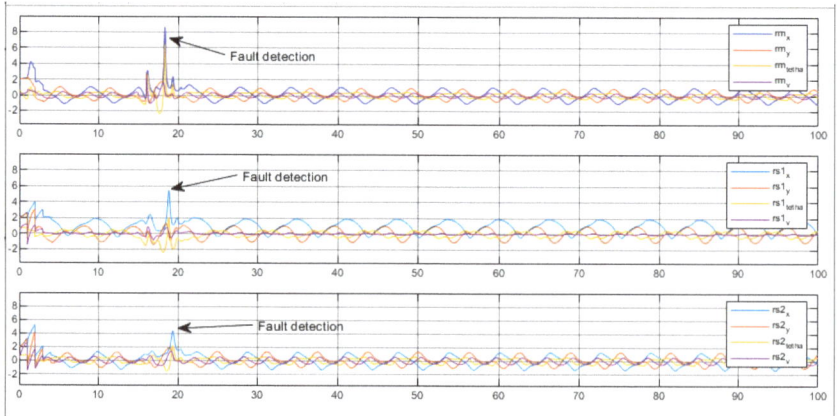

Figure 28. The residual of the three robots when a fault is applied on the master.

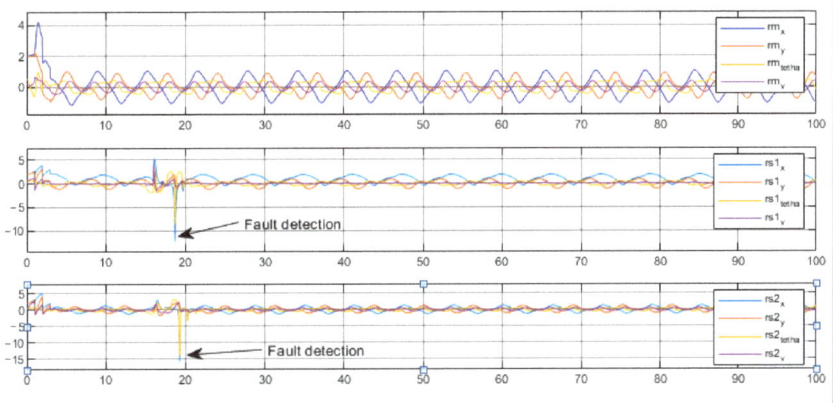

Figure 29. The residual of the three robots when a fault is applied on the first slave.

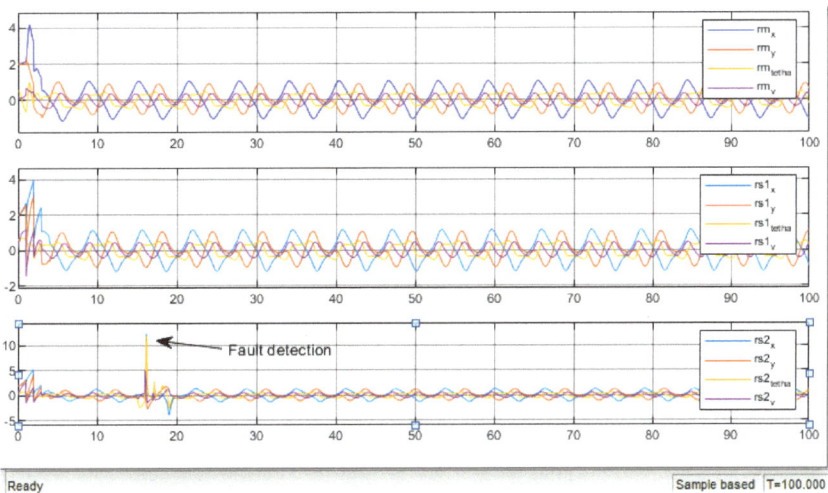

Figure 30. The residual of the three robots when the a fault is applied on the second slave.

Table 1. Residual signatures under fault.

Fault on	Residue of Master Robot	Residue of First Slave Robot	Residue of Second Slave Robot
Master robot	1	1	1
First slave robot	0	1	1
Second slave robot	0	0	1

5. Conclusions

In this paper, we consider the architecture of a network of three robots, with a master and two slaves working in consensus. The main task is assigned to the master, and the two slaves follow. Graph theory is used to model the whole system. Next, in order to continuously estimate the state of each robot, an extended Kalman filter is assigned to each robot. The residual generator compares the current state of the robot with its estimated state. It deduces whether a fault has occurred or not. We noticed that faults propagate from the master to the slaves according to the orientation of the arcs in the graph. In analyzing the fault signature table and the adjacency matrix (A), this observation allowed us to isolate faulty robots. The problem becomes more complex when the graph is not a directed graph. In this case, a temporal analysis of the moment of fault occurrence is mandatory. This is our perspective for future work.

Author Contributions: Conceptualization, A.J. and B.B.; methodology, A.J. and B.B.; software, A.Z.; validation, B.B., A.Z. and M.N.A.; formal analysis, A.J.; investigation, A.J. and B.B.; resources, A.J.; data curation, B.B.; writing—original draft preparation, A.J. and B.B.; writing—review and editing, A.J. and B.B.; visualization, B.B.; supervision, A.Z.; project administration, M.N.A.; funding acquisition A.J. All authors have read and agreed to the published version of the manuscript.

Funding: This research received no external funding.

Data Availability Statement: The data presented in this study are available on request from the corresponding author due to (specify the reason for the restriction).

Conflicts of Interest: The authors declare no conflicts of interest.

References

1. Ribeiro, A.F.A.; Lopes, A.C.C.; Ribeiro, T.A.; Pereira, N.S.S.M.; Lopes, G.T.; Ribeiro, A.F.M. Probability-Based Strategy for a Football Multi-Agent Autonomous Robot System. *Robotics* **2024**, *13*, 5. [CrossRef]
2. Guo, X.G.; Xu, W.D.; Wang, J.L.; Park, J.H. Distributed neuroadaptive fault-tolerant sliding-mode control for 2-D plane vehicular platoon systems with spacing constraints and unknown direction faults. *Automatica* **2021**, *129*, 109675. [CrossRef]
3. Ruch, C.; Gachter, J.; Hakenberg, J.; Frazzoli, E. The +1 Method: Model-Free Adaptive Repositioning Policies for Robotic Multi-Agent Systems. *IEEE Trans. Netw. Sci. Eng.* **2020**, *7*, 3171–3184. [CrossRef]
4. Engelmann, D.C.; Ferrando, A.; Panisson, A.R.; Ancona, D.; Bordini, R.H.; Mascardi, V. RV4JaCamdash;Towards Runtime Verification of Multi-Agent Systems and Robotic Applications. *Robotics* **2023**, *12*, 49. [CrossRef]
5. Jouili, A.; Boussaid, B.; Zouinkhi, A.; Abdelkrim, M. Finite time consensus for multi-tricycle systems under graph theory. In Proceedings of the 2020 20th International Conference on Sciences and Techniques of Automatic Control and Computer Engineering (STA), Sfax, Tunisia, 20–22 December 2020; pp. 24–29. [CrossRef]
6. Sayed-Mouchaweh, M.; Billaudel, P. Abrupt and Drift-Like Fault Diagnosis of Concurrent Discrete Event Systems. In Proceedings of the 2012 11th International Conference on Machine Learning and Applications, Boca Raton, FL, USA, 12–15 December 2012; Volume 2, pp. 434–439. [CrossRef]
7. Khan, A.Q. Observer-Based Fault Detection in Nonlinear Systems. Ph.D. Thesis, University of Duisburg Essen, Essen, Germany, 2010.
8. Sallem, F. Détection et Isolation de Défauts Actionneurs Basées sur un Modèle de L'organe de Commande. Ph.D. Thesis, Université Paul Sabatier—Toulouse III, Toulouse, France, 2013.
9. Mouzakitis, A. Classification of fault diagnosis methods for control systems. *Meas. Control.* **2013**, *46*, 303–308. [CrossRef]
10. Fourlas, G.K.; Karras, G.C.; Kyriakopoulos, K.J. Sensors fault diagnosis in autonomous mobile robots using observer—Based technique. In Proceedings of the 2015 International Conference on Control, Automation and Robotics, Singapore, 20–22 May 2015; pp. 49–54. [CrossRef]
11. Yutian, L.; Jungan, C. Integrated Fault Diagnosis Method of Mobile Robot. In *Theoretical and Mathematical Foundations of Computer Science*; Springer: Berlin/Heidelberg, Germany, 2011; pp. 372–379.
12. Rhouma, T.; Keller, J.Y.; Abdelkrim, M.N. A Kalman filter with intermittent observations and reconstruction of data losses. *Int. J. Appl. Math. Comput. Sci.* **2022**, *32*, 241–253. [CrossRef]
13. Ducard, G. Smac–Fdi: A Single Model Active Fault Detection and Isolation System for Unmanned Aircraft. *Int. J. Appl. Math. Comput. Sci.* **2015**, *25*, 189–201. [CrossRef]
14. Jouili, A.; Boussaid, B.; Zouinkhi, A.; Abdelkrim, M. Fault detection in wheeled mobile robot based on extended kalman filter. In Proceedings of the 2022 IEEE 21st international Ccnference on Sciences and Techniques of Automatic Control and Computer Engineering (STA), Sousse, Tunisia, 19–21 December 2022; pp. 237–242.
15. Jouili, A.; Boussaid, B.; Zouinkhi, A.; Abdelkrim, M.N. FDI based extended kalman filter for multi robot system. In Proceedings of the 2023 20th International Multi-Conference on Systems, Signals & Devices (SSD), Mahdia, Tunisia, 20–23 February 2023; pp. 675–680. [CrossRef]
16. Kia, S.S.; Van Scoy, B.; Cortes, J.; Freeman, R.A.; Lynch, K.M.; Martinez, S. Tutorial on Dynamic Average Consensus: The Problem, Its Applications, and the Algorithms. *IEEE Control Syst. Mag.* **2019**, *39*, 40–72. [CrossRef]
17. Boizot, N.; Busvelle, E.; Jean-Paul, G. An adaptive high-gain observer for nonlinear systems. *Automatica* **2010**, *46*, 1483–1488. [CrossRef]

18. Bokor, J.; Szabo, Z. Fault detection and isolation in nonlinear systems. *Annu. Rev. Control* **2009**, *33*, 113–123. [CrossRef]
19. Samia, M.; Graton, G.; El Mostafa, E.; Ouladsine, M.; Planchais, A. On fault detection and isolation applied on unicycle mobile robot sensors and actuators. In Proceedings of the 2018 7th International Conference on Systems and Control (ICSC), Valencia, Spain, 24–26 October 2018; pp. 148–153.

Disclaimer/Publisher's Note: The statements, opinions and data contained in all publications are solely those of the individual author(s) and contributor(s) and not of MDPI and/or the editor(s). MDPI and/or the editor(s) disclaim responsibility for any injury to people or property resulting from any ideas, methods, instructions or products referred to in the content.

Article

Comparative Study of Methods for Robot Control with Flexible Joints

Ranko Zotovic-Stanisic [1,*], Rodrigo Perez-Ubeda [2] and Angel Perles [3]

[1] Institute of Industrial Control Systems and Computing (ai2), Universitat Politècnica de València, 46022 Valencia, Spain
[2] Mechanical Engineering Department, Universidad de Antofagasta, Antofagasta 1200000, Chile; rodrigo.perez.ubeda@uantof.cl
[3] ITACA Institute, Universitat Politècnica de València, 46022 Valencia, Spain; aperles@upv.es
* Correspondence: rzotovic@isa.upv.es

Abstract: Robots with flexible joints are gaining importance in areas such as collaborative robots (cobots), exoskeletons, and prostheses. They are meant to directly interact with humans, and the emphasis in their construction is not on precision but rather on weight reduction and soft interaction with humans. Well-known rigid robot control strategies are not valid in this area, so new control methods have been proposed to deal with the complexity introduced by elasticity. Some of these methods are seldom used and are unknown to most of the academic community. After selecting the methods, we carried out a comprehensive comparative study of algorithms: simple gravity compensation (Sgc), the singular perturbation method (Spm), the passivity-based approach (Pba), backstepping control design (Bcd), and exact gravity cancellation (Egc). We modeled these algorithms using MATLAB and simulated them for different stiffness levels. Furthermore, their practical implementation was analyzed from the perspective of the magnitudes to be measured and the computational costs of their implementation. In conclusion, the Sgc method is a fast and affordable solution if joint stiffness is relatively high. If good performance is necessary, the Pba is the best option.

Keywords: control; robot flexible joints; backstepping; passivity

Citation: Zotovic-Stanisic, R.; Perez-Ubeda, R.; Perles, A. Comparative Study of Methods for Robot Control with Flexible Joints. *Actuators* **2024**, *13*, 299. https://doi.org/10.3390/act13080299

Academic Editor: Giulia Scalet

Received: 3 June 2024
Revised: 18 July 2024
Accepted: 3 August 2024
Published: 6 August 2024

Copyright: © 2024 by the authors. Licensee MDPI, Basel, Switzerland. This article is an open access article distributed under the terms and conditions of the Creative Commons Attribution (CC BY) license (https:// creativecommons.org/licenses/by/ 4.0/).

1. Introduction

Robots with flexible joints are becoming increasingly relevant. New types of robots are gaining importance on the market, such as collaborative robots (cobots), exoskeletons, and prostheses. They are meant to directly interact with humans. In this new generation of robots, the emphasis in their construction is not on precision (such as for rigid robot counterparts) but rather on weight reduction (collaborative robots) and/or soft interaction with humans (exoskeletons and prostheses). Thus, these new robots use more elastic mechanical transmissions.

Cobots typically have harmonic drive transmissions instead of classical gears [1] due to their light weight, high reduction ratio, and relatively good back-driveability. Wearable robotics mostly use series elastic actuators (SEAs) [2,3] for transmission. SEAs are added to some cobots to increase the compliance of their harmonic drives [4], such as those produced by the Rethink company.

Before the advent of flexible robots, most robots were rigid to achieve high precision. Controlling rigid manipulators is well covered and included in robotics textbooks [5–7]. In these cases, the best performance is obtained using inverse dynamics control methods, also called computed torque. This involves compensating all nonlinear forces that act on the robot, such as gravity, inertia, and centrifugal and Coriolis forces.

When approaching the problem of controlling flexible robots, the first idea that comes to mind is adapting the well-known inverse dynamics method. However, in a rigid robot, the actuators are directly connected to the links, compensating for external forces. In a robot with flexible links, the motor acts on an elastic element, causing its torsion, which causes the link to move. Thus, the dynamic between the actuator and the link does not directly compensate for external forces.

Many applications involve a wide range of compliances in their joints. According to [8], stiffnesses may vary from 5 to 10 kNm/rad down to 0.2 to 1 kNm/rad. This wide elasticity range complicates control considerably. In addition, stability analysis is much more difficult.

For example, oscillations may occur, possibly prohibiting many robotics tasks.

To achieve a task, the trajectory of the link ($q, \dot{q}$, etc.) must be controlled, but it is only possible to act on the motor ($\theta, \dot{\theta}$, etc.).

Another complication in flexible robots vs. rigid ones is the higher order of the system. While the former is second-order, the latter is fourth [9,10]. Thus, it may be necessary to measure and include higher-order derivatives.

Several control strategies have been proposed to deal with this wide range of elasticity. The late 1980s and early 1990s were prolific regarding contributions in this field; researchers aimed to control motor position and velocity to achieve good trajectory tracking with links.

In [11,12], some less conventional control methods, like the *singular perturbation method* (Spm) or *backstepping control design* (Bcd), were proposed. Tomei [13] introduced an extremely simple PD with the *simple gravity compensation* (Sgc) method and demonstrated its stability criteria. The authors of [14] improved the previous method, proposing *exact gravity cancellation* (Egc) while introducing less restrictive criteria with better trajectory tracking. The authors of [15,16] introduced the *passivity-based approach* (Pba) to determine the control action.

For each case, it is difficult to decide which method is appropriate and which constraints to use for its practical application, such as computational costs and expensive sensor requirements. Although some of these methods have been described in previous work [11,12,17,18], this study models and simulates a selection of methods to provide a clearer picture of the performance of each for different stiffness levels.

This study is dedicated to applying "classical" methods to control robots with flexible joints. A few recent strategies have not been included since they have several versions. Their analysis would be very extensive and has been left for future work. However, they are briefly mentioned below.

One approach is model predictive control (MPC) [19–21]. This method includes constraints such as maximum motor torques and velocities in the controller design.

Another strategy is sliding mode control [22–24]. It achieves good and robust trajectory tracking, but it may need a very fast sampling period.

Several authors have dedicated their research to robustly controlling robots with elastic joints [25–27]. This is a wide area, and there are many very different contributions.

This paper is organized as follows: the Section 2 presents the approach used to model the selected control algorithms. It then briefly describes the basis of each control algorithm and, finally, the simulation parameters. Then, Section 3 presents the output of the simulations for different stiffness levels. Next, Section 4 provides an interpretation of the simulation results, the requirements of each method for its practical implementation, and the pros and cons. Finally, Section 5 discusses the advantages and disadvantages of each controller. At the end of the article, Appendix A describes the first and second derivatives of the inertia, gravity, centrifugal, and Coriolis matrices.

2. Materials and Methods

2.1. Approaches for Modeling Robots with Flexible Joints

The dynamic model of the rigid robot is well known and can be found in textbooks [5–7]. It can be represented by the following expression:

$$\tau = M(q)\ddot{q} + C(q, \dot{q})\dot{q} + G(q) \tag{1}$$

where τ is the vector of the motor torque; q, $\dot{q}$, and $\ddot{q}$ are the vectors of the motor position, velocity, and acceleration, respectively; $M(q)$ is the inertia matrix of the robot; $C(q, \dot{q})$ is the matrix of the centrifugal and Coriolis forces; and $G(q)$ is the vector of gravity torques on the motors.

The following subsections describe the two possible ways to model the dynamics of robots with elastic joints: conventional modeling and the singularly perturbed model.

2.1.1. Conventional Elastic Modeling

The main difference between modeling a rigid robot and a flexible robot is an elastic element between the motor rotor and the link (see Figure 1).

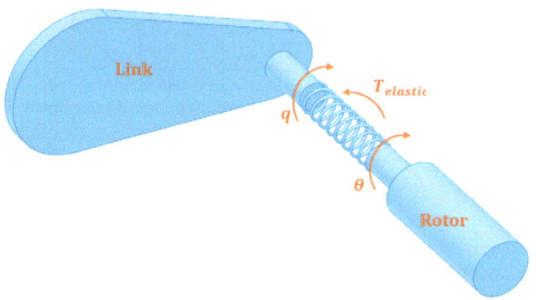

Figure 1. Schema of an elastic joint. θ is the position of the motor rotor, q is the position of the link, and $\tau_{elastic} = K(\theta - q)$ is the elastic torque. K is the stiffness of the joint in this figure.

The dynamics can be separated into two parts: the motor side and the link side. We can directly actuate the former, but we need to control the latter to achieve tasks, for example, as in [17,18]. This fact can be determined by assuming three conditions:

- A1: Joint deflections are small, so flexibility effects are limited to the linear elasticity domain.
- A2: Actuator rotors are modeled as uniform bodies with their centers of mass on the rotation axis.
- A3: Each motor is located in the robot arm before the driven link. This can be generalized to the case of multiple motors simultaneously driving multiple distal links.

In this case, the complete model can be represented by the following expression:

$$\begin{bmatrix} M(q) & S(q) \\ S^T(q) & J_m \end{bmatrix} \begin{bmatrix} \ddot{q} \\ \ddot{\theta} \end{bmatrix} + \begin{bmatrix} c(q,\dot{q}) + c_1(q,\dot{q},\dot{\theta}) \\ c_2(q,\dot{q}) \end{bmatrix} + \begin{bmatrix} G(q) + K(q-\theta) \\ K(\theta - q) \end{bmatrix} = \begin{bmatrix} 0 \\ \tau \end{bmatrix} \quad (2)$$

where τ is the vector of the motor torque; q, $\dot{q}$, and $\ddot{q}$ are the vectors of the link position, velocity, and acceleration, respectively; θ, $\dot{\theta}$, and $\ddot{\theta}$ are the vectors of the rotor position, velocity, and acceleration, respectively; $M(q)$ is the inertia matrix of the robot; $c(q,\dot{q})$, $c_1(q,\dot{q},\dot{\theta})$ and $c_2(q,\dot{q})$ are the matrices of the centrifugal and Coriolis forces; $G(q)$ is the vector of gravity torque on the motors; and $K(\theta - q)$ is the elastic torque. The matrix, S, represents the inertial coupling between the rotors and the links.

S is smaller than the other terms and is neglected by most authors, as are the $c_1(q,\dot{q},\dot{\theta})$ and $c_2(q,\dot{q})$ components, providing a reduced model:

$$\begin{bmatrix} M(q) & 0 \\ 0 & J_m \end{bmatrix} \begin{bmatrix} \ddot{q} \\ \ddot{\theta} \end{bmatrix} + \begin{bmatrix} c(q,\dot{q}) \\ 0 \end{bmatrix} + \begin{bmatrix} G(q) + K(q-\theta) \\ K(\theta - q) \end{bmatrix} = \begin{bmatrix} 0 \\ \tau \end{bmatrix} \quad (3)$$

This model is used for stability analysis in all of the studies mentioned in this article and, in general, by most authors. It will also be used in this study.

2.1.2. Singularly Perturbed Model

Another approach is using a singular perturbation model, that is, to refer to a situation in which a system exhibits two or more distinct time scales of motion. In these systems, one of the time scales is much slower than the others, separating the fast and slow dynamics. It was used in [6,11–13].

With a flexible joint, the elastic torque is much faster than the link. This separates the fast dynamics (elastic torque) from the slow dynamics (motion of the link). A singularly perturbated model can be obtained using a new coordinate space:

$$\begin{bmatrix} q \\ z \end{bmatrix} = \begin{bmatrix} I & 0 \\ -K & K \end{bmatrix} \begin{bmatrix} q \\ \theta \end{bmatrix} = \begin{bmatrix} q \\ K(q-\theta) \end{bmatrix} \qquad (4)$$

where $z = K(q - \theta)$ is the elastic torque.

From Equation (3),

$$\ddot{\theta} = J_m^{-1}(\tau + z) \qquad (5)$$

and

$$\ddot{q} = M(q)^{-1}(-c(q,\dot{q}) - G(q) - z) \qquad (6)$$

From Equation (4),

$$\ddot{z} = K(\ddot{\theta} - \ddot{q}) = K(J_m^{-1}(\tau + z) - M(q)^{-1}(-c(q,\dot{q}) - G(q) - z)) \qquad (7)$$

$$\ddot{z} = K((J_m + M(q)^{-1})z + J_m^{-1}\tau + M(q)^{-1}(c(q,\dot{q}) + G(q))) \qquad (8)$$

If we assume that the matrix, K, has large and similar elements, it is possible to extract a large common scale factor, $\hat{K} \gg 1$, from K: $K = \frac{1}{\epsilon^2}\hat{K} = \frac{1}{\epsilon^2}diag\{\hat{k}_1, \hat{k}_2, \ldots, \hat{k}_n\}, 0 < \epsilon \ll 1$.

Thus, Equation (8) can be rewritten as

$$\epsilon^2 \ddot{z} = \hat{K}((J_m + M(q)^{-1})z) + J_m^{-1}\tau + M(q)^{-1}(c(q,\dot{q}) + G(q)) \qquad (9)$$

Higher stiffness values mean lower ε values.

2.2. Control Strategies

This subsection briefly explains the control methods used in this study.

2.2.1. Singular Perturbation Method

The singular perturbation method [28] control strategy is used for processes that have one part that is much faster than the other. This method treats the slow and the fast parts separately, making control much easier. Two control actions are generated: one for the slow part and another one for the fast one.

The output of the slow loop is used as the input for the fast loop. To obtain the final control action, slow and fast control actions are added.

For the slow part, the control action (torque) can be generated according to the laws of control for rigid robots, which have been known for decades, for example, the inverse dynamic method provided by Equation (4).

The fast control receives the slow control action as a reference value and must ensure that it will be tracked. According to [18], a possible control law is

$$\tau_{fast} = K_{p\tau}(\tau_{slow} - \tau_{elastic}) - \epsilon K_{d\tau}\dot{\tau}_{elastic} \qquad (10)$$

This is a PD control law for the elastic torque, and $K_{p\tau}$ and $K_{d\tau}$ are the proportional and derivative constants, respectively.

The final motor torque should be

$$\tau = \tau_{fast} + \tau_{slow} \qquad (11)$$

Notably, the stability criteria for this control method are established according to the Tikhonov theorem [28]. This states that if both the slow and fast loops are separately asymptotically stable and ϵ tends toward zero, their respective errors also tend toward zero. However, since $\epsilon = \frac{1}{K^2}$, it is greater than 0. Consequently, the convergence and stability cannot be determined analytically.

Since the convergence criteria assume that ε tends toward zero, the singular perturbation control will work better for robots with stiffer joints than robots with elastic ones.

This unclear stability criteria definition limits the singular perturbation method. It cannot be used for applications such as robust or adaptive control.

2.2.2. Backstepping Control Design

Backstepping control design [29] is a control technique that stabilizes systems with nonlinear dynamics. It involves transforming the nonlinear dynamics into a series of intermediate systems with linear or linearizable dynamics and then applying a sequence of feedback controllers to each intermediate system, from top to bottom. The goal is to design a feedback control law that drives the system to its desired trajectory.

This system must be expressed so that each state variable derivative depends on this state, the next, and the previous ones. Only the last state derivative depends on the control action and all the previous states:

$$\begin{aligned}
\dot{x}_1 &= f_1(x_1) + g_1(x_1, x_2) x_2 \\
\dot{x}_2 &= f_2(x_1, x_2) x_2 + g_2(x_1, x_2, x_3) x_3 \\
&\vdots \\
\dot{x}_n &= f_n(x_1, x_2, \ldots, x_n) x_2 + g_n(x_1, x_2, \ldots, x_n) u
\end{aligned} \quad (12)$$

x_2 is a virtual input to guarantee the stability of x_1. Then, x_3 is used as a virtual input to guarantee the stability of x_2. This is repeated iteratively until the last state, which is stabilized by the control action, u.

This control method was first used to control elastic joints in [11]. As will be demonstrated in simulations, this method works well. Nonetheless, it needs higher derivatives for the link position, and the system must be represented in a chained form, as in Equation (14).

2.2.3. Simple Gravity Compensation

Simple gravity compensation [13] proposes a PD controller with gravity compensation. The control law is

$$\tau = K_p(\theta_d - \theta) - K_d \dot{\theta} + G(q_d) \quad (13)$$

where

$$\theta_d = q_d + K^{-1} G(q_d) \quad (14)$$

q_d is the reference position of the link.

Asymptotic stability is demonstrated for this case if $\lambda_{min}\left(\begin{bmatrix} K & -K \\ -K & K + K_p \end{bmatrix}\right) > \alpha$, where α is a number that fulfills the following condition for the given robot: $\|G(q_1) - G(q_2)\| \leq \alpha \|q_1 - q_2\|$.

This method is very simple. The reference position of the motors is necessary to compensate for the gravity torque of the links; it does not need the feedback of the link position. Only the motor position and velocity are used in the loop, helping to assure its stability.

2.2.4. Exact Gravity Cancellation

Exact gravity cancellation was proposed in [14]. Its control action consists of two parts:

$$\tau_m = \tau_g + \tau_0 \quad (15)$$

The first component dynamically compensates for gravity:

$$\tau_g = G(q) + JK^{-1}\ddot{G}(q) \tag{16}$$

The second component is a PD-type law:

$$\tau_0 = K_p(q_d - \theta + K^{-1}G(q)) - K_d(\dot{\theta} - K^{-1}\dot{G}(q)) \tag{17}$$

The global asymptotical stability can be shown via Lyapunov analysis. There are no constraints on the proportional constant.

This method is an improvement over the previous one.

2.2.5. Passivity-Based Approach

The passivity-based approach was proposed by the German Aerospace Center group and the Kuka company [15,16].

The final control law can be expressed as

$$\tau_m = JJ_\theta^{-1}u + (I - JJ_\theta^{-1})\tau_{elastic} \tag{18}$$

$$u = J_\theta\ddot{\theta}_{ref} + K(\theta_{ref} - q_{ref}) - K_\theta\tilde{\theta} - K_{D\theta}\dot{\tilde{\theta}} \tag{19}$$

$$\theta_{ref} = q_{ref} + K^{-1}\left(M(q)\ddot{q}_{ref} + C(q,\dot{q})\dot{q}_{ref} + G(q)\right) \tag{20}$$

J_θ is introduced for inertia shaping of the rotor since control is easier if the rotor and link inertias are similar orders of magnitude. The passivity of the system is thus demonstrated.

However, to obtain θ_{ref}, this method must compensate for the elastic torque and the feedback of the link velocity and acceleration to compute the inertia and centrifugal matrices. Regarding the elastic torque, the authors of [16] proposed a lowpass filter with a cut-off frequency of 250 Hz.

Thus, it is necessary to compute up to the second derivatives of the inertia, centrifugal, and gravity matrices to obtain $\ddot{\theta}_{ref}$, which has a very high computational cost.

2.3. Modeling Robot Dynamics

The described control methods were modeled with MATLAB. The model assumes a two-degrees-of-freedom robot with revolute joints. The MATLAB files needed for the simulations are included in the Supplementary Materials. There are five files, one for each controller. There is also a file called gentray5 that contains the fifth-order trajectory generator used by the other files.

The dynamics equations were obtained from [17,18]. For simplicity, the S inertia coupling matrix was set to 0 and the gear ratios to 1.

The links were modeled as uniform thin rods, with the following characteristics according to suggestions from experts in the field:

- Their lengths are $L_1 = L_2 = 0.5$ m.
- Their masses are $m_1 = 10$ kg and $m_2 = 0.5$ kg.
- The distances of the centers of gravity from the rotation axes are both $d_1 = d_2 = 0.25$ m.
- Moments of inertia: $I_1 = \frac{m_1 L_1^2}{12}$ kgm^2 and $I_2 = \frac{m_2 L_2^2}{12}$ kgm^2.
- Gear ratios: $r_1 = r_2 = 1$.
- Weight of the rotor of the second motor: $m_{r2} = 2$ kg.
- Inertia carried by the second motor: $J_{m2} = (I_2 + m_2 d_2^2)/r_2^2$.
- Inertia carried by the first motor: $J_{m1} = (I_1 + m_1 d_1^2 + m_{r2} L_1^2)/r_2^2$.
- The stiffnesses are set to $K_1 = K_2 = 200$, $K_1 = K_2 = 10^3$, and $K_1 = K_2 = 10^4$ Nm/rad in different simulations.

Some intermediate variables were introduced:

$$a_1 = I_1 + m_1 d_1^2 + (m_{r2} + m_2) L_1^2 + I_2 + m_2 d_2^2$$
$$a_2 = I_2 + m_2 d_2^2 \tag{21}$$
$$a_3 = m_2 L_1 d_2$$

For the dynamics expressed in Equation (2), the following matrix values were obtained:

$$B(q) = \begin{bmatrix} a_1 + a_3 \cos(q_2) & a_2 + a_3 \cos(q_2) \\ a_2 + a_3 \cos(q_2) & a_2 \end{bmatrix} \tag{22}$$

$$J = \begin{bmatrix} J_{m1} & 0 \\ 0 & J_{m2} \end{bmatrix} \tag{23}$$

$$S = \begin{bmatrix} 0 & 0 \\ 0 & 0 \end{bmatrix} \tag{24}$$

$$M = \begin{bmatrix} B & S \\ S^T & J \end{bmatrix} \tag{25}$$

$$c(q, \dot{q}) = \begin{bmatrix} -a_3 \sin(q_2)\left(\dot{q}_1 \dot{q}_2 + \dot{q}_2^2\right) \\ a_3 \sin(q_2) \dot{q}_1^2 \end{bmatrix} \tag{26}$$

$$c_1\left(q, \dot{q}, \dot{\theta}\right) = c_2(q, \dot{q}) = \begin{bmatrix} 0 \\ 0 \end{bmatrix} \tag{27}$$

$$G(q) = \begin{bmatrix} m_1 g d_1 \cos(q_1) + m_{r2} g L_1 \cos(q_1) + m_2 g (L_1 \cos(q_1) + d_2 \cos(q_1 + q_2)) \\ m_2 g d_2 \cos(q_1 + q_2) \end{bmatrix} \tag{28}$$

2.4. Adjusting the Gains for the Controllers

All the controllers use some sort of feedback, typically proportional–derivative. Their performance will depend on their gains.

To control a single joint [6], there are generally several (or infinite) combinations of proportional and derivative constants that work very well. They are computed based on the desired dynamics of the system, i.e., the natural frequency and damping ratio. To compute the proportional and derivative constants, it is necessary to know the inertia moment and the viscous friction coefficient of the system. Generally, better trajectory tracking is achieved with higher proportional and derivative gains. However, there is a point when increasing the gains practically does not improve the controller.

A multiple-degrees-of-freedom robot is much more complicated. The inertia carried by a motor is variable. Furthermore, centrifugal and Coriolis forces and gravity act on the links.

Most robot controllers (for rigid robots) compensate for the external forces and add a proportional–derivative controller for feedback [5–7]. If all the dynamics (inertia, gravity, centrifugal forces) is compensated, the values of the proportional and derivative gains may be computed for the required natural frequency and damping ratio. However, when, for example, only gravity is compensated, the optimal values of the gains vary as the robot moves.

Usually, authors do not explain how these gains are obtained. One option is to adjust them through trial and error. Another is computing the value of the gains for each motor in real time, as in the case of a single joint, for the desired natural frequency and damping ratio of the system. However, this is time-consuming and not frequently used. Another method [5] is gain scheduling. This involves reading the best gains for the actual robot configuration from a database in every sampling period.

In this study, the trial-and-error method was used. For every controller, many combinations were simulated. The simulations stopped when no more important improvements could be obtained.

3. Results

The simulations were conducted for a fifth-order polynomial trajectory generator. The first joint went from 0 to 2π and the second from 0 to $-\pi$ in five seconds.

The simulation was repeated for stiffnesses of $K_1 = K_2 = 200$, $K_1 = K_2 = 10^3$, and $K_1 = K_2 = 10^4$ Nm/rad for both joints. The first two values are typical for elastic mechanical transmissions like harmonic drives. A value of 200 is almost the most elastic found in the bibliographic research we conducted for this article [30].

Before comparing the different controllers, simulations were conducted, controlling the robot as if it was rigid, i.e., directly compensating for the inertia, gravity, centrifugal, and Coriolis terms. For cases $K_1 = K_2 = 200$ and $K_1 = K_2 = 10^3$, the system became unstable. For case $K_1 = K_2 = 10^4$, it worked acceptably. Of course, this result also depends on the other dynamic parameters of the robot, such as its mass and moments of inertia. Figure 2 shows the results of the simulation for $K_1 = K_2 = 10^4$. The mean quadratic errors for both joints were 0.0006 and 0.0036.

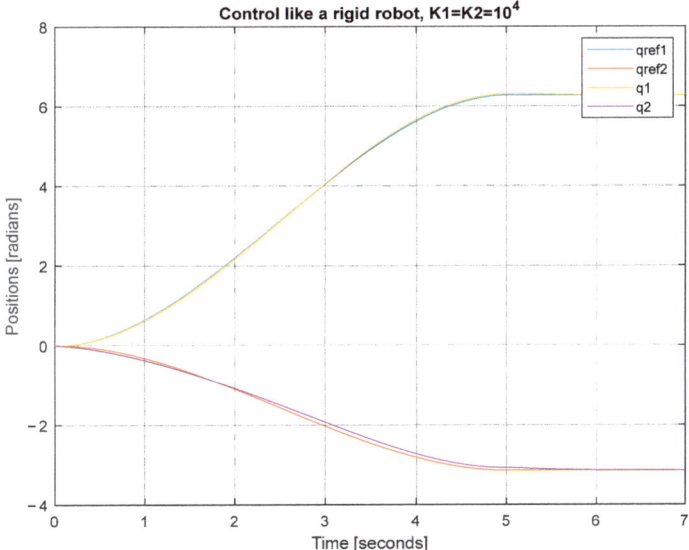

Figure 2. The positions of the links when both stiffnesses are $K_1 = K_2 = 10^4$. The blue (first joint) and red (second joint) lines represent the reference positions (first link in blue and second link in red), while the yellow (first joint) and purple (second joint) lines represent the real positions.

Then, the simulations were run for the different control strategies and stiffnesses. The simulation results for stiffness $K_1 = K_2 = 200$ are shown in Figure 3, and the mean quadratic errors are shown in Table 1.

Table 1. Mean quadratic errors with a stiffness of K = 200.

Model	Error J1	Error J2
Simple gravity compensation	0.68959863	0.05578526
Singular perturbation method	0.03245501	0.00112681
Passivity-based approach	0.00005066	0.00093868
Backstepping control design	0.00002342	0.00001048
Exact gravity cancellation	0.05286061	0.02621246

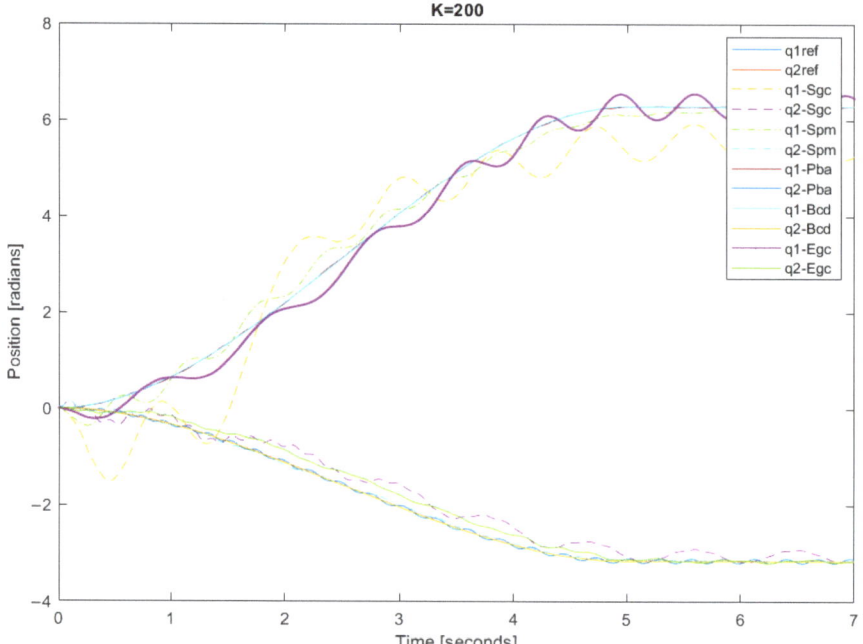

Figure 3. The positions of the links when both stiffnesses are $K_1 = K_2 = 200$. The blue (first joint) and red (second joint) lines represent the reference positions, while the other lines represent the real positions for Sgc, Spm, Pba, Bcd, and Egc.

The simulation results for stiffness $K_1 = K_2 = 10^3$ are shown in Figure 4, and the mean quadratic errors are shown in Table 2.

Table 2. The mean quadratic errors with a stiffness of K = 1000.

Model	Error J1	Error J2
Simple gravity compensation	0.030661799	0.004333034
Singular perturbation method	0.001062728	0.000887938
Passivity-based approach	0.000003148	0.000021080
Backstepping control design	0.000023420	0.000010484
Exact gravity cancellation	0.003362791	0.003530729

Finally, the simulation results for stiffness $K_1 = K_2 = 10^4$ are shown in Figure 5, and the mean quadratic errors are shown in Table 3.

Table 3. The mean quadratic errors with a stiffness of K = 10,000.

Model	Error J1	Error J2
Simple gravity compensation	0.003610014	0.003408527
Singular perturbation method	0.000083979	0.000830958
Passivity-based approach	0.000000127	0.000000020
Backstepping control design	0.000023419	0.000010488
Exact gravity cancellation	0.003358803	0.003387955

Figures 6 and 7 summarize the mean quadratic errors of the different methods for joint 1 and joint 2 for the three stiffness values.

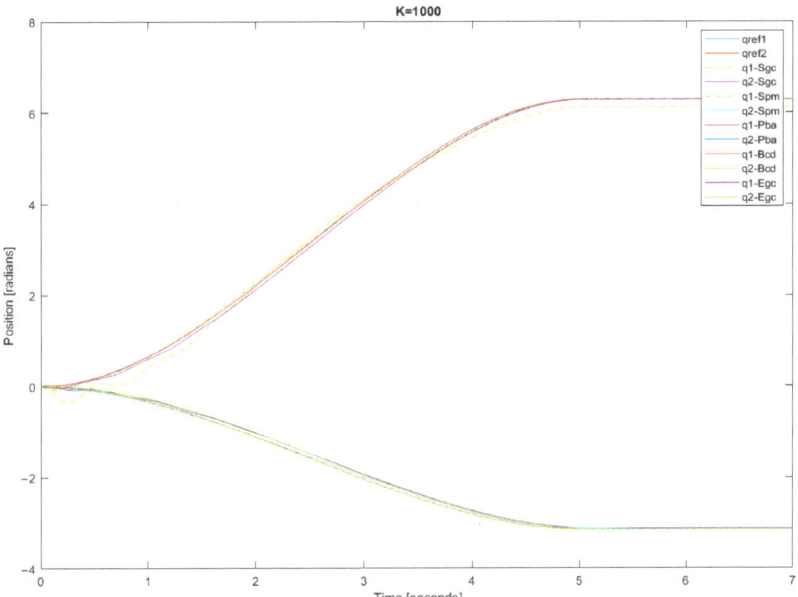

Figure 4. The positions of the links when both stiffnesses are $K_1 = K_2 = 10^3$. The blue (first joint) and red (second joint) lines represent the reference positions (first link in blue and second link in red), while the other lines represent the real positions for Sgc, Spm, Pba, Bcd, and Egc.

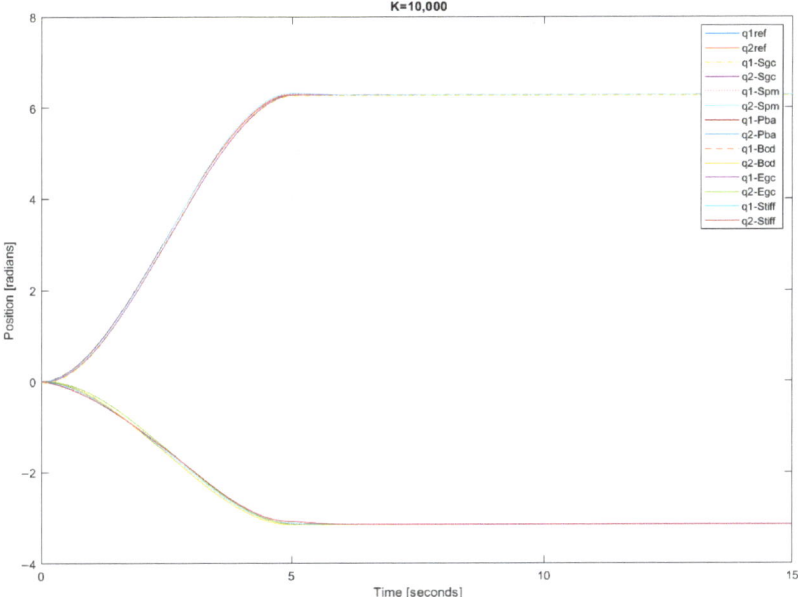

Figure 5. The positions of the links when both stiffnesses are $K_1 = K_2 = 10^4$. The blue (first joint) and red (second joint) lines represent the reference positions (first link in blue and second link in red), while the other lines represent the real positions for Sgc, Spm, Pba, Bcd, and Egc.

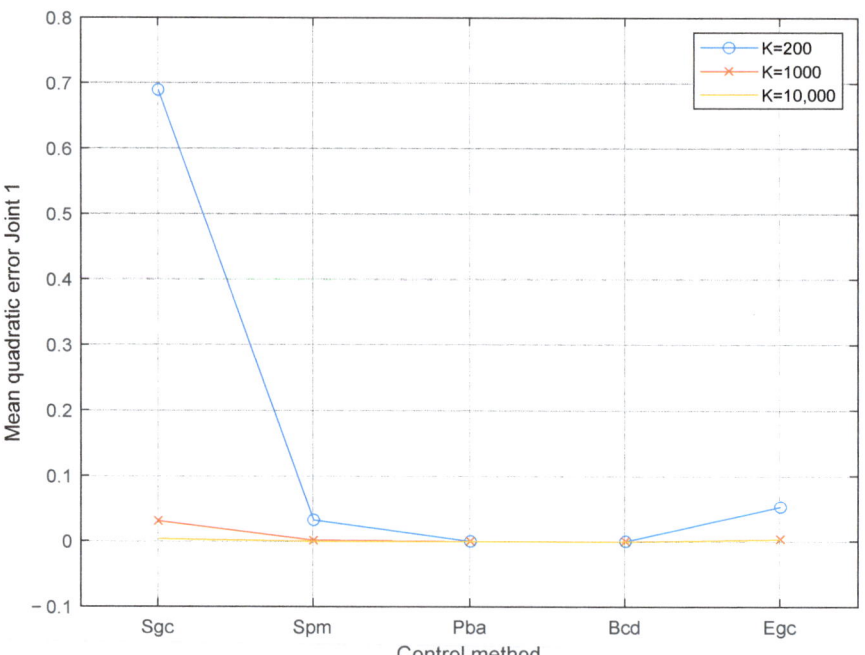

Figure 6. The mean quadratic error of joint 1 when the stiffnesses are 200, 1000, and 10,000.

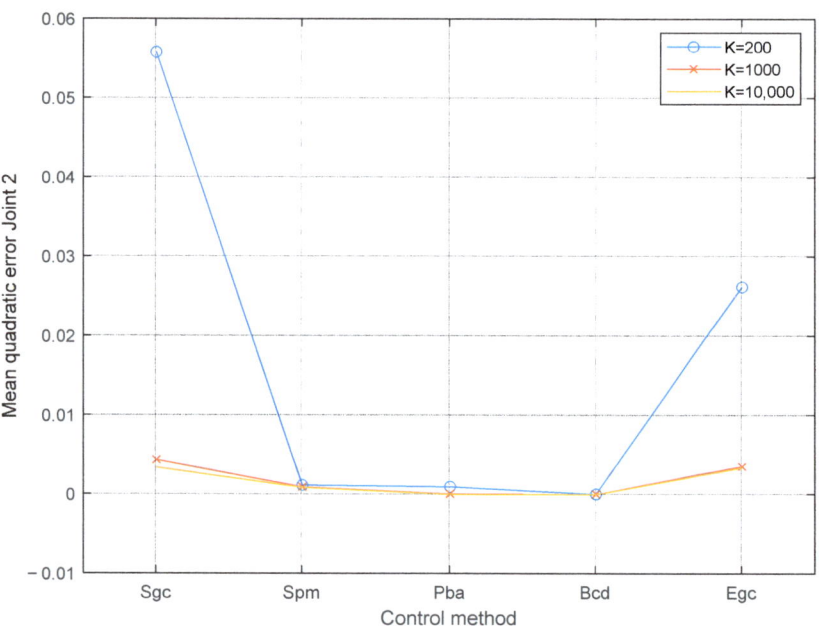

Figure 7. The mean quadratic error of joint 2 when the stiffnesses are 200, 1000, and 10,000.

The results show that the simple gravity control method presents the highest position errors with many oscillations when the rigidity is 200. A specific analysis was carried out for this control method: First, the position error was evaluated for various levels of rigidity with values between 200 and 1000. In turn, the proportionality and derivative gains of this controller were modified to observe their influence. Figures 8 and 9 show the position errors of each joint for different stiffness and controller gain levels.

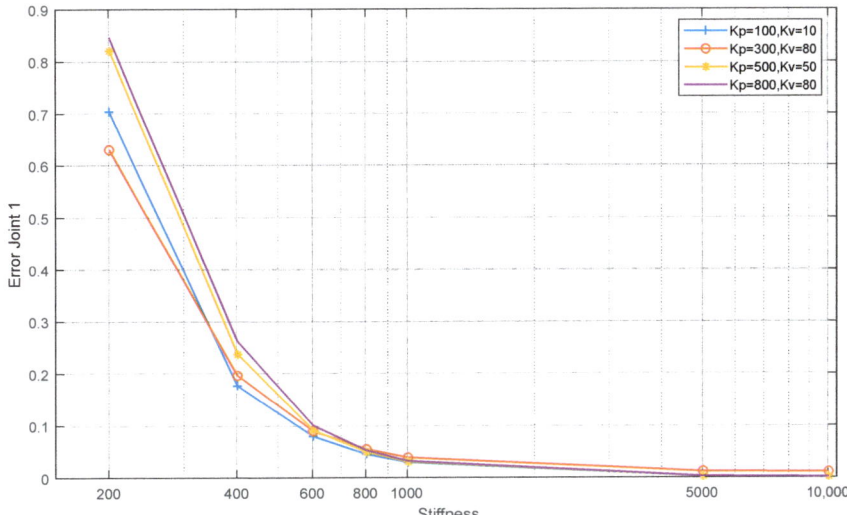

Figure 8. The mean quadratic error of joint 1 with simple gravity compensation with various stiffness values and control gains.

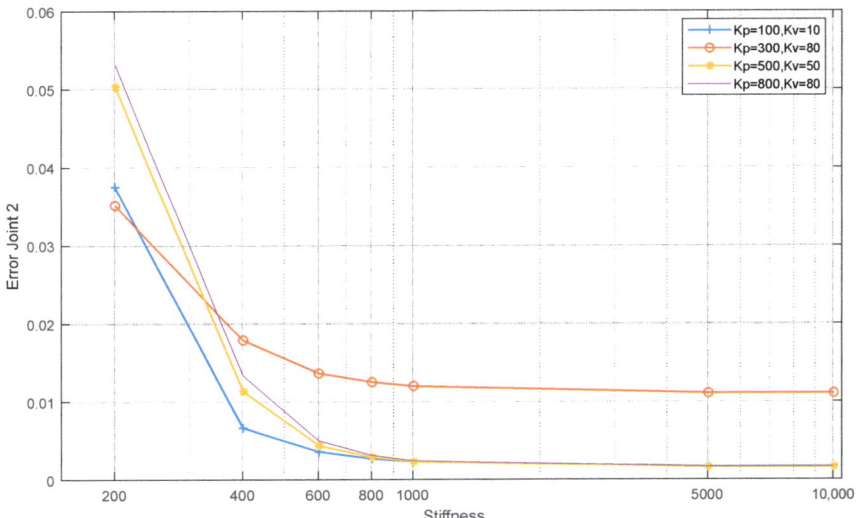

Figure 9. The mean quadratic error of joint 2 with simple gravity compensation with various stiffness values and control gains.

In joint 1, the decreased error is more significant when stiffness increases. The controller gains may provide a minor error, but this is insignificant. For all controllers with low stiffnesses, the system oscillates. When the stiffness value reaches 600–800, the oscillations begin to disappear.

In joint 2, like joint 1, the error decreases as the stiffness increases; however, when low proportional gain and high derivative gain are used, a steady state error occurs with stiffnesses greater than 600.

4. Discussion

As expected, the simulations show that the errors are higher when the joint stiffness is lower. In addition, oscillations appear with low stiffness values for the Sgc, Spm, and Egc controllers.

Regarding mean quadratic errors, the Pba exhibits better results than the other controllers independently of the stiffness value.

The error in the first joint is one order of magnitude higher than that in the second joint, probably because the higher load carried by the first motor increases the nonlinearities. The error is much higher for K = 200 than for the other cases. The worst results are obtained with Sgc and then Egc. Spb is comparable to the Pba and Bcd for the second joint but not for the first one.

Although some controllers work very well in simulations, they need to measure or estimate certain magnitudes, such as the high derivatives of the position or torque. For example, position and velocity may be fed back using low-cost sensors and computer interfaces. However, many authors are reluctant to feed back the acceleration because of the significant effect of the noise. Few have used the first and second derivatives of acceleration (jerk and snap, respectively). Thus, the feasibility of these controllers in the real world is doubtful. The same problem occurs with torque feedback because torque measurements are noisy, and its derivative may be impossible to determine.

To summarize the requirements of each method, Table 4 enumerates the necessary sensors and dynamic parameters.

Table 4. Magnitudes necessary to be measured and parameters to be known for each controller.

Method	Magnitudes to Be Measured	Dynamic Parameters to Be Known
Sgc	Joint positions and velocities.	Stiffnesses, masses, and positions of the c.o.g.
Spm	Elastic torques and their first derivatives, positions, and velocities of the links.	All the inertia moments (links and motors), masses, and centers of gravity of the links.
Pba	Position and velocity of the rotor, up to the third derivative of the link position.	All
Bcd	Up to the fourth derivative of the position.	All
Egc	Position and velocity of the rotor, up to the second derivative of the link.	Stiffness, rotor inertia, mass, and position of the c.o.g.

More than one factor influences the complexity of the controller. One aspect is the necessary amount of computation. Another is the set of dynamic parameters that must be known. Some of these factors are not easy to identify, like moments of inertia. The dynamical parameters may vary from one robot to another even if they are the same model.

Another point to be considered is the necessary sampling period. A too-short sampling time may cause problems with real-time calculus and necessitate a more powerful computer. The necessary sampling time depends on the rate of change in the measured magnitude. Thus, typically, methods (e.g., Pba) that need to control the elastic torque need faster sampling than those that require only the positions and their derivatives.

From the cost perspective, position sensors are cheap, and they are necessary for all the control methods described in this article. However, adding torque sensors greatly increases the cost of the system.

The Spm works well for the stiffest case; however, it worsens as elasticity increases. For a stiffness of 200 Nm/rad, oscillations appear. This is logical since the initial supposition of this strategy is that the fast part is much faster than the slow one. On the other hand, the stability of this technique is determined by Tikhonov's theorem [28], which does not provide exact criteria for stability. This affects the robustness of the controller. In addition, the singular perturbation method requires a torque sensor and the first derivative of the elastic torque. Generally, this method is the third best regarding trajectory tracking.

The Bcd method has the second-best performance regarding trajectory tracking in simulations. However, it is hard to implement it in real applications since it requires feeding back higher-order derivatives.

The Sgc method is extremely simple and cheap (only sensors for the rotor position are required). It has the worst trajectory following, and its performance worsens as joint elasticity increases. For stiffnesses of 200 Nm/rad and 1000 Nm/rad, oscillations appear.

The Egc method is the second worst regarding trajectory tracking. It requires measuring the acceleration of the link to fully compensate the gravity. For a stiffness of 200 Nm/rad, oscillations appear.

The Pba has good performance regardless of joint stiffness. However, it requires an expensive torque sensor for each joint. The sampling period must be faster.

Since adjusting gains is an important part of controller design, a few words will be dedicated to this topic.

The Sgc and Egc methods have no feedback for the link position end velocity—just the motor side. Thus, the link works in an open loop. Varying the gains on the motor side cannot control the link side well for robots with relatively high elasticity.

The Spm has two sets of proportional–derivative gains: one for the fast part and another for the slow part. The fast part is very sensitive, and system stability can be easily lost with small variations in gains.

Regarding the Pba and Bcd methods, all relevant variables are fed back. These controllers are not approximative but exact methods. For these reasons, good trajectory tracking may be achieved with several gain combinations.

5. Conclusions

All the methods performed well for joints with small elasticity; however, oscillations appeared in Sgc for medium and high elasticity and Egc and the Spm for low stiffness.

Considering all the drawbacks and the advantages of the Spm, it is not the most advisable.

Egc is an improvement over simple gravity compensation. However, its small trajectory tracking upgrade does not justify the high derivative requirement or the increased computational cost.

The backstepping method has very good performance in simulations. However, its implementation in the real world is problematic.

In conclusion, the Sgc method is a fast and affordable solution if joint stiffness is relatively high. If good performance is necessary, the Pba is the best option.

Supplementary Materials: The following supporting information can be downloaded at: https://www.mdpi.com/article/10.3390/act13080299/s1, MATLAB files; SingularPerturbationMethod.m, SimpleGravityCompensation.m, BacksteppingControlDesign.m, PassivityBasedApproach.m, ExactGravityCancellation.m and gentray5.m.

Author Contributions: Conceptualization, R.Z.-S.; methodology, R.Z.-S. and A.P.; software, R.Z.-S. and R.P.-U.; validation, R.Z.-S., R.P.-U., and A.P.; formal analysis, R.Z.-S. and R.P.-U.; investigation, R.Z.-S.; resources, R.Z.-S., R.P.-U. and A.P.; data curation, R.Z.-S. and R.P.-U.; writing—original draft preparation, R.Z.-S.; writing—review and editing, R.Z.-S., R.P.-U. and A.P.; visualization, R.Z.-S., R.P.-U. and A.P.; supervision, R.Z.-S., R.P.-U. and A.P.; project administration, R.Z.-S.; funding acquisition, A.P. All authors have read and agreed to the published version of the manuscript.

Funding: Grant PID2020-117713RB-I00 funded by MCIN/AEI/10.13039/501100011033.

Data Availability Statement: The data presented in this study are available on request from the corresponding author.

Conflicts of Interest: The authors declare no conflicts of interest.

Appendix A

Appendix A.1. Determining the Derivatives of the Dynamic Model

According to Equation (2), this term can be expressed as

$$\tau_{inertia} = \begin{bmatrix} B(q) & S(q) \\ S^T(q) & J_m \end{bmatrix} \begin{bmatrix} \ddot{q} \\ \ddot{\theta} \end{bmatrix} \tag{A1}$$

Since J and S are constant, their derivatives are zero.
Deriving Equation (A1) gives

$$\dot{\tau}_{inertia} = \begin{bmatrix} \dot{B}(q) & 0 \\ 0 & 0 \end{bmatrix} \begin{bmatrix} \ddot{q} \\ \ddot{\theta} \end{bmatrix} + \begin{bmatrix} B(q) & 0 \\ 0 & 0 \end{bmatrix} \begin{bmatrix} \dddot{q} \\ \dddot{\theta} \end{bmatrix} \tag{A2}$$

Deriving it again gives

$$\ddot{\tau}_{inertia} = \begin{bmatrix} \ddot{B}(q) & 0 \\ 0 & 0 \end{bmatrix} \begin{bmatrix} \ddot{q} \\ \ddot{\theta} \end{bmatrix} + 2 \begin{bmatrix} \dot{B}(q) & 0 \\ 0 & 0 \end{bmatrix} \begin{bmatrix} \dddot{q} \\ \dddot{\theta} \end{bmatrix} + \begin{bmatrix} B(q) & 0 \\ 0 & 0 \end{bmatrix} \begin{bmatrix} q^{(4)} \\ \theta^{(4)} \end{bmatrix} \tag{A3}$$

The first and second derivatives of the matrix can be obtained for Equation (22):

$$\dot{B}(q) = \begin{bmatrix} -2a_3 \sin(q_2)\dot{q}_2 & -a_3 \sin(q_2)\dot{q}_2 \\ -a_3 \sin(q_2)\dot{q}_2 & 0 \end{bmatrix} \tag{A4}$$

$$\ddot{B}(q) = \begin{bmatrix} -2a_3\left(\cos(q_2)\dot{q}_2 + \sin(q_2)\ddot{q}_2\right) & -a_3\left(\cos(q_2)\dot{q}_2 + \sin(q_2)\ddot{q}_2\right) \\ -a_3\left(\cos(q_2)\dot{q}_2 + \sin(q_2)\ddot{q}_2\right) & 0 \end{bmatrix} \tag{A5}$$

In summary, the first derivative of the inertia matrix depends on the positions and velocities of the joints. Its second derivative also depends on acceleration. The total inertia torques depend on up to the fourth derivative of the position.

Appendix A.2. The Centrifugal and Coriolis Terms

Given Equations (21) and (22), for a two-degrees-of-freedom robot, the torque related to centrifugal and Coriolis forces can be expressed as

$$\tau_C = c(q, \dot{q})\dot{q} \tag{A6}$$

By deriving, we obtain

$$\dot{\tau}_C = \dot{c}(q, \dot{q}, \ddot{q})\dot{q} + c(q, \dot{q})\ddot{q} \tag{A7}$$

The first derivative of the matrix c is

$$\dot{c} = \begin{bmatrix} -a_3\left(\cos(q_2)\dot{q}_2^2 + \sin(q_2)\ddot{q}_2\right) & -a_3\left(\cos(q_2)\dot{q}_2^2 + \sin(q_2)\ddot{q}_2\right) \\ a_3\left(\cos(q_2)\dot{q}_1\dot{q}_2 + \sin(q_2)\ddot{q}_1\right) & 0 \end{bmatrix} \tag{A8}$$

We then introduce the following:

$$aux1 = \left(-\sin(q_2)\dot{q}_2^3 + 3\cos(q_2)\dot{q}_2\ddot{q}_2 + \sin(q_2)\dddot{q}_2\right) \tag{A9}$$

$$aux2 = \left(\sin(q_2)\dot{q}_1\dot{q}_2^2 - \cos(q_2)(\dot{q}_1\ddot{q}_2 + \dot{q}_2\ddot{q}_1) - \cos(q_2)\dot{q}_2\ddot{q}_1 + \sin(q_2)\dddot{q}_1\right) \tag{A10}$$

$$aux2 = \left(\sin(q_2)\dot{q}_1\dot{q}_2^2 - \cos(q_2)(\dot{q}_1\ddot{q}_2 + 2\dot{q}_2\ddot{q}_1) + \sin(q_2)\dddot{q}_1\right) \tag{A11}$$

By deriving (A8) and introducing (A9) into (A11), we obtain

$$\ddot{c} = -a3 * \begin{bmatrix} aux1 & aux1 \\ aux2 & 0 \end{bmatrix} \quad (A12)$$

The first derivative of the centrifugal and Coriolis terms depends on the positions, velocities, and accelerations of the joints. Its second derivative depends on the jerks.

Appendix A.3. The Gravity Term

The first derivative is obtained by deriving the gravity term using Equation (28):

$$\dot{G} = g \begin{bmatrix} -m_1 d_1 \sin(q_1)\dot{q}_1 - m_{r2} L_1 \sin(q_1)\dot{q}_1 - m_2(L_1 \sin(q_1)\dot{q}_1 - d_2 \sin(q_1+q_2)(\dot{q}_1+\dot{q}_2)) \\ -m_2 d_2 \sin(q_1+q_2)(\dot{q}_1+\dot{q}_2) \\ 0 \\ 0 \end{bmatrix} \quad (A13)$$

We then introduce the intermediate variables:

$$\begin{aligned} g_{11} &= (m_1 d_1 + m_{r2} L_1)(\sin(q_1)\ddot{q}_1 + \cos(q_1)\dot{q}_1^2) \\ g_{12} &= m_2 L_1 (\sin(q_1)\ddot{q}_1 + \cos(q_1)\dot{q}_1^2) \\ g_{13} &= m_2 d_2 \bigl(\sin(q_1+q_2)(\ddot{q}_1+\ddot{q}_2) + \cos(q_1+q_2)(\dot{q}_1^2 + \dot{q}_2^2 + 2\dot{q}_1 \dot{q}_2) \bigr) \end{aligned} \quad (A14)$$

The second derivative of the gravity torque is obtained by deriving Equation (A13) and substituting with Equation (A14):

$$\ddot{G} = g \begin{bmatrix} -g_{11} - g_{12} - g_{13} \\ -g_{13} \\ 0 \\ 0 \end{bmatrix} \quad (A15)$$

The first derivative of the gravity term depends on the positions and the velocities of the joints. Its second derivative depends on acceleration.

References

1. Harmonic Drive SE. Robotics, Handling & Automation, Brochure. Available online: https://harmonicdrive.de/en/applications/robotics-handling-automation (accessed on 2 June 2024).
2. Pratt, G.A.; Williamson, M.M. Series elastic actuators. In Proceedings of the IEEE International Conference on Intelligent Robots and Systems, Pittsburgh, PA, USA, 5–9 August 1995; Volume 1, pp. 399–406.
3. Pratt, J.E.; Krupp, B.T. Series elastic actuators for legged robots. *SPIE* **2004**, *5422*, 135–144. [CrossRef]
4. Guizzo, E.; Ackerman, E. How Rethink Robotics Built Its New Baxter Robot Worker. IEEE Spectrum. Available online: https://spectrum.ieee.org/rethink-robotics-baxter-robot-factory-worker (accessed on 2 June 2024).
5. Barrientos, A.; Peñín, L.F.; Balaguer, C.; Aracil, R. *Fundamentos de Robótica*, 2nd ed.; McGraw-Hill: Madrid, Spain, 1997.
6. Spong, M.W.; Hutchinson, S.; Vidyasagar, M. *Robot Modeling and Control*, 1st ed.; John Wiley & Sons: Hoboken, NJ, USA, 2006.
7. Siciliano, B.; Sciavicco, L.; Villani, L.; Oriolo, G. *Robotics, Modelling, Planning and Control*, 1st ed.; Springer: London, UK, 2009.
8. De Luca, A. A review on the control of flexible joint manipulators. In *IROS: Workshop on Soft Robotic Modeling and Control, Proceedings of the IEEE/RSJ International Conference on Intelligent Robots and Systems, Madrid, Spain, 1–5 October 2018*; Sapienza Università di Roma: Roma, Italy, 2018.
9. Vallery, H.; Veneman, J.; Van Asseldonk, E.; Ekkelenkamp, R.; Buss, M.; Van Der Kooij, H. Compliant Actuation of Rehabilitation Robots. *IEEE Robot. Autom. Mag.* **2008**, *15*, 60–69. [CrossRef]
10. Hyun, D.J.; Lim, H.; Park, S.; Nam, S. Singular Wire-Driven Series Elastic Actuation with Force Control for a Waist Assistive Exoskeleton, H-WEXv2. *IEEE/ASME Trans. Mechatron.* **2020**, *25*, 1026–1035. [CrossRef]
11. Spong, M.W. Modeling and Control of Elastic Joint Robots. *ASME J. Dyn. Syst. Meas. Control* **1987**, *109*, 310–319. [CrossRef]
12. Spong, M.W. *Control of Robots with Flexible Joints: A Survey*; Coordinated Science Laboratory Report no. UILU-ENG-90-2203, DC-116; University of Illinois: Urbana, IL, USA, 1990.
13. Tomei, P. A simple PD controller for robots with elastic joints. *IEEE Trans. Autom. Control* **1991**, *36*, 1208–1213. [CrossRef]
14. De Luca, A.; Flacco, F. A PD-type regulator with exact gravity cancellation for robots with flexible joints. In Proceedings of the IEEE International Conference on Robotics and Automation, Shanghai, China, 9–13 May 2011.

15. Albu-Schaffer, A.; Ott, C.; Hirzinger, G. A unified passivity-based control framework for position, torque and impedance control of flexible joint robots. *Int. J. Robot. Res.* **2007**, *24*, 23–39. [CrossRef]
16. Ott, C. *Cartesian Impedance Control of Redundant and Flexible-Joint Robots*; Springer: Berlin, Germany, 2008.
17. De Luca, A. Elastic joints. In *Theory of Robot Control*; de Witt, C.C., Siciliano, B., Bastin, G., Eds.; Springer: London, UK, 1996; pp. 179–218.
18. De Luca, A. Elastic joints. In *Springer Handbook of Robotics*; Siciliano, B., Khatib, O., Eds.; Springer: Berlin, Germany, 2016; pp. 243–263.
19. Ghahrmani, N.O.; Towhidkhah, F. Constrained incremental predictive controller design for a flexible joint robot. *ISA Trans.* **2009**, *48*, 321–326. [CrossRef]
20. Iskandar, M.; Van Ommeren, C.; Wu, X.; Albu-Schäffer, A.; Dietrich, A. Model Predictive Control Applied to Different Time-Scale Dynamics of Flexible Joint Robots. *IEEE Robot. Autom. Lett.* **2023**, *8*, 672–679. [CrossRef]
21. Ott, C.; Beck, F.; Keppler, M. An Experimental Study on MPC based Joint Torque Control for Flexible Joint Robots. In Proceedings of the 13th IFAC Symposium on Robot Control (SYROCO), Matsumoto, Japan, 17–20 October 2022. [CrossRef]
22. Zaare, S.; Soltanpour, M.R.; Moattari, M. Voltage based sliding mode control of flexible joint robot manipulators in presence of uncertainties. *Robot. Auton. Syst.* **2019**, *118*, 204–219. [CrossRef]
23. Rsetam, K.; Cao, Z.; Man, Z. Cascaded-extended-state-observer-based sliding-mode control for underactuated flexible joint robot. *IEEE Trans. Ind. Electron.* **2020**, *67*, 10822–10832. [CrossRef]
24. Tuan, H.M.; Sanfilippo, F.; Hao, N.V. A novel adaptive sliding mode controller for a 2-DOF elastic robotic arm. *Robotics* **2022**, *11*, 47. [CrossRef]
25. Fateh, M.M. Robust control of flexible-joint robots using voltage control strategy. *Nonlinear Dyn.* **2012**, *67*, 1525–1537. [CrossRef]
26. Izadbakhsh, A. Robust control design for rigid-link flexible-joint electrically driven robot subjected to constraint: Theory and experimental verification. *Nonlinear Dyn.* **2016**, *85*, 751–765. [CrossRef]
27. Ullah, H.; Malik, F.M.; Raza, A.; Mazhar, N.; Khan, R.; Saeed, A.; Ahmad, I. Robust Output Feedback Control of Single-Link Flexible-Joint Robot Manipulator with Matched Disturbances Using High Gain Observer. *Sensors* **2021**, *21*, 3252. [CrossRef] [PubMed]
28. Kokotovic, P.; Khalil, H.K.; O'Reilly, J. *Singular Perturbation Methods in Control: Analysis and Design*; Society for Industrial and Applied Mathematics: Philadelphia, PA, USA, 1999; Reprint edition.
29. Vaidyanathan, S.; Azar, A.T. An introduction to backstepping control. In *Backstepping Control of Nonlinear Dynamical Systems*; Academic Press: Cambridge, MA, USA, 2021; pp. 1–32. [CrossRef]
30. De Luca, A.; Siciliano, B.; Zollo, L. PD control with on-line gravity compensation for robots with elastic joints: Theory and experiments. *Automatica* **2005**, *41*, 809–1819. [CrossRef]

Disclaimer/Publisher's Note: The statements, opinions and data contained in all publications are solely those of the individual author(s) and contributor(s) and not of MDPI and/or the editor(s). MDPI and/or the editor(s) disclaim responsibility for any injury to people or property resulting from any ideas, methods, instructions or products referred to in the content.

Article

Bionic Walking Control of a Biped Robot Based on CPG Using an Improved Particle Swarm Algorithm

Yao Wu [1,*], Biao Tang [1,2], Shuo Qiao [1,*] and Xiaobing Pang [1]

1 School of Mechatronic Engineering, Changsha University, Changsha 410022, China; tangbiao@mails.guet.edu.cn (B.T.); pangxiaobing55@ccsu.edu.cn (X.P.)
2 School of Mechanical and Electrical Engineering, Guilin University of Electronic Technology, Guilin 541004, China
* Correspondence: wuyao@whu.edu.cn (Y.W.); z20190623@ccsu.edu.cn (S.Q.)

Abstract: In the domain of bionic walking control for biped robots, optimizing the parameters of the central pattern generator (CPG) presents a formidable challenge due to its high-dimensional and nonlinear characteristics. The traditional particle swarm optimization (PSO) algorithm often converges to local optima, particularly when addressing CPG parameter optimization issues. To address these challenges, one improved particle swarm optimization algorithm aimed at enhancing the stability of the walking control of biped robots was proposed in this paper. The improved PSO algorithm incorporates a spiral function to generate better particles, alongside optimized inertia weight factors and learning factors. Evaluation results between the proposed algorithm and comparative PSO algorithms were provided, focusing on fitness, computational dimensions, convergence rates, and other metrics. The biped robot walking validation simulations, based on CPG control, were implemented through the integration of the V-REP (V4.1.0) and MATLAB (R2022b) platforms. Results demonstrate that compared with the traditional PSO algorithm and chaotic PSO algorithms, the performance of the proposed algorithm is improved by about 45% (two-dimensional model) and 54% (four-dimensional model), particularly excelling in high-dimensional computations. The novel algorithm exhibits a reduced complexity and improved optimization efficiency, thereby offering an effective strategy to enhance the walking stability of biped robots.

Keywords: biped robot; central pattern generator; PSO; bionic control

Citation: Wu, Y.; Tang, B.; Qiao, S.; Pang, X. Bionic Walking Control of a Biped Robot Based on CPG Using an Improved Particle Swarm Algorithm. *Actuators* **2024**, *13*, 393. https://doi.org/10.3390/act13100393

Received: 13 August 2024
Revised: 11 September 2024
Accepted: 1 October 2024
Published: 2 October 2024

Copyright: © 2024 by the authors. Licensee MDPI, Basel, Switzerland. This article is an open access article distributed under the terms and conditions of the Creative Commons Attribution (CC BY) license (https://creativecommons.org/licenses/by/4.0/).

1. Introduction

The humanoid biped robot has attracted the attention of researchers because it can simulate the walking characteristics of human beings and shows potential to walk in complex terrains. In daily life and industrial production, biped robots have broad application prospects. The superior mobility and environmental adaptability of biped robots enable them to perform diverse tasks efficiently, thereby increasing production efficiency and reducing labor costs. The continuous progress in this field not only promotes the development of robot technology, but also provides important support for the construction of an intelligent society in the future [1]. However, the biped robot is a multi-rigid-body and nonlinear under-actuated system, and the stability control of its walking is a very challenging task [2].

In recent years, inspired by the principle of bionics, more and more researchers have added the central pattern generator (CPG) [3] to the study of walking control of bipedal or multi-legged robots. Based on the CPG model, Sun et al. [4] proposed a bionic control method based on the human–exoskeleton coupling dynamic model by using the human–exoskeleton interaction model. Li et al. [5] proposed a method combining the reinforcement of learning with CPG to enhance the terrain adaptability of hexapod robots in walking planning. CPG can generate self-sustaining multidimensional rhythm signals without any external input, thereby controlling the coordinated periodic motion of organisms [6]. The biped robot under the control of a CPG network has good anti-interference ability and adaptability to varying environments. However, the CPG network is usually controlled by multiple parameters. The adjustment of these parameters does not have uniform rules, and it is necessary to introduce an optimization algorithm to adjust them. The output control target of the CPG neural oscillator can be the joint torque of the two swing phases of the biped robot, which involves a large number of variables. The intelligent optimization algorithm can quickly find the appropriate oscillator parameters and provide preconditions for stable walking of the biped robot. The focus of this paper is to design a new intelligent optimization algorithm to optimize the parameters of the CPG network for biped walking control. It is expected that the optimized CPG network parameters can implement the stable walking of biped robots.

Among intelligent algorithms, PSO is simpler than the genetic algorithm (GA) [7], ant colony algorithm [8], neural network algorithm [9], tree structure encoding [10], fuzzy logic [11,12], gravitational search algorithm [13], grey wolf optimization (GWO) [14], and other intelligent algorithms. One of the main advantages of the PSO algorithm is the fast convergence speed. However, in the process of optimization, the PSO algorithm is prone to falling into the local optimal solution, and because the search accuracy of the global optimal solution is not high, the algorithm will stagnate for a long time. In order to solve the above problems, many scholars have improved the PSO algorithm. Zaman et al. [15] proposed to combine the backtracking search optimization algorithm with a particle swarm optimization algorithm and introduced a new mutation operator to improve the global search ability of the particle swarm optimization algorithm. Das et al. [16] used two evolutionary operators to improve the particle swarm optimization algorithm, which helped improve the convergence of the algorithm and remove the local optimal solution. Yuan et al. [17] used a differential evolution algorithm to improve the particle swarm optimization algorithm, which solved the limitations of traditional particle swarm optimization. Zhao et al. [18] proposed an adaptive weight adjustment strategy to improve the search ability of the algorithm. Shao et al. [19] applied it to the path planning of aerial robots based on the comprehensively improved particle swarm optimization algorithm. Song et al. [20] used the method of a continuous high-order Bessel curve to optimize the PSO algorithm and plan the smooth path of a mobile robot. Li et al. [21] proposed an improved hybrid algorithm based on the PSO algorithm and the GA algorithm, which can shorten the robot planning path and accelerate the convergence speed.

Tao et al. [22] proposed a walking optimization method based on the parallel comprehensive learning particle swarm optimizer (PCLPSO), which improved the fast and stable walking ability of humanoid robots. Although PCLPSO enhances the global search ability through multi-group parallel operation and information exchange between the master and slave groups, its efficiency may be affected when dealing with multimodal functions and high-dimensional space. Sahu et al. [23] designed an adaptive particle swarm optimization algorithm (APSO) and used this algorithm to plan the path for a biped robot. However, this method mainly focuses on the optimization of parameters such as the learning factors in a traditional particle swarm optimization algorithm, and as such, the improvement of the algorithm structure is limited.

In summary, although the traditional particle swarm optimization algorithm has been improved and optimized to a certain extent, in the face of high-dimensional and nonlinear models, such as biped robot gait planning and walking control, there are still problems, such as low computational efficiency, low optimization accuracy, and slow search speed, which need further improvement. In our previous research work [24–26], we optimized the structural parameters of the biped robot and designed the walking controller. These studies provided new ideas for the design of the bionic walking controller for a biped robot. In this paper, one improved particle swarm optimization algorithm was proposed, inspired by the idea of spiral function improvement, to provide a solution for the problem of the traditional PSO being prone to falling into the local optimal solution. Aiming at the parameter optimization problem of the CPG control network of biped robots, a new fitness function was designed and the improved particle swarm optimization algorithm was used to find the optimal parameters for the CPG network. This method helped to improve the stability and robustness of biped robot walking and promotes the performance of population intelligent algorithms in practical applications.

2. Biped Bionic Walking Control Based on CPG

2.1. Structural Design of the Biped Robot

The structural design of the biped robot is mainly divided into three parts: hip joint, knee joint, and ankle joint. The overall design height of the biped robot is more than 0.85 m, and the number of degrees of freedom for the joints is 6. The overall structural design of the robot is shown in Figure 1:

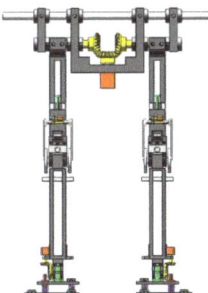

Figure 1. The mechanical structure of the biped robot.

The main physical parameters of the biped robot are shown in Table 1. In the subsequent biped robot walking simulations, the physical parameters in Table 1 will be used as the input conditions.

Table 1. The mechanical structure parameters of the biped robot.

Parameter Name	Sign	Values [Unit]
leg length	$L = l_1 + l_2$	0.9062 [m]
waist mass	m_H	11 [kg]
leg mass	m_1	2.2 [kg]
thigh mass	m_2	11 [kg]
centroid position of calf	a_1/l_1	0.614
centroid position of thigh	a_2/l_2	0.468
acceleration of gravity	g	9.8 [m s^{-2}]

2.2. Design of the Bionic Walking Control of the Biped Based on CPG

The walking generation of the biped robot based on CPG mimics the neural control mechanism of a biologically rhythmic motion. By establishing a CPG control network, the output of the CPG neural oscillator is used to control the joint angle or torque, thereby generating a stable biped robot during walking. By optimizing the parameters of the CPG control network through learning and training, the stability of the bipedal walking can be further improved.

Since the Hopf oscillator has benefits such as good stability and ease of generating periodic motion as a limit cycle, it is widely used in the walking planning of legged robots and has good stability. In this paper, the bionic walking control of the biped robot based on the Hopf oscillator is proposed. The following model is established:

$$\begin{cases} \dot{x} = \alpha(u - x^2 - y^2)x - \omega_H y \\ \dot{y} = \beta(u - x^2 - y^2)y + \omega_H x \end{cases} \quad (1)$$

where x and y are two stated variables of nonlinear differential equations, which are self-oscillating functions with respect to time; α and β can control the convergence speed of the oscillator; $\sqrt{u}$ is the amplitude of the oscillator; and ω_H is the vibration frequency of the oscillator.

In order to control the torque of the biped robot joint, the Hopf output mapping function should be established:

$$\begin{cases} X = X_a + X_R \cdot x \\ Y = Y_a + Y_R \cdot y \end{cases} \quad (2)$$

where X and Y are the output of the Hopf oscillator, X_a and Y_a are the offsets of the self-excited oscillation function relative to the initial position, and X_R and Y_R are the amplitudes of the self-excited oscillation function curve.

The CPG walking controller established in this paper is the result of a multi-factor interaction. It is necessary to use the optimization algorithm to determine the parameters of α, β, ω_H, u, X_a, Y_a, X_R, Y_R and so on.

3. Improved Design of the Particle Swarm Optimization Algorithm

3.1. Overview of the Traditional Particle Swarm Optimization Algorithm

The particle swarm optimization algorithm is an efficient algorithm that imitates the foraging behavior of birds. It was proposed by Kennedy et al. in 1995 [27]. The algorithm finds the global optimal solution through the coordination and sharing of information among individuals in the bionic bird population. The main advantages of the traditional PSO algorithm consist of self-organizing ability, evolutionary ability and memory function, strong overall optimization ability, and a fast self-optimization speed. In the D-dimensional space, the particle swarm is composed of N particles. Then, the position vector of the i-th particle is as follows:

$$X_i = (x_{i1}, x_{i2}, \cdots, x_{iD}), i = 1, 2, \cdots, N \quad (3)$$

In the k-th iteration, the velocity update formula of the i-th particle is as follows:

$$v_i^k = \omega v_i^{k-1} + c_1 r_1 \left(Pbest_i^k - x_i^k \right) + c_2 r_2 \left(Gbest_i^k - x_i^k \right) \quad (4)$$

where ω is the velocity inertia weight, $Pbest_i^k$ is the best position for the i-th particle to pass by until the k-th iteration. $Gbest_i^k$ is the best position for all particles to pass through by the k-th iteration. c_1 is the individual learning factor of the particle, c_2 is the social learning factor of the particle, and r_1, r_2 is a random number in the range of $[0, 1]$.

The position update formula of the i-th particle is as follows:

$$x_i^{k+1} = x_i^k + v_i^k t \quad (5)$$

3.2. The Improved PSO Algorithm

In order to solve the problems that the traditional PSO algorithm has, which includes being prone to falling into the local optimum and having difficulties in escaping the search process, as well as optimization problems in high-dimensional, nonlinear, and other complex models, one improved PSO algorithm was proposed: a spiral function in the optimization strategy was used to improve the search ability and convergence speed of the algorithm.

The spiral function formula is as follows:

$$\begin{cases} x_s = a \times e^{b \times \theta_s} \times \cos(\theta_s) \\ y_s = a \times e^{b \times \theta_s} \times \sin(\theta_s) \end{cases} \quad (6)$$

where a is the compression coefficient of the spiral, b is the rotation coefficient of the spiral, and θs is the angle of the spiral function.

The spiral function is used to improve the optimization strategy of the traditional particle swarm optimization algorithm, which can effectively improve the global search ability of the particle swarm. The optimization process of this method is shown in Figure 2. When the particle update stagnation occurs, the spiral function is calculated with the current optimal particle as the center. With an increasing number of iterations, the spatial size of the spiral function distribution and the number of sampling points are gradually changed. The trend is that the dispersion space will become smaller and the number of sampling points will become greater, but the value of the dispersion space and the number of sampling points needs to be limited.

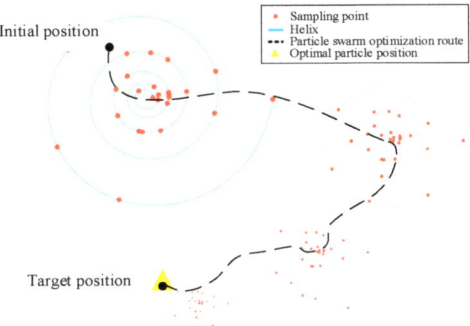

Figure 2. A diagram of the improved particle swarm optimization algorithm.

On the spiral function, the points are randomly selected via normal distribution, as shown in Figure 3. The red dots in Figure 3 are the points taken, with the starting and ending positions on the helix as the boundary and the middle part randomly selected.

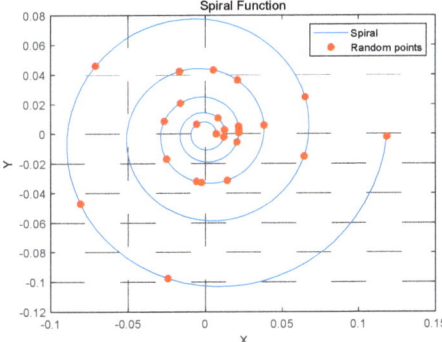

Figure 3. Sampling plot of the normal distribution of the spiral function.

If d points are randomly sampled on the spiral function, then in n-dimensional space, the position of the i-th particle is as follows:

$$Z_{i \times n} = Gbest + \begin{bmatrix} \underbrace{x_i, y_i}_{\text{group 1}}, \cdots, \underbrace{x_i, y_i}_{\text{group n/2}} \end{bmatrix}, i = 1, 2, 3, \cdots d \qquad (7)$$

where $Gbest$ is the global optimal value in traditional PSO and $[x_i, y_i]$ is the output particle value after taking a point on the spiral function. If the dimension is odd, x_i or y_i are added separately.

The fitness values of these obtained particles $Z_{d\times n}$ are calculated and compared with the fitness values of the particles *Gbest*. When a better particle than the particle *Gbest* appears, some particles are randomly taken from the spiral function to replace some of the particles in the particle swarm. The update is completed in this way every time.

In the process of iteration, the PSO algorithm will have a long time for the iteration without updating the fitness value, indicating that the calculation of the particle swarm in this interval is of little significance or that the calculation results have not progressed. When this problem occurs, the improved particle swarm optimization algorithm is based on whether the same fitness value appears in the search. Before entering the next step of calculation, the spiral function centered on the global optimal value appearing in the current iteration number is preferentially calculated. Random sampling points are randomly selected on the generated spiral function in the form of a normal distribution, and the fitness values of these sampling points are calculated step by step. If there are better particles than the global optimal particle position in the current iteration, the algorithm will randomly replace the particles in the current iteration. Thus, a part of the particle population is updated quickly to improve the efficiency of the next iteration, and the newly generated particles can avoid falling near the local optimal solution. A flowchart of the improved particle swarm optimization algorithm based on the spiral function is shown in Figure 4. The pseudocode for the improved PSO algorithm is described as Algorithm 1.

Algorithm 1. The particle swarm optimization algorithm based on the spiral function

Input: The optimization space of each CPG network parameter.
Output: Optimal CPG network parameters.
Step 1: Set the number of particles N, the number of iterations k, and then randomly set the initial position x_i and velocity v_i of the particles within a limited range.
Step 2: Calculate the fitness value $F\left(x_i^k\right)$ of the current particle.

If $F\left(x_i^k\right) < F\left(x_i^{k-1}\right)$, Individual optimal particle $Pbest_i^k = x_i^k$.
If $F\left(Pbest_i^k\right) < \left[F\left(Pbest_1^k\right)\dots F\left(Pbest_N^k\right)\right]$, Global optimal particle $Gbest_i^k = x_i^k$.

Step 3: Using the spiral function to update the particles.

If $Gbest_i^k == Gbest_i^{k-1}$, the spiral function is generated with $Gbest_i^k$ particles as the center. The particles are sampled by normal distribution on the spiral line, and the particles in the k-th iteration are randomly replaced by the sampled particles.

Step 4: Iterate steps 2 and 3 until the maximum number of iterations k is reached.
Return the minimum fitness particle swarm

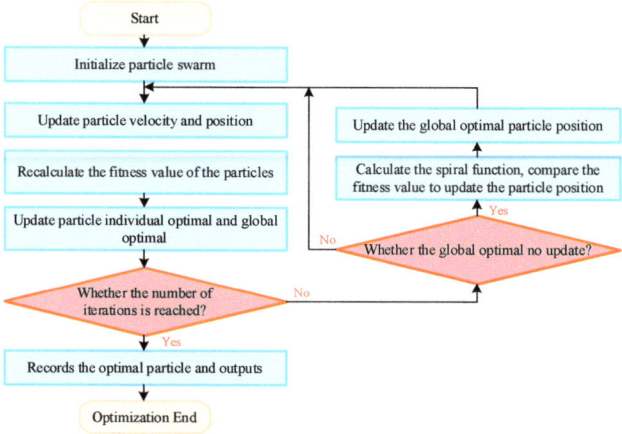

Figure 4. The improved particle swarm algorithm flow chart.

4. The Biped Robot Walking Controller Optimization

The improved particle swarm optimization algorithm and the comparative optimization algorithms are used to optimize the parameters of the biped robot network based on CPG control. Based on the CPG network parameters obtained by optimization, the limit cycle of walking can be obtained by substituting the optimized parameters into the bipedal walking controller, so that the biped robot can achieve stable walking.

In order to fully reflect the efficient optimization ability of the particle swarm optimization algorithm based on the spiral function, this paper will present results from two-dimensional, four-dimensional, and ten-dimensional models, and other parameter optimization tests from low to high dimensions. The basic parameters of the algorithm are shown in Table 2.

Table 2. The basic parameters of the algorithm.

Parameter Name	Sign	Values
velocity inertia weight	ω	0.3~0.9
particle individual learning factor	c_1	$0.95 + 0.1 \times \text{rand}[0, 1]$
particle social learning factor	c_2	$0.95 + 0.1 \times \text{rand}[0, 1]$
random number 1	r_1	[0, 1]
random number 2	r_2	[0, 1]
the compression coefficient of spiral line	a	[0.007, 0.18]
rotation coefficient of spiral line	b	0.09
angle range of spiral function	θ_s	$[0, 10\pi]$

The initial value of the biped robot walking control is set as follows:

$$x_0 = [0.1925\ -0.3919\ -0.3921\ -1.0746\ 1.5121\ 1.3789] \tag{8}$$

According to the characteristics of biped robot walking, the fitness function formula is set as follows:

$$\begin{cases} A = x_0(1,1) \\ B = (L \times cos(x_0(1,1)) + L \times cos(x_0(2,1))) \div cos(\theta) \\ C = B - B_0 \end{cases}$$

$$F = 0.4 \times \sqrt{\frac{\sum(A_i - \overline{A})^2}{n}} + 0.3 \times \sqrt{\frac{\sum(B_i - \overline{B})^2}{n}} + 0.3 \times \sqrt{\frac{\sum(C_i - \overline{C})^2}{n}} \tag{9}$$

where x_0 is the initial value of the passive walking of the biped robot, $x_0[1, 1]$ is the position angle of the support leg under the generalized coordinate, L is the total length of the robot leg, θ is the angle between the slope and the ground, $x_0[2, 1]$ is the position angle of the swing leg under the generalized coordinate, B is the step size of the biped robot, and C is the step size difference of the biped robot.

The fitness function is a measure of the stability index of the biped robot during walking, and it is also an important part of the particle swarm optimization algorithm. The smaller the implicated fitness function value, the higher the walking stability of the biped robot and the better the control effect.

In order to evaluate the performance of the algorithm optimization, an evaluation index is proposed. According to the relationship between the fitness change of the algorithm and the number of iterations, the search efficiency (SE) of the algorithm can be obtained. The formula is as follows:

$$SE = \frac{F_s - F_e}{I_e} \times 100\% \tag{10}$$

where F_s is the fitness value at the beginning, F_e is the fitness value at the end, and I_e is the number of iterations of the algorithm.

When optimizing the parameters of different dimensions, this paper tests multiple comparative algorithms. These algorithms include the improved particle swarm optimization algorithm (IPSO), the traditional particle swarm optimization algorithm (TPSO), the chaotic particle swarm optimization algorithm 1 (CPSO1), and the chaotic particle swarm optimization algorithm 2 (CPSO2). Random numbers r_1 and r_2 were replaced by chaotic sequences in CPSO1 [28]. Additionally, a chaotic search was conducted as the inactive particles were randomly generated and incorporated in the new population in CPSO2 [29].

4.1. The Two-Dimensional Comparison Test

Some parameters in CPG $[X_{a1}, X_{R1}]$ were introduced into the particle swarm optimization algorithm and the improved algorithm, respectively. The number of particles was 50 and the number of iterations was 80. In MATLAB (R2022b), several cases were run for the TPSO algorithm and the IPSO algorithm, respectively, and the optimization results are shown in Figure 5.

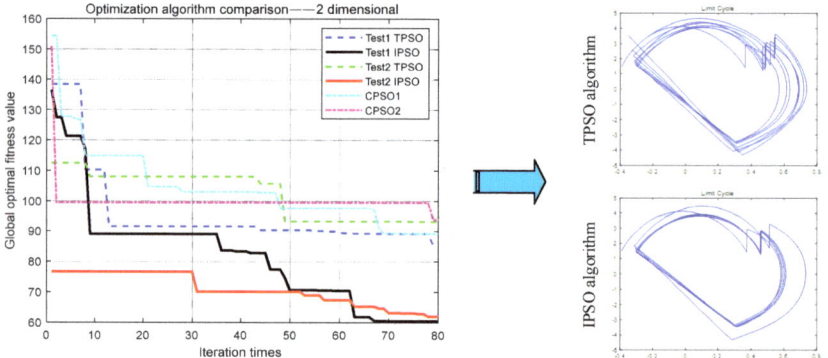

Figure 5. The two-dimensional comparison test for the optimization algorithms.

The results from Figure 5 showed that the TPSO algorithm had about 15 iterations, and began to have a longer number of stagnation updates, with less updates until the 80th update. The IPSO algorithm had a strong update ability in the later stage and had a wide range of updates from the 30th time. Better results were gained via IPSO than the chaotic PSO algorithms. From the comparison of the IPSO and TPSO algorithms with an initial global optimal fitness value of about 140, the IPSO algorithm was about 45% more efficient than the TPSO algorithm. The calculation is as follows:

$$[(140-60)-(140-85)] \div (140-85) = 45\% \qquad (11)$$

Using the calculated global optimal value, the limit cycle of biped robot walking can be obtained by substituting it into the mathematical model of biped robot hybrid dynamics. From the comparison of the two limit cycles, shown on the right side of Figure 5, the CPG parameters found by the IPSO algorithm were better, and the walking stability of the biped robot was higher. For low-dimensional parameter optimization, the IPSO algorithm had greater advantages than the TPSO algorithm.

4.2. The Four-Dimensional Comparison Test

Some parameters $[X_{a1}, X_{R1}, Y_{a1}, Y_{R1}]$ in CPG were introduced into the particle swarm optimization algorithm and the improved algorithm, respectively. The number of particles was 50 and the number of iterations was 30. In MATLAB, the TPSO algorithm and the IPSO algorithm were run several times, and the optimization results are shown in Figure 6.

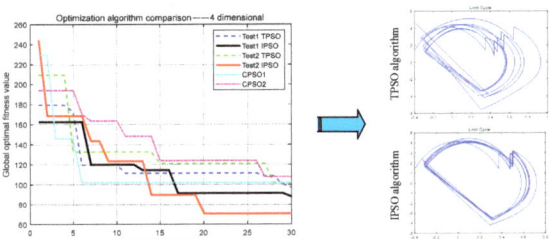

Figure 6. The four-dimensional comparison test for the optimization algorithms.

The results from the graph show that the TPSO algorithm had about seven iterations and began to have a longer number of stagnation updates, with less updates until the 30th update. The IPSO algorithm still had a strong update ability in the later period, and it also had an update state after the 13th time. From the comparison of the IPSO and TPSO algorithms with the initial global optimal fitness value of about 210, the IPSO algorithm was about 54% higher than the traditional algorithm. The calculation is as follows:

$$[(240 - 70) - (210 - 100)] \div (210 - 100) = 54\% \tag{12}$$

Using the calculated global optimal value, the limit cycle of biped robot walking can be obtained by substituting it into the mathematical model of biped robot hybrid dynamics. From the comparison of the two limit cycles, shown on the right side of Figure 6, the IPSO algorithm had the strongest results in efficient optimization and better convergence of the limit cycle.

4.3. The Ten-Dimensional Comparison Test

Some parameters $[X_{a1}, X_{R1}, Y_{a1}, Y_{R1}, Z_{a1}, Z_{R1}, X_{a2}, X_{R2}, Y_{a2}, Y_{R2}]$ in CPG were introduced into the particle swarm optimization algorithm and the improved algorithm, respectively. The number of particles was 50 and the number of iterations was 50. In MATLAB, the TPSO algorithm and the IPSO algorithm were run several times, and the optimization results are shown in Figure 7.

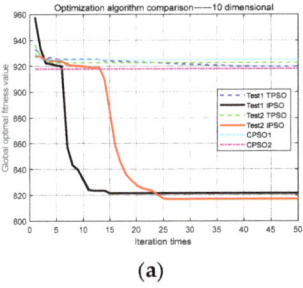
(a)

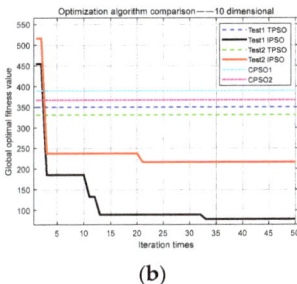
(b)

Figure 7. The ten-dimensional comparison test for the optimization algorithms: (**a**) higher fitness value position; (**b**) lower fitness value position.

The results from Figure 7 show that under different initial positions, the TPSO algorithm had a very low optimization efficiency, very few iterative update fitness values, and the image tended to be horizontal, especially in the lower initial fitness values. It is possible that the TPSO algorithm needed more particles and iterations to continue to update the fitness value. However, the IPSO algorithm had a strong ability to update, and the update span was very large. Compared with TPSO and the chaotic PSO algorithms, the IPSO algorithm had a greater impact on the optimization efficiency whether the initial fitness value position was large or small.

The calculated global optimal value was substituted into the hybrid dynamic mathematical model of the biped robot, and the limit cycle of the biped robot was obtained, as shown in Figure 8.

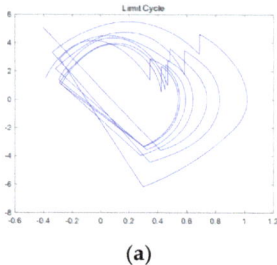
(a)

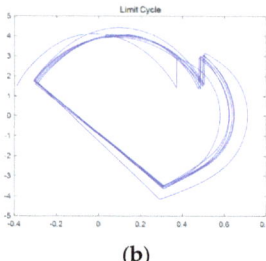
(b)

Figure 8. Diagrams from the ten-dimensional limit cycle comparison: (**a**) TPSO algorithm; (**b**) IPSO algorithm.

Based on the above test data, the optimization performance of the algorithm in different dimensions was evaluated. For each algorithm test, the optimal calculation results were selected to calculate the SE value from Equation (10). The evaluation results are shown in Table 3. From the comparison results, the improved particle swarm optimization algorithm had a higher optimization ability, especially IPSO, which had better results in high-dimensional optimization.

Table 3. Performance evaluation of the different algorithms.

Dimensional	Iteration Times	Algorithm	SE
2-dimensions	80	TPSO	66%
		IPSO	95.6%
		CPSO1	81.5%
		CPSO2	71.5%
4-dimensions	30	TPSO	275%
		IPSO	579.3%
		CPSO1	424.3%
		CPSO2	287.6%
10-dimensions (higher fitness value position)	50	TPSO	28%
		IPSO	278.6%
		CPSO1	11.2%
		CPSO2	<1%

The IPSO algorithm proposed in this paper does not produce very complex mathematical calculations. When the global optimal fitness value is stagnant and updated, the calculation of the spiral function starts to provide a new particle position for the next iteration. The computational complexity of the IPSO algorithm is proportional to the dimension of the input parameters and the number of stagnation updates of the fitness value. The proposed algorithm is more suitable for high-dimensional parameter optimization occasions, such as parameter optimization of the PID controller, time optimization of manipulator space planning, etc.

5. Walking Control Results of the Biped Robot

Using the ten-dimensional calculation results obtained by the IPSO algorithm, these were substituted into the CPG control network, and the walking simulation of the biped robot was performed using V-REP [30] and MATLAB. The parameters of the CPG walking controller, based on the improved PSO algorithm, are as follows:

$$\begin{cases} X_{a1} = 5.2762, \ X_{R1} = 3.8630; \quad Y_{a1} = -0.2491, \ Y_{R1} = 2.2230; \\ Z_{a1} = 0.3565, \ Z_{R1} = 0.8835; \\ X_{a2} = -83.8180, \ X_{R2} = 33.4561; \quad Y_{a2} = -2.4189, \ Y_{R2} = 0.9675; \\ \alpha = 1, \ \beta = 1, \ u = 3, \ \omega_H = 7\pi \end{cases} \quad (13)$$

For the above parameters, the corresponding CPG oscillator output results are shown in Figure 9.

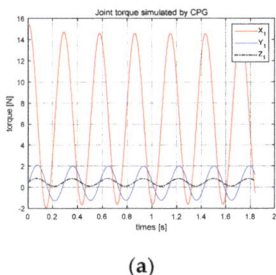
(a)

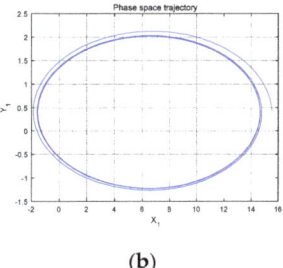
(b)

Figure 9. The CPG oscillator output results: (**a**) joint torque simulated by CPG; (**b**) phase space trajectory.

The data in Formula (13) were brought into the hybrid dynamic model of the biped robot [24]. In MATLAB, the joint angle and joint angular velocity of the biped robot were obtained by calculation, as shown in Figure 10.

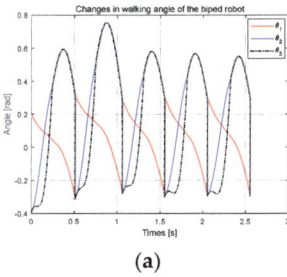
(a)

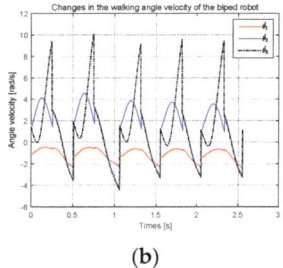
(b)

Figure 10. The walking results of the biped robot: (**a**) changes in the walking angle of the biped robot; (**b**) changes in the walking angle velocity of the biped robot.

Using the joint data from Figure 10, the stick figure of the biped robot walking in a flat environment was obtained. As shown in Figure 11, during the walking process, the robot continued to walk stably and the walking state switched smoothly.

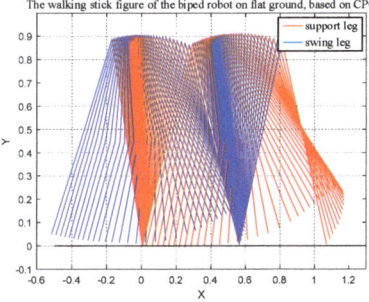

Figure 11. The walking stick figure of the biped robot on flat ground, based on CPG.

In order to simulate the biped robot walking on the flat ground, the robot's mechanical model was imported into the cross-platform, open-source simulation software V-REP (V4.1.0). Additionally, MATLAB (R2022b) was used to conduct the optimization process and control the biped robot to walk.

The virtual prototype model of the biped robot was established in V-REP, as shown in Figure 12.

Figure 12. The biped robot's virtual prototype model.

In order to evaluate the performance of the CPG parameter optimization, two walking simulation tests were conducted for the biped robots in flat ground scenarios. Specifically, one walking simulation was used as a baseline, where the robot was controlled under the CPG network with unoptimized parameters, and the other was the experiment, where robot was simulated under CPG control with the optimized parameters gained by the improved PSO algorithm.

In the flat ground scene of V-REP, the one-step walking gait of the biped robot based on the unoptimized CPG parameters is shown in Figure 13. The robot could not maintain balance and easily fell.

Figure 13. The biped robot's gait simulation under the unoptimized CPG parameters.

In the flat ground scene of V-REP, the one-step walking gait of the biped robot based on the optimized CPG parameters is shown in Figure 14. It was seen that stable bipedal locomotion could be gained under the CPG optimized parameters by using the proposed IPSO algorithm. The step length of the bipedal locomotion was about 0.56 m and the single step walking time was about 0.5 s.

6. Ijspeert, A.J. Central pattern generators for locomotion control in animals and robots: A review. *Neural Netw.* **2008**, *21*, 642–653. [CrossRef]
7. Urraca, R.; Sodupe-Ortega, E.; Antonanzas, J.; Antonanzas-Torres, F.; Martinez-de-Pison, F.J. Evaluation of a novel GA-based methodology for model structure selection: The GA-PARSIMONY. *Neurocomputing* **2018**, *271*, 9–17. [CrossRef]
8. Wu, Y.; Gong, M.; Ma, W.; Wang, S. High-order graph matching based on ant colony optimization. *Neurocomputing* **2019**, *328*, 97–104. [CrossRef]
9. Madani, T.; Daachi, B.; Benallegue, A. Adaptive variable structure controller of redundant robots with mobile/fixed obstacles avoidance. *Robot. Auton. Syst.* **2013**, *61*, 555–564. [CrossRef]
10. Ju, M.Y.; Wang, S.E.; Guo, J.H. Path planning using a hybrid evolutionary algorithm based on tree structure encoding. *Sci. World J.* **2014**, *2014*, 746260. [CrossRef]
11. Son, C. Intelligent rule-based sequence planning algorithm with fuzzy optimization for robot manipulation tasks in partially dynamic environments. *Inf. Sci.* **2016**, *342*, 209–221. [CrossRef]
12. El Ferik, S.; Nasir, M.T.; Baroudi, U. A Behavioral Adaptive Fuzzy controller of multi robots in a cluster space. *Appl. Soft Comput.* **2016**, *44*, 117–127. [CrossRef]
13. Purcaru, C.; Precup, R.E.; Iercan, D.; Fedorovici, L.O.; David, R.C.; Dragan, F. Optimal robot path planning using gravitational search algorithm. *Int. J. Artif. Intell.* **2013**, *10*, 1–20.
14. Liu, E.; Yao, X.; Liu, M.; Jin, H. AGV path planning based on improved grey wolf optimization algorithm and its implementation prototype platform. *Comput. Integr. Manuf. Syst.* **2018**, *24*, 2779–2791.
15. Zaman, H.R.R.; Gharehchopogh, F.S. An improved particle swarm optimization with backtracking search optimization algorithm for solving continuous optimization problems. *Eng. Comput.* **2022**, *38*, 2797–2831. [CrossRef]
16. Das, P.K.; Jena, P.K. Multi-robot path planning using improved particle swarm optimization algorithm through novel evolutionary operators. *Appl. Soft Comput.* **2020**, *92*, 106312. [CrossRef]
17. Yuan, Q.; Sun, R.; Du, X. Path planning of mobile robots based on an improved particle swarm optimization algorithm. *Processes* **2022**, *11*, 26. [CrossRef]
18. Zhao, G.; Jiang, D.; Liu, X.; Tong, X.; Sun, Y.; Tao, B.; Kong, J.; Yun, J.; Liu, Y.; Fang, Z. A tandem robotic arm inverse kinematic solution based on an improved particle swarm algorithm. *Front. Bioeng. Biotechnol.* **2022**, *10*, 832829. [CrossRef]
19. Shao, S.; Peng, Y.; He, C.; Du, Y. Efficient path planning for UAV formation via comprehensively improved particle swarm optimization. *ISA Trans.* **2020**, *97*, 415–430. [CrossRef]
20. Song, B.; Wang, Z.; Zou, L. An improved PSO algorithm for smooth path planning of mobile robots using continuous high-degree Bezier curve. *Appl. Soft Comput.* **2021**, *100*, 106960. [CrossRef]
21. Li, X.; Tian, B.; Hou, S.; Li, X.; Li, Y.; Liu, C.; Li, J. Path planning for mount robot based on improved particle swarm optimization algorithm. *Electronics* **2023**, *12*, 3289. [CrossRef]
22. Tao, C.; Xue, J.; Zhang, Z.; Cao, F.; Li, C.; Gao, H. Gait optimization method for humanoid robots based on parallel comprehensive learning particle swarm optimizer algorithm. *Front. Neurorobotics* **2021**, *14*, 600885. [CrossRef] [PubMed]
23. Sahu, C.; Parhi, D.R. Navigational strategy of a biped robot using regression-adaptive PSO approach. *Soft Comput.* **2022**, *26*, 12317–12341. [CrossRef]
24. Wu, Y.; Qiao, S.; Yao, D. A hybrid chaotic controller integrating hip stiffness modulation and reinforcement learning-based torque control to stabilize passive dynamic walking. *Proc. Inst. Mech. Eng. Part C J. Mech. Eng. Sci.* **2023**, *237*, 673–691. [CrossRef]
25. Wu, Y.; Yao, D.; Xiao, X.; Guo, Z. Intelligent controller for passivity-based biped robot using deep Q network. *J. Intell. Fuzzy Syst.* **2019**, *36*, 731–745. [CrossRef]
26. Wu, Y.; Yao, D.; Xiao, X. Optimal design for flexible passive biped walker based on chaotic particle swarm optimization. *J. Electr. Eng. Technol.* **2018**, *13*, 2493–2503.
27. Kennedy, J.; Eberhart, R. Particle Swarm Optimization. In Proceedings of the IEEE International Conference on Neural Networks, Perth, WA, Australia, 27 November–1 December 1995; IEEE Press: Piscataway, NJ, USA, 1995; pp. 1942–1947.
28. Yang, C.H.; Tsai, S.W.; Chuang, L.Y.; Yang, C.H. An improved particle swarm optimization with double-bottom chaotic maps for numerical optimization. *Appl. Math. Comput.* **2012**, *219*, 260–279. [CrossRef]
29. Meng, H.J.; Zheng, P.; Wu, R.Y.; Hao, X.J.; Xie, Z. A hybrid particle swarm algorithm with embedded chaotic search. In Proceedings of the IEEE Conference on Cybernetics and Intelligent Systems, Singapore, 1–3 December 2004; pp. 367–371.
30. Ao, T.; Li, M.; Liu, M.; Wang, H. Control Simulation of Dual-Arm Robot Based on Sliding Mode Controller. *Process Autom. Instrum.* **2019**, *40*, 34–38.

Disclaimer/Publisher's Note: The statements, opinions and data contained in all publications are solely those of the individual author(s) and contributor(s) and not of MDPI and/or the editor(s). MDPI and/or the editor(s) disclaim responsibility for any injury to people or property resulting from any ideas, methods, instructions or products referred to in the content.

Figure 14. The biped robot's gait simulation under the optimized CPG parameters.

6. Conclusions and Future Work

The improved particle swarm optimization algorithm based on a spiral function that was proposed in this paper is effective in solving the problem of CPG parameter optimization. The improved particle swarm optimization algorithm had benefits such as a lower likelihood of falling into the local optimum and a high optimization efficiency.

The IPSO algorithm was compared with the TPSO algorithm, the CPSO1 algorithm, and the CPSO2 algorithm in the two-dimensional, four-dimensional, and ten-dimensional CPG parameter optimization. The optimization results showed that the efficiency of the IPSO algorithm was about 45% (for two-dimensional optimization) and 54% (for four-dimensional optimization) higher than that of the TPSO algorithm near the same initial position. The IPSO algorithm had a better optimization efficiency and faster convergence speed. The improved algorithm had a good performance, especially for high-dimensional optimization.

Through the joint simulation of V-REP and MATLAB, these results show that a biped robot based on CPG control using the IPSO algorithm can walk stably on flat ground. These optimized parameter results will be used to design real biped robot prototype experiments in the future.

Author Contributions: All authors contributed to this work. Conceptualization, Y.W. and S.Q.; methodology, Y.W. and B.T.; software, B.T.; validation, Y.W. and X.P.; writing—original draft preparation, B.T.; writing—review and editing, B.T. and Y.W.; funding acquisition, Y.W., S.Q. and X.P. All authors have read and agreed to the published version of the manuscript.

Funding: This research was funded by the Research Foundation of Education Bureau of Hunan Province, China (22B0826, 22A0600), and the Science and Technology Innovation Program of Hunan Province (2022RC1138, 2023JJ30079).

Data Availability Statement: The data that support the findings of this study are available on request from the corresponding author. The data are not publicly available due to privacy and ethical restrictions.

Conflicts of Interest: The authors declare no conflicts of interest.

References

1. Katayama, S.; Murooka, M.; Tazaki, Y. Model predictive control of legged and humanoid robots: Models and algorithms. *Adv. Robot.* **2023**, *37*, 298–315. [CrossRef]
2. Wang, J.; Lu, S.; Chen, J. A CPG-based gait planning method for bipedal robots. *Artif. Life Robot.* **2024**, *29*, 340–348.
3. Matsubara, T.; Morimoto, J.; Nakanishi, J.; Sato, M.A.; Doya, K. Learning CPG-based biped locomotion with a policy gradient method. *Robot. Auton. Syst.* **2006**, *54*, 911–920. [CrossRef]
4. Sun, T.; Zhang, S.; Li, R.; Yan, Y. A Bionic Control Method for Human–Exoskeleton Coupling Based on CPG Model. *Actuators* **2023**, *12*, 321. [CrossRef]
5. Li, D.; Wei, W.; Qiu, Z. Combined Reinforcement Learning and CPG Algorithm to Generate Terrain-Adaptive Gait of Hexapod Robots. *Actuators* **2023**, *12*, 157. [CrossRef]

MDPI AG
Grosspeteranlage 5
4052 Basel
Switzerland
Tel.: +41 61 683 77 34

Actuators Editorial Office
E-mail: actuators@mdpi.com
www.mdpi.com/journal/actuators

Disclaimer/Publisher's Note: The title and front matter of this reprint are at the discretion of the Guest Editor. The publisher is not responsible for their content or any associated concerns. The statements, opinions and data contained in all individual articles are solely those of the individual Editor and contributors and not of MDPI. MDPI disclaims responsibility for any injury to people or property resulting from any ideas, methods, instructions or products referred to in the content.

www.ingramcontent.com/pod-product-compliance
Lightning Source LLC
LaVergne TN
LVHW072340090526
838202LV00019B/2446